Acclaim for

LEON LEDERMAN and *THE GOD PARTICLE*

"You'll laugh so hard you won't realize how much you are learning." —*San Francisco Examiner*

"Rewarding . . . What Lederman offers is the chance for humans, insignificant specks of self-replicating matter on a small planet circling an undistinguished star in an ordinary galaxy, to deduce the fundamental laws that made it all happen."

—*Washington Post Book World*

"A cross between Stephen Hawking and Garrison Keillor."

—*Dallas Observer*

"Lederman will leave you laughing." —*Newsday*

"Lederman's unfailing good humor and knack for storytelling are especially welcome." —*Chicago Tribune*

"The book proves that a monumental intellect, complete with requisite mussed white hair and a Nobel Prize, can relate theory and experimentation without being a monumental bore."

—*Sunday Oklahoman*

"As funny and irreverent as he is brilliant and articulate . . . Lederman puts his own witty personal spin on quantum physics."

—*Booklist*

"A delight to read and absorb, far more accessible than most books about contemporary physics."

—*Library Journal*

"A rollicking tour of mankind's two-thousand-year search for the nature of matter and . . . what grand discoveries lie ahead. A highly knowledgeable insider's look at the world of physics."

—*Kirkus Reviews*

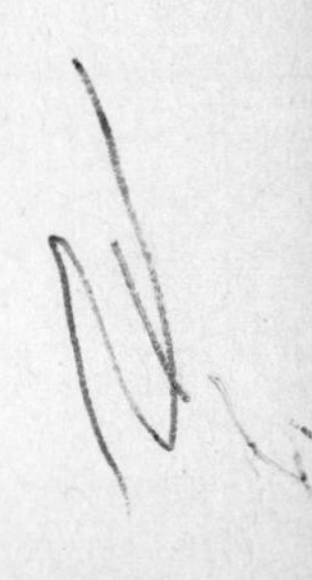

BOOKS BY LEON LEDERMAN

From Quarks to the Cosmos: Tools of Discovery
(with David N. Schramm)

Portraits of Great American Scientists
(with Judith A. Scheppler)

Symmetry and the Beautiful Universe
(with C. T. Hill)

BOOKS BY DICK TERESI

Omni's Continuum: Dramatic Phenomena from the New Frontiers of Science

Laser: Light of a Million Uses
(with Jeff Hecht)

The Three-Pound Universe: Revolutionary Discoveries About the Brain from the Chemistry of Mind to the New Frontiers of the Soul
(with Judith Hooper)

Lost Discoveries: The Ancient Roots of Modern Science—from the Babylonians to the Maya

THE GOD PARTICLE

•

If the Universe Is the Answer,

What Is the Question?

•

LEON LEDERMAN

WITH DICK TERESI

A MARINER BOOK

Houghton Mifflin Company

BOSTON NEW YORK

FOR EVAN AND JAYNA

First Mariner Books edition 2006

Visit our Web site: www.houghtonmifflinbooks.com.

Library of Congress Cataloging-in-Publication Data

Lederman, Leon M.
The god particle : if the universe is the answer, what is the question? / Leon Lederman ; with Dick Teresi.
p. cm.
Includes index.
Originally published: 1993.
ISBN-13: 978-0-618-71168-0
ISBN-10: 0-618-71168-6
1. Higgs bosons. 2. Particles (Nuclear physics)—Philosophy.
3. Matter—Constitution. 4. Science—Philosophy. I. Teresi, Dick. II. Title.
QC793.5.B62L43 2006
539.7'21—dc22 2006009243

Printed in the United States of America
Drawings by Mary Reilly
MV 10 9 8 7 6 5 4 3

I like relativity and quantum theories
because I don't understand them
and they make me feel as if space shifted
  about like a swan that can't settle,
refusing to sit still and be measured;
and as if the atom were an impulsive thing
always changing its mind.

—D. H. Lawrence

CONTENTS

PREFACE

A FUNNY THING happened to me on the way to Waxahachie . . .

It is awkward writing a preface to a new edition of a 1993 book that was originally based on a false premise. It wasn't the major premise, but a premise nonetheless. And the title of the book, *The God Particle,* problematic to begin with, was based on this misguided assumption.

I had assumed back then that the world of science was on the brink of a series of exciting new discoveries that would bring us closer to understanding how the universe works and the identities of the building blocks that make it possible. We were so close to a major epiphany in 1993 as we looked forward to a brand-new instrument, the Superconducting Super Collider (or SSC), then under construction in Waxahachie, Texas. It was to be the most powerful particle accelerator, or "atom smasher," ever built, designed to answer our most serious questions. But the unexpected intervened.

Before I get into that, however, let me review the main thrust of the book, a thrust that was valid then and remains valid today. *The God Particle* is a history of particle physics that began in about 600 B.C. with the philosopher Thales in the Greek colony of Miletus, as Thales asked himself whether all the varied objects in the universe could be traced back to a single, basic substance, and a simple, overarching principle. The approach of Thales and his followers is still with us today—a belief in ultimate simplicity, still with us in spite of the apparent complexity of our universe revealed in the research of the past 2600 years. Our story paused at Democritus (450 B.C.), who coined the term *atomos* ("too small to see and that which cannot be cut") and proceeded through the centuries and into modern times to explore the accomplishments of Albert Einstein, Enrico Fermi, Richard Feynman, Murray Gell-Mann, Sheldon Glashow, T. D. Lee, Steven Weinberg, C. N. Yang, and many other heroes of particle physics. Although I name only theorists, it was my fellow experimentalists who really did all the heavy lifting.

In 1993 we were justified, I think, in being optimistic about our chances to forge what my colleague Steven Weinberg calls "a final theory." Late in the nineteenth century only one *atomos* elementary particle, the electron, had been experimentally discovered. The ensuing decades saw us corral the rest: five more leptons (cousins of the electron), the six quarks, and the essential bosons, the photon, the W and Z, and the gluons, all force-carrying particles. One important particle had eluded us, though: the Higgs boson, a particle that would finally illuminate many of the mysteries of matter. The SSC's primary mission was to find the Higgs.

We were sanguine about the future. The SSC's construction was 20 percent complete. Our pleas for this machine began under the presidency of Ronald Reagan, construction started in 1990, and we thought we were home free until Congress canceled the project in 1993. Einstein said a physicist's job was to "read the mind of God." But how do you read the mind of a U.S. congressman? Albert, you had it so easy! Junking the SSC would free up $11 billion that would fund a cornucopia of other physics experiments, plug up the deficit, eradicate the national debt, banish poverty, cure acne, and bring us peace in our time. (How did that work out, by the way?) But I digress.

Here's the good news. *The God Particle* was ahead of its time. There is now a brand-new machine about to come online. It's called the Large Hadron Collider (LHC). Its first beams are expected in 2007 and 2008, and it is advertised to find the Higgs, to discover supersymmetry (so read the book!), and to explore several new outrageous, if not totally crazy, ideas that have emerged since that black day in 1993. So I was smarter than I thought, just writing in the wrong decade. This new instrument will not be surrounded by the friendly folks of Waxahachie, but will be located in Geneva, Switzerland, which has fewer good rib restaurants but more fondue, and is easier to spell and pronounce. One of the ideas to be explored by the LHC that turns normally phlegmatic theoretical physicists incoherent with excitement is the idea of "extra dimensions." Hidden dimensions adding to our up-down, left-right, and to-and-fro dimensions (or x-y-z) would reveal a new kind of universe in which we live and play. This is not only important to help underpin exciting "theories of everything," but, as the experimenter Henry Frisch says, "It will help us find all those missing socks."

Now, as for the title, *The God Particle,* my coauthor, Dick Teresi, has agreed to accept the blame (I paid him off). I mentioned the phrase as a joke once in a speech, and he remembered it and used it as the working title of the book. "Don't worry," he said, "no publisher ever uses the working title on the final book." The rest is history. The title ended up offending two groups: 1) those who believe in God, and 2) those who do not. We were warmly received by those in the middle.

But we are stuck with it. Some of the physics community has picked up the phrase, and both the *Los Angeles Times* and the *Christian Science Monitor* have referred to the Higgs boson as "The God Particle." This may advance our hopes for a movie version. After all, this time we are certainly on the verge of finding the Higgs and exposing a simpler and more elegant universe hitherto hidden from our sight. It's all in the book.

Have I ever lied to you?

—Leon Lederman, 2006

DRAMATIS PERSONAE

Atomos or a-tom: Theoretical particle invented by Democritus. The a-tom, invisible and indivisible, is the smallest unit of matter. Not to be confused with the so-called chemical atom, which is merely the smallest unit of each of the elements (hydrogen, carbon, oxygen, and so on).

Electron: The first a-tom discovered, in 1898. Like all modern a-toms, the electron is believed to have the curious property of "zero radius." It is a member of the lepton family of a-toms.

Quark: One of the a-toms. There are six quarks—five discovered, one still sought after (in 1993). Each of the six quarks comes in three colors. Only two of the six, the up and the down quark, exist naturally in today's universe.

Neutrino: Another a-tom in the lepton family. There are three different kinds. Neutrinos are not used to build matter, but they are essential to certain reactions. They win the minimalist contest: zero charge, zero radius, and very possibly zero mass.

Muon and tau: These leptons are cousins of the electron, only much heavier.

Photon, graviton, the W^+, W^-, and Z^0 family, and gluons: These are particles, but not matter particles like quarks and leptons. They transmit the electromagnetic, gravitational, weak, and strong forces, respectively. Only the graviton has not yet been detected.

The void: Nothingness. Also invented by Democritus. A place that a-toms can move around in. Today's theorists have littered the void with a potpourri of virtual particles and other debris. Modern terms: the vacuum and, from time to time, the aether (see below).

The aether: Invented by Isaac Newton, reinvented by James Clerk Maxwell. This is the stuff that fills up the empty space of the universe. Discredited and discarded by Einstein, the aether is now making a Nixonian comeback. It's really the vacuum, but burdened by theoretical, ghostly particles.

Accelerator: A device for increasing the energy of particles. Since $E = mc^2$, an accelerator makes these particles heavier.

Experimenter: A physicist who does experiments.

Theorist: A physicist who doesn't do experiments.

And introducing . . .

The God Particle

(also known as the Higgs particle, a.k.a. the Higgs boson, a.k.a. the Higgs scalar boson)

· 1 ·

THE INVISIBLE SOCCER BALL

> Nothing exists except atoms and empty space; everything else is opinion.
>
> — Democritus of Abdera

IN THE VERY BEGINNING there was a void — a curious form of vacuum — a nothingness containing no space, no time, no matter, no light, no sound. Yet the laws of nature were in place, and this curious vacuum held potential. Like a giant boulder perched at the edge of a towering cliff . . .

Wait a minute.

Before the boulder falls, I should explain that I really don't know what I'm talking about. A story logically begins at the beginning. But this story is about the universe, and unfortunately there are *no data* for the Very Beginning. None, zero. We don't know anything about the universe until it reaches the mature age of a billionth of a trillionth of a second — that is, some very short time after creation in the Big Bang. When you read or hear anything about the birth of the universe, someone is making it up. We are in the realm of philosophy. Only God knows what happened at the Very Beginning (and so far She hasn't let on).

Now, where were we? Oh yes . . .

Like a giant boulder perched at the edge of a towering cliff, the void's balance was so exquisite that only whim was needed to produce a change, a change that created the universe. And it happened. The nothingness exploded. In this initial incandescence, space and time were created.

Out of this energy, matter emerged — a dense plasma of particles that dissolved into radiation and back to matter. (Now we're working with at least a few facts and some speculative theory in hand.) Particles collided and gave birth to new particles. Space and time boiled and foamed as black holes formed and dissolved. What a scene!

As the universe expanded and cooled and grew less dense, particles coalesced, and forces differentiated. Protons and neutrons formed, then nuclei and atoms and huge clouds of dust, which, still expanding, condensed locally here and there to form stars, galaxies, and planets. On one planet — a most ordinary planet, orbiting a mediocre star, one speck on the spiral arm of a standard galaxy — surging continents and roiling oceans organized themselves, and out of the oceans an ooze of organic molecules reacted and built proteins, and life began. Plants and animals evolved out of simple organisms, and eventually human beings arrived.

The human beings were different primarily because they were the only species intensely curious about their surroundings. In time, mutations occurred, and an odd subset of humans began roaming the land. They were arrogant. They were not content to enjoy the magnificence of the universe. They asked "How?" How was the universe created? How can the "stuff" of the universe be responsible for the incredible variety in our world: stars, planets, sea otters, oceans, coral, sunlight, the human brain? The mutants had posed a question that could be answered — but only with the labor of millennia and with a dedication handed down from master to student for a hundred generations. The question also inspired a great number of wrong and embarrassing answers. Fortunately, these mutants were born without a sense of embarrassment. They were called physicists.

Now, after examining this question for more than two thousand years — a mere flicker on the scale of cosmological time — we are beginning to glimpse the entire story of creation. In our telescopes and microscopes, in our observatories and laboratories — and on our notepads — we begin to perceive the outlines of the pristine beauty and symmetry that governed in the first moments of the universe. We can almost see it. But the picture is not yet clear, and we sense that something is obscuring our vision — a dark force that blurs, hides, obfuscates the intrinsic simplicity of our world.

HOW DOES THE UNIVERSE WORK?

This book is devoted to one problem, a problem that has confounded science since antiquity. What are the ultimate building blocks of matter? The Greek philosopher Democritus called the smallest unit the *atomos* (literally "not able to be cut"). This a-tom is not the atom you learned about in high school science courses, like hydrogen, helium, lithium, and proceeding all the way to uranium and beyond. Those are big, klunky, complicated entities by today's standards (or by Democritus's standards, for that matter). To a physicist, or even a chemist, such atoms are veritable garbage cans of smaller particles — electrons, protons, and neutrons — and the protons and neutrons in turn are buckets full of still smaller guys. We need to know the most primitive objects there are, and we need to understand the forces that control the social behavior of these objects. It is Democritus's a-tom, not your chemistry teacher's atom, that is the key to matter.

The matter we see around us today is complex. There are about a hundred chemical atoms. The number of useful combinations of atoms can be calculated, and it is huge: billions and billions. Nature uses these combinations, called molecules, to build planets, suns, viruses, mountains, paychecks, Valium, literary agents, and other useful items. It was not always so. During the earliest moments after the creation of the universe in the Big Bang, there was no complex matter as we know it today. No nuclei, no atoms, nothing that was made of simpler pieces. This is because the searing heat of the early universe did not allow the formation of composite objects; such objects, if formed by transient collisions, would be instantly decomposed into their most primitive constituents. There was perhaps one kind of particle and one force — or even a unified particle/force — and the laws of physics. Within this primordial entity were contained the seeds of the complex world in which humans evolved, perhaps primarily to think about these things. You might find the primordial universe boring, but to a particle physicist, those were the days! Such simplicity, such beauty, however mistily visualized in our speculations.

THE BEGINNING OF SCIENCE

Even before my hero Democritus, there were Greek philosophers who dared to try to explain the world using rational arguments and rigorously excluding superstition, myth, and the intervention of gods. These had served as valuable assets in accommodating to a world full of fearsome and seemingly arbitrary phenomena. But the Greeks were impressed too by regularities, by the alternation of day and night, the seasons, the action of fire and wind and water. By the year 650 B.C. a formidable technology had arisen in the Mediterranean basin. The people there knew how to survey land and navigate by the stars; they had a sophisticated metallurgy and a detailed knowledge of the positions of stars and planets for making calendars and assorted predictions. They made elegant tools, fine textiles, and elaborately formed and decorated pottery. And in one of the colonies of the Greek empire, the bustling town of Miletus on the west coast of what is now modern Turkey, the belief was articulated that the seemingly complex world was intrinsically simple — and that this simplicity could be discovered through logical reasoning. About two hundred years later, Democritus of Abdera proposed a-toms as the key to a simple universe, and the search was on.

The genesis of physics was astronomy because the earliest philosophers looked up in awe at the night sky and sought logical models for the patterns of stars, the motions of planets, the rising and setting of the sun. Over time, scientists turned their eyes earthward: phenomena taking place at the surface of the earth — apples falling from trees, the flight of an arrow, the regular motion of a pendulum, winds, and tides — gave rise to a set of "laws of physics." Physics blossomed during the Renaissance, becoming a separate, distinct discipline by about 1500. As the centuries rolled by, and as our powers of observation sharpened with the invention of microscopes, telescopes, vacuum pumps, clocks, and so on, more and more phenomena were uncovered that could be described meticulously by recording numbers in notebooks, by constructing tables and drawing graphs, and then by triumphantly noting conformity to mathematical behavior.

By the early part of the twentieth century atoms had become the frontier of physics; in the 1940s, nuclei became the focus of research. Progressively, more and more domains became subject to observation. With the development of instruments of ever-increasing power, we looked more and more closely at things smaller and smaller. The observations and measurements were followed inevitably by syntheses, compact summaries of our understanding. With each major advance, the field divided; some scientists followed the "reductionist" road toward the nuclear and subnuclear domain, while others followed the path to a greater understanding of atoms (atomic physics), molecules (molecular physics and chemistry), nuclear physics, and so on.

THE ENTRAPMENT OF LEON

I started out as a molecules kid. In high school and early college I loved chemistry, but I gradually shifted toward physics, which seemed cleaner — odorless, in fact. I was strongly influenced, too, by the kids in physics, who were funnier and played better basketball. The giant of our group was Isaac Halpern, now a professor of physics at the University of Washington. He claimed that the only reason he went to see his posted grades was to determine whether the A had a "flat top or a pointy top." Naturally, we all loved him. He could also broad-jump farther than any of us.

I became intrigued with the issues in physics because of their crisp logic and clear experimental consequences. In my senior year in college, my best friend from high school, Martin Klein, the now eminent Einstein scholar at Yale, harangued me on the splendors of physics during a long evening over many beers. That did it. I entered the U.S. Army with a B.S. in chemistry and a determination to be a physicist if I could only survive basic training and World War II.

I was born at last into the world of physics in 1948, when I began my Ph.D. research working with the world's most powerful particle accelerator of its time, the synchrocyclotron at Columbia University. Dwight Eisenhower, president of Columbia, cut the ribbon dedicating the machine in June of 1950. Having helped Ike win the war, I was obviously much appreciated by the Columbia authorities, who paid me almost $4,000 for just one year of ninety-hour weeks. These were heady times. In the 1950s, the synchrocyclotron and other powerful new devices created the new discipline of particle physics.

To the outsider, perhaps the most salient characteristic of particle physics is the equipment, the instruments. I joined the quest just as particle accelerators were coming of age. They dominated physics for the next four decades, and still do. The first "atom smasher" was a few inches in diameter. Today the world's most powerful accelerator is housed at Fermi National Accelerator Laboratory (Fermilab) in Batavia, Illinois. Fermilab's machine, called the Tevatron, is four miles around, and smashes protons and antiprotons together with unprecedented energies. By the year 2000 or so, the Tevatron's monopoly of the energy frontier will be broken. The Superconducting Super Collider (SSC), the mother of all accelerators, presently being built in Texas, will be fifty-four miles around.

Sometimes we ask ourselves: have we taken a wrong turn somewhere? Have we become obsessed with the equipment? Is particle physics some sort of arcane "cyber science," with huge groups of researchers and megalithic machines dealing with phenomena so abstract that even She is not sure what happens when particles collide at high energies? We can gain confidence and inspiration by viewing the process as following a chronological Road, one that could plausibly have started in the Greek colony of Miletus in 650 B.C. The Road's destination is a city where all is understood — where the sanitation workers and even the mayor know how the universe works. Many have followed The Road: Democritus, Archimedes, Copernicus, Kepler, Galileo, Newton, Faraday, all the way to Einstein, Fermi, and my contemporaries.

The Road narrows and broadens; it passes long stretches of nothing (like Route 80 through Nebraska) and curvy sections of intense activity. There are tempting side streets labeled "electrical engineering," "chemistry," "radio communications," or "condensed matter." Those who have taken the side streets have changed the way people live on this planet. But those who stay with The Road find that it is clearly marked all the way with the same sign: "How does the universe work?" It is on this Road that we find the accelerators of the 1990s.

I got on The Road at Broadway and 120th Street in New York City. In those days the scientific problems seemed very clear and very important. They had to do with the properties of what's called the strong nuclear force and some theoretically predicted particles called pi mesons, or pions. Columbia's accelerator was designed to produce lots of pions by bombarding innocent targets with protons. The instrumentation was rather simple at the time, simple enough for a graduate student to understand.

Columbia was a hotbed of physics in the 1950s. Charles Townes would soon discover the laser and win the Nobel Prize. James Rainwater would win the Prize for his nuclear model, and Willis Lamb for measuring the tiny shift in hydrogen's spectral lines. Nobel laureate Isadore Rabi, who inspired all of us, headed up a team that included Norman Ramsey and Polykarp Kusch, both to become Nobel winners in due course. T. D. Lee shared the Nobel for his theory of parity violation. The density of professors who had been anointed with Swedish holy water was both exhilarating and depressing. As young faculty, some of us wore lapel buttons that read "Not Yet."

For me the Big Bang of professional recognition took place in the period 1959–1962 when two of my Columbia colleagues and I carried out the first-ever measurement of high-energy neutrino collisions. Neutrinos are my favorite particles. A neutrino has almost no properties: no mass (or very little), no electric charge, and no radius — and, adding insult to injury, no strong force acts on it. The euphemism used to describe a neutrino is "elusive." It is barely a fact, and it can pass through millions of miles of solid lead with only a tiny chance of being involved in a measurable collision.

Our 1961 experiment provided the cornerstone for what came to be known in the 1970s as the "standard model" of particle physics. In 1988 the experiment was recognized by the Royal Swedish Academy of Science with the Nobel Prize. (Everybody asks, why did they wait twenty-seven years? I don't really know. I used to give my family the facetious excuse that the Academy was dragging its feet because they couldn't decide which of my great achievements to honor.) Winning the Prize was of course a great thrill. But that thrill does not really compare with the incredible excitement that gripped us at the moment when we realized our experiment was a success.

Physicists today feel the same emotions that scientists have felt for centuries. The life of a physicist is filled with anxiety, pain, hardship, tension, attacks of hopelessness, depression, and discouragement. But these are punctuated by flashes of exhilaration, laughter, joy, and exultation. These epiphanies come at unpredictable times. Often they are generated simply by the sudden understanding of something new and important, something beautiful, that someone else has revealed. However, if you are mortal, like most of the scientists I know, the far sweeter moments come when you yourself discover some new fact about the universe. It's astonishing how often this happens at 3 A.M., when you are alone in the lab and you have learned something profound, and you realize that not one of the other five billion people on earth knows what you now know. Or so you hope. You will, of course, hasten to tell them as soon as possible. This is known as "publishing."

This is a book about a string of infinitely sweet moments that scientists have had over the past 2,500 years. These sweet moments add up to our present knowledge about what the universe is and how it works. The pain and depression are part of the story, too. Often it is the obstinacy, the stubbornness, the pure orneriness of nature that gets in the way of the "Eureka" moment.

The scientist, however, cannot depend on Eureka moments to make his life fulfilling. There must be some joy in day-to-day activities. For me, this joy is in designing and building apparatus that will teach us about this extraordinarily abstract subject. When I was an impressionable graduate student at Columbia, I helped a world-famous professor visiting from Rome build a particle counter. I was the virgin in this and he a past master. Together we turned the brass tube on the lathe (it was after 5 P.M. and the machinists had all gone home). We soldered on the glass-tipped end caps and strung a gold wire through the short, insulated metal straw penetrating the glass. Then we soldered some more. We flushed the special gas through the counter for a few hours while hooking an oscilloscope to the wire, protected from a 1,000-volt power supply by a special capacitor. My professor friend — let's call him Gilberto, because that was his name — kept peering at the green trace of the oscilloscope while lecturing me in faultlessly broken English on the history and evolution of particle counters. Suddenly Gilberto went stark, raving wild. "Mamma mia! Regardo incredibilo! Primo secourso!" (Or something like that.) He shouted, pointed, lifted me up in the air — even though I was six inches taller and fifty pounds heavier than he — and danced me around the room.

"What happened?" I stammered.

"Mufiletto!" he replied. "Izza counting. Izza counting!"

He was probably putting some of this on for my benefit, but he was genuinely excited that we had, with our hands, eyes, and brains, fashioned a device that detected the passage of cosmic ray particles, registered them by small blips in the sweep of the oscilloscope. Although he must have seen this phenomenon thousands of times, he never got over the thrill. That one of these particles may just possibly have started its voyage to 120th Street and Broadway, tenth floor, light-years ago in a distant galaxy was only part of the excitement. Gilberto's seemingly never-ending enthusiasm was contagious.

THE LIBRARY OF MATTER

When explaining the physics of fundamental particles, I often borrow (and embellish on) a lovely metaphor from the Roman poet-philosopher Lucretius. Suppose we are given the task of discovering the most basic elements of a library. What would we do? First we might think of books in their various subject categories: history, science, biography. Or perhaps we would organize them by size: thick, thin, tall, short. After considering many such divisions we realize that books are complex objects that can be readily subdivided. So we look inside. Chapters, paragraphs, and sentences are quickly dismissed as inelegant and complex constituents. Words! Here we recall that on a table near the entrance there is a fat catalogue of all the words in the library —the dictionary. By following certain rules of behavior, which we call grammar, we can use the dictionary words to compose all the books in the library. The same words are used over and over again, fitted together in different ways.

But there are so many words. Further reflection would lead us to letters, since words are "cuttable." Now we have it! Twenty-six letters can make the tens of thousands of words, and they can in turn make the millions (billions?) of books. Now we must introduce an additional set of rules: spelling, to constrain the combinations of letters. Without the intercession of a very young critic we might publish our discovery prematurely. The young critic would say, smugly no doubt, "You don't need twenty-six letters, Grandpa. All you need is a zero and a one." Children today grow up playing with digital crib toys and are comfortable with computer algorithms that convert zeroes and ones to the letters of the alphabet. If you are too old for this, perhaps you are old enough to remember Morse code, composed of dots and dashes. In either case we now have the sequence: 0 or 1 (or dot and dash) with appropriate code to make the twenty-six letters; spelling to make all the words in the dictionary; grammar to compose the words into sentences, paragraphs, chapters, and, finally, books. And the books make the library.

Now, if it makes no sense to take apart the 0 or the 1, we have discovered the primordial, a-tomic components of the library. In the metaphor, imperfect as it is, the universe is the library, the forces of nature are the grammar, spelling, and algorithm, and the 0 and 1 are what we call quarks and leptons, our current candidates for Democritus's a-toms. All of these objects, of course, are invisible.

QUARKS AND THE POPE

The lady in the audience was stubborn. "Have you ever *seen* an atom?" she insisted. It is an understandable if irritating question to a scientist who has long lived with the objective reality of atoms. I can visualize their internal structure. I can call up mental pictures of cloudlike blurs of electron "presence" surrounding the tiny dot nucleus that draws the misty electron cloud toward it. This mental picture is never precisely the same for two different scientists because both are constructing these images from equations. Such written prescriptions are not user-friendly when it comes to humoring the scientist's human need for a visual image. Yet we can "see" atoms and protons and, yes, quarks.

My attempts to answer this thorny question always begin with trying to generalize the word "see." Do you "see" this page if you are wearing glasses? If you are looking at a microfilm version? If you are looking at a photocopy (thereby robbing me of my royalty)? If you are reading the text on a computer screen? Finally, in desperation, I ask, "Have you ever seen the pope?"

"Well, of course," is the usual response. "I saw him on television." Oh, really? What she saw was an electron beam striking phosphorus painted on the inside of a glass screen. My evidence for the atom, or the quark, is just as good.

What is that evidence? Tracks of particles in a bubble chamber. In the Fermilab accelerator, the "debris" from a collision between a proton and an antiproton is captured electronically by a three-story-tall, $60 million detector. Here the "evidence," the "seeing," is tens of thousands of sensors that develop an electrical impulse as a particle passes. All of these impulses are fed through hundreds of thousands of wires to electronic data processors. Ultimately a record is made on spools of magnetic tape, encoded by zeroes and ones. This tape records the hot collisions of proton against antiproton, which can generate as many as seventy particles that fly apart into the various sections of the detector.

Science, especially particle physics, gains confidence in its conclusions by duplication — that is, an experiment in California is confirmed by a different style of accelerator operating in Geneva. Also by building into each experiment checks and tests confirming that the apparatus is functioning as designed. It is a long and involved process, the result of decades of experiments.

Still, particle physics remains unfathomable to many people. That stubborn lady in the audience isn't the only one mystified by a bunch of scientists chasing after tiny invisible objects. So let's try another metaphor . . .

THE INVISIBLE SOCCER BALL

Imagine an intelligent race of beings from the planet Twilo. They look more or less like us, they talk like us, they do everything like humans

— except for one thing. They have a fluke in their visual apparatus. They can't see objects with sharp juxtapositions of black and white. They can't see zebras, for example. Or shirts on NFL referees. Or soccer balls. This is not such a bizarre fluke, by the way. Earthlings are even stranger. We have two literal blind spots in the center of our field of vision. The reason we don't see these holes is because our brain extrapolates from the information in the rest of the field to guess what *should* be in these holes, then fills it in for us. Humans routinely drive 100 miles per hour on the autobahn, perform brain surgery, and juggle flaming torches, even though a portion of what they see is merely a good guess.

Let's say this contingent from the planet Twilo comes to earth on a goodwill mission. To give them a taste of our culture, we take them to see one of the most popular cultural events on the planet: a World Cup soccer match. We, of course, don't know that they can't see the black-and-white soccer ball. So they sit there watching the match with polite but confused looks on their faces. As far as the Twiloans are concerned, a bunch of short-pantsed people are running up and down the field kicking their legs pointlessly in the air, banging into each other, and falling down. At times an official blows a whistle, a player runs to the sideline, stands there, and extends both his arms over his head while the other players watch him. Once in a great while the goalie inexplicably falls to the ground, a great cheer goes up, and one point is awarded to the opposite team.

The Twiloans spend about fifteen minutes being totally mystified. Then, to pass the time, they attempt to understand the game. Some use classification techniques. They deduce, partially because of the clothing, that there are two teams in conflict with one another. They chart the movements of the various players, discovering that each player appears to remain more or less within a certain geographical territory on the field. They discover that different players display different physical motions. The Twiloans, as humans would do, clarify their search for meaning in World Cup soccer by giving names to the different positions played by each footballer. The positions are categorized, compared, and contrasted. The qualities and limitations of each position are listed on a giant chart. A major break comes when the Twiloans discover that *symmetry* is at work. For each position on Team A, there is a counterpart position on Team B.

With two minutes remaining in the game, the Twiloans have composed dozens of charts, hundreds of tables and formulas, and scores of complicated rules about soccer matches. And though the rules might all be, in a limited way, correct, none would really capture the essence of the game. Then one young pipsqueak of a Twiloan, silent until now, speaks his mind. "Let's postulate," he ventures nervously, "the existence of an invisible ball."

"Say what?" reply the elder Twiloans.

While his elders were monitoring what appeared to be the core of the game, the comings and goings of the various players and the demarcations of the field, the pipsqueak was keeping his eyes peeled for rare events. And he found one. Immediately before the referee announced a score, and a split second before the crowd cheered wildly, the young Twiloan noticed the momentary appearance of a bulge in the back of the goal net. Soccer is a low-scoring game, so there were few bulges to observe, and each was very short-lived. Even so, there were enough events for the pipsqueak to note that the shape of each bulge was hemispherical. Hence his wild conclusion that the game of soccer is dependent upon the existence of an invisible ball (invisible, at least, to the Twiloans).

The rest of the contingent from Twilo listen to this theory and, weak as the empirical evidence is, after much arguing, they conclude that the youngster has a point. An elder statesman in the group — a physicist, it turns out — observes that a few rare events are sometimes more illuminating than a thousand mundane events. But the real clincher is the simple fact that there *must* be a ball. Posit the existence of a ball, which for some reason the Twiloans cannot see, and suddenly everything works. The game makes sense. Not only that, but all the theories, charts, and diagrams compiled over the past afternoon remain valid. The ball simply gives meaning to the rules.

This is an extended metaphor for many puzzles in physics, and it is especially relevant to particle physics. We can't understand the rules (the laws of nature) without knowing the objects (the ball) and, without a belief in a logical set of laws, we would never deduce the existence of all the particles.

THE PYRAMID OF SCIENCE

We're talking about science and physics here, so before we proceed, let's define some terms. What is a physicist? And where does this job description fit in the grand scheme of science?

A discernible hierarchy exists, though it is not a hierarchy of social value or even of intellectual prowess. Frederick Turner, a University of Texas humanist, put it more eloquently. There exists, he said, a science pyramid. The base of the pyramid is mathematics, not because math is more abstract or more groovy, but because mathematics does not rest upon or need any of the other disciplines, whereas physics, the next layer of the pyramid, relies on mathematics. Above physics sits chemistry, which requires the discipline of physics; in this admittedly simplistic separation, physics is not concerned with the laws of chemistry. For example, chemists are concerned with how atoms combine to form molecules and how molecules behave when in close proximity. The forces between atoms are complex, but ultimately they have to do with the law of attraction and repulsion of electrically charged particles — in other words, physics. Then comes biology, which rests on an understanding of both chemistry and physics. The upper tiers of the pyramid become increasingly blurred and less definable: as we reach physiology, medicine, psychology, the pristine hierarchy becomes confused. At the interfaces are the hyphenated or compound subjects: mathematical physics, physical chemistry, biophysics. I have to squeeze astronomy into physics, of course, and I don't know what to do with geophysics or, for that matter, neurophysiology.

The pyramid may be disrespectfully summed up by an old saying: the physicists defer only to the mathematicians, and the mathematicians defer only to God (though you may be hard pressed to find a mathematician that modest).

EXPERIMENTERS AND THEORISTS: FARMERS, PIGS, AND TRUFFLES

Within the discipline of particle physics there are theorists and experimenters. I am of the latter persuasion. Physics in general progresses because of the interplay of these two divisions. In the eternal love-hate relation between theory and experiment, there is a kind of scorekeeping. How many important experimental discoveries were predicted by theory? How many were complete surprises? For example, the positive electron (positron) was anticipated by theory, as were the pion, the antiproton, and the neutrino. The muon, tau lepton, and upsilon were surprises. A more thorough study indicates rough equality in this silly debate. But who's counting?

Experiment means observing and measuring. It involves the construction of special conditions under which observations and measurements are most fruitful. The ancient Greeks and modern astronomers share a common problem. They did not, and do not, manipulate the objects they are observing. The early Greeks either could not or would not; they were satisfied to merely observe. The astronomers would dearly love to bash two suns together — or better, two galaxies — but they have yet to develop this capability, and must be content with improving the quality of their observations. But in España we have 1,003 ways of studying the properties of our particles.

Using accelerators, we can design experiments to search for the existence of new particles. We can organize particles to impinge on atomic nuclei, and read the details of the subsequent deflections the way Mycenaean scholars read Linear B — if we crack the code. We produce particles; then we "watch" them to see how long they live.

A new particle is predicted when a synthesis of existing data by a perceptive theorist seems to demand its existence. More often than not, it doesn't exist, and that particular theory suffers. Whether it succumbs or not depends on the resilience of the theorist. The point is that both kinds of experiments are carried out: those designed to test a theory and those designed to explore a new domain. Of course, it is often much more fun to *disprove* a theory. As Thomas Huxley wrote, "The great tragedy of science — the slaying of a beautiful hypothesis by an ugly fact." Good theories explain what is already known and predict the results of new experiments. The interaction of theory and experiment is one of the joys of particle physics.

Of the prominent experimentalists in history, some — including Galileo, Kirchhoff, Faraday, Ampère, Hertz, the Thomsons (both J. J. and G. P.), and Rutherford — were fairly competent theorists as well. The experimenter-theorist is a vanishing breed. In our time Enrico Fermi was an outstanding exception. I. I. Rabi expressed his concern about the widening gap by commenting that European experimentalists could not add a column of figures, and theorists couldn't tie their own shoelaces. Today we have two groups of physicists both with the common aim of understanding the universe but with a large difference in cultural outlook, skills, and work habits. Theorists tend to come in late to work, attend grueling symposiums on Greek islands or Swiss mountaintops, take real vacations, and are at home to take out the garbage much more frequently. They tend to worry about insomnia. One theorist, it is said, went to the lab physician with great concern: "Doctor, you have to help me! I sleep well all night, and the mornings aren't bad, but all afternoon I toss and turn." This behavior gives rise to the unfair characterization of *The Leisure of the Theory Class* as a takeoff on Thorstein Veblen's bestseller.

Experimenters don't come in late — they never went home. During an intense period of lab work, the outside world vanishes and the

obsession is total. Sleep is when you can curl up on
floor for an hour. A theoretical physicist can spend hi
missing the intellectual challenge of experimental wo
none of the thrills and dangers — the overhead crane
load, the flashing skull and crossbones and DANGER,
signs. A theorist's only real hazard is stabbing himse
while attacking a bug that crawls out of his calculations. My attitude toward theorists is a blend of envy and fear but also respect and affection. Theorists write all the popular books on science: Heinz Pagels, Frank Wilczek, Stephen Hawking, Richard Feynman, et al. And why not? They have all that spare time. Theorists tend to be arrogant. During my reign at Fermilab I solemnly cautioned our theory group against arrogance. At least one of them took me seriously. I'll never forget the prayer I overheard emanating from his office: "Dear Lord, forgive me the sin of arrogance, and Lord, by arrogance I mean the following . . ."

Theorists, like many other scientists, can be fiercely, sometimes absurdly, competitive. But some theorists are serene, way above the battles that mere mortals engage in. Enrico Fermi is a classic example. At least outwardly, the great Italian physicist never even hinted that competition was a relevant concept. Whereas the common physicist might say, "We did it first!" Fermi only wanted to know the details. However, at a beach near Brookhaven Laboratory on Long Island one summer day, I showed him how one can sculpt realistic structures in the moist sand. He immediately insisted that we compete to see who would make the best reclining nude. (I decline to reveal the results of that competition here. It depends on whether you're partial to the Mediterranean or the Pelham Bay school of nude sculpting.)

Once, at a conference, I found myself on the lunch line next to Fermi. Awed to be in the presence of the great man, I asked him what his opinion was of the evidence we had just listened to, for a particle named the K-zero-two. He stared at me for a while, then said, "Young man, if I could remember the names of these particles I would have been a botanist." This story has been told by many physicists, but the impressionable young researcher was *me*.

Theorists can be warm, enthusiastic human beings with whom experimentalists (mere plumbers and electricians we) love to converse and learn. It has been my good fortune to enjoy long conversations with some of the outstanding theorists of our times: the late Richard Feynman, his Cal Tech colleague Murray Gell-Mann, the arch Texan Steven Weinberg, and my rival comic Shelly Glashow. James Bjorken,

Martinus Veltman, Mary Gaillard, and T. D. Lee are other great ones who have been fun to be with, to learn from, and to tweak. A significant fraction of my experiments emerged from the papers of, and discussions with, these savants. Some theorists are much less enjoyable, their brilliance marred by a curious insecurity, reminiscent perhaps of Salieri's view of the young Mozart in the movie *Amadeus:* "Why, Lord, did you encapsulate so transcendent a composer in the body of an asshole?"

Theorists tend to peak at an early age; the creative juices tend to gush very early and start drying up past the age of fifteen — or so it seems. They need to know just enough; when they're young they haven't accumulated useless intellectual baggage.

Of course, theorists tend to receive an undue share of the credit for discoveries. The sequence of theorist, experimenter, and discovery has occasionally been compared to the sequence of farmer, pig, truffle. The farmer leads the pig to an area where there might be truffles. The pig searches diligently for the truffles. Finally, he locates one, and just as he is about to devour it, the farmer snatches it away.

GUYS WHO STAYED UP LATE

In the following chapters I approach the history and future of matter as seen through the eyes of discoverers, stressing — not, I hope, out of proportion — the experimenters. Think of Galileo, schlepping up to the top of the Leaning Tower of Pisa and dropping two unequal weights onto a wooden stage so he could listen for two impacts or one. Think of Fermi and his colleagues establishing the first sustained chain reaction beneath the football stadium of the University of Chicago.

When I talk about the pain and hardship of a scientist's life, I'm speaking of more than existential angst. Galileo's work was condemned by the Church; Madame Curie paid with her life, a victim of leukemia wrought by radiation poisoning. Too many of us develop cataracts. None of us gets enough sleep. Most of what we know about the universe we know thanks to a lot of guys (and ladies) who stayed up late at night.

The story of the a-tom, of course, includes theorists. They help us through what Steven Weinberg calls "the dark times between experimental breakthroughs," leading us, as he says, "almost imperceptibly to changes in previous beliefs." Weinberg's book *The First Three Min-*

utes was one of the best, though now dated, popular accounts of the birth of the universe. (I always thought the book sold so well because people thought it was a sex manual.) My emphasis will be on the crucial measurements we have made in the atom. But you cannot talk about data without touching on theory. What do all these measurements mean?

UH-OH, MATH

We're going to have to talk a bit about math. Even experimenters cannot make it through life without some equations and numbers. To avoid mathematics entirely would be like playing the role of an anthropologist who avoids examining the language of the culture that is being studied, or like a Shakespearean scholar who hasn't learned English. Mathematics is such an intricate part of the weave of science — especially physics — that to dismiss it is to leave out much of the beauty, much of the aptness of expression, much of the ritualistic costuming of the subject. On a practical level, math makes it easier to explain how ideas developed, how devices work, how the whole thing is woven together. You find a number here; you find the same number there — maybe they're related.

But take heart. I'm not going to do calculations. And there won't be any math on the final. In a course I taught for nonscience majors at the University of Chicago (called "Quantum Mechanics for Poets"), I straddled the issue by pointing at the mathematics and talking about it without actually *doing it*, God forbid, in front of the whole class. Even so, I find that abstract symbols on the blackboard automatically stimulate the organ that secretes eye-glaze juice. If, for instance, I write $x = vt$ (read: ex equals vee times tee), a gasp arises in the lecture hall. It isn't that these brilliant children of parents paying $20,000 tuition per year cannot deal with $x = vt$. Give them numbers for x and t and ask them to solve for v, and 48 percent would get it right, 15 percent would refuse to answer on advice of counsel, and 5 percent would vote present. (Yes, I know that doesn't add up to 100. But I'm an experimenter, not a theorist. Besides, dumb mistakes give my class confidence.) What freaks out the students is that they know I'm going to talk about it. Talking about math is new to them and brings about extreme anxiety.

To regain my students' respect and affection I immediately switch to a more familiar and comfortable subject. Examine the following:

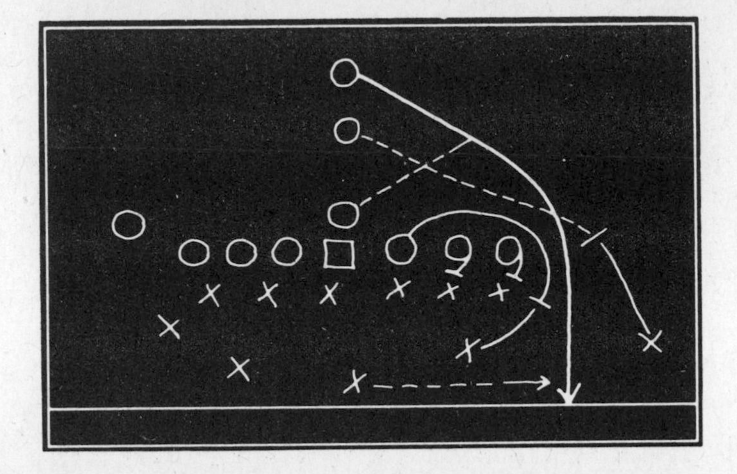

Think of a Martian staring at this diagram, trying to understand it. Tears spray out of his belly button. But your average high-school-dropout football fan yelps, "That's the Washington Redskins' goal-line 'Blast'!" Is this representation of a fullback off-tackle run that much simpler than $x = vt$? Actually, it's just as abstract, and certainly more arcane. The equation $x = vt$ works anywhere in the universe. The Redskins' short-yardage play might score a touchdown in Detroit or Buffalo, but never against the Bears.

So think of equations as having a real-world meaning, just as diagrams of football plays — overcomplicated and inelegant as they are — have a real-world meaning on the gridiron. In fact, it's not all that important to manipulate the equation $x = vt$. It's more important to be able to read it, to understand it as a statement about the universe in which we live. To understand $x = vt$ is to have power. You will be able to predict the future and to read the past. It is both the Ouija board and the Rosetta Stone. So what does it mean?

The x tells where the thing is. The thing can be Harry cruising along the interstate in his Porsche or an electron zipping out of an accelerator. When $x = 16$ units, for example, we mean that Harry or the electron is located 16 units away from a place we call zero. The v is how fast Harry or the electron is moving — such as Harry tooling along at 80 mph or the electron dawdling at 1 million meters per second. The t represents the time elapsed after someone yells "go." Now we can predict where the thing will be at any time, whether $t =$ 3 seconds or 16 hours or 100,000 years. We can also tell where the thing was, whether $t = -7$ seconds (7 seconds before $t = 0$) or $t =$ -1 million years. In other words, if Harry starts out from your driveway and drives due east for one hour at a speed of 80 mph, then obviously he will be 80 miles east of your driveway an hour from "go." In reverse, you can also calculate where Harry was an hour ago (-1 hour), assuming his velocity was always v and that v is known — a critical assumption, because if Harry is a lush, he may have stopped at Joe's Bar an hour ago.

Richard Feynman presents the subtlety of the equation another way. In his version a cop stops a woman in a station wagon, sidles up to her window, and snarls, "Did you know you were going eighty miles an hour?"

"Don't be ridiculous," the woman replies. "I only left the house fifteen minutes ago." Feynman, thinking he had invented a humorous entrée to differential calculus, was shocked when he was accused of being a sexist for telling this story, so I won't tell it here.

The point of our little excursion into the land of math is that equations have solutions, and these solutions can be compared to the "real world" of measurement and observation. If the outcome of this confrontation is succcessful, one's confidence in the original law is increased. We'll see from time to time that the solutions do not always agree with observation and measurement, in which case, after due checking and rechecking, the "law" from which the solution emerged is relegated to the dustbin of history. Occasionally the solutions to the equations expressing a law of nature are completely unexpected and bizarre, and therefore bring the theory under suspicion. If subsequent observations show that it was right after all, we rejoice. Whatever the outcome, we know that the overarching truths about the universe as well as the functioning of a resonant electronic circuit or the vibrations of a structural steel beam are all expressed in the language of mathematics.

THE UNIVERSE IS ONLY SECONDS OLD (10^{18} OF THEM)

One more thing about numbers. Our subject often switches from the world of the very tiny to the world of the enormous. Thus we will be dealing with numbers that are often very, very large or very, very small. So, for the most part, I shall write them using scientific notation. For instance, instead of writing one million as 1,000,000, I write it like this: 10^6. That means 10 raised to the sixth power, which is a 1 followed by six zeroes, which is the approximate cost, in dollars, of running the U.S. government for about 20 seconds. Big numbers that don't conveniently start with a 1 can also be written in scientific notation. For instance, 5,500,000 is written 5.5×10^6. With tiny numbers, we just insert a minus sign. One millionth (1/1,000,000) is written like this: 10^{-6}, which means a 1 that is six places to the right of a decimal point, or .000001.

What's important is to grasp the scale of these numbers. One of the disadvantages of scientific notation is that it hides the true immensity of numbers (or their smallness). The span of scientifically relevant times is mind-boggling: 10^{-1} seconds is an eye blink; 10^{-6} seconds is the lifetime of the muon particle, and 10^{-23} seconds is the time it takes a photon, a particle of light, to cross the nucleus. Keep in mind that going up by powers of ten escalates the stakes tremendously. Thus 10^7 seconds is equal to a bit more than four months, and 10^9 seconds is thirty years. But 10^{18} seconds is roughly the age of the universe, the amount of time that has transpired since the Big Bang. Physicists measure it in seconds — just a lot of them.

Time isn't the only quantity that ranges from the unimaginably infinitesimal to the endless. The smallest distance that is relevant to measurement today is something like 10^{-17} centimeters, which is how far a thing called the Z^0 (zee zero) can travel before it departs our world. Theorists sometimes deal in much smaller space concepts; for instance, when they talk about superstrings, a trendy but very abstract and very hypothetical theory of particles, they say that the size of a superstring is 10^{-35} centimeters, real small. At the other extreme, the largest distance is the radius of the observable universe, somewhat under 10^{28} centimeters.

A TALE OF TWO PARTICLES AND THE ULTIMATE T-SHIRT

When I was ten years old, I came down with the measles, and to cheer me up my father bought me a book with big print called *The Story of Relativity,* by Albert Einstein and Leopold Infeld. I'll never forget the beginning of Einstein and Infeld's book. It talked about detective stories, about how every detective story has a mystery, clues, and a detective. The detective tries to solve the mystery by using the clues.

There are essentially two mysteries to be solved in the following story. Both manifest themselves as particles. The first is the long-sought a-tom, the invisible, indivisible particle of matter first postulated by Democritus. The a-tom lies at the heart of the basic questions of particle physics.

We've struggled to solve this first mystery for 2,500 years. It has thousands of clues, each uncovered with painstaking labor. In the first few chapters, we'll see how our predecessors have attempted to put the puzzle together. You'll be surprised to see how many "modern" ideas were embraced in the sixteenth and seventeenth centuries, and even centuries before Christ. By the end, we'll be back to the present and chasing a second, perhaps even greater mystery, one represented by the particle that I believe orchestrates the cosmic symphony. And you will see through the course of the book the natural kinship between a sixteenth-century mathematician dropping weights from a tower in Pisa and a present-day particle physicist freezing his fingers off in a hut on the cold, wind-swept prairie of Illinois as he checks the data flowing in from a half-billion-dollar accelerator buried beneath the frozen ground. Both asked the same questions. What is the basic structure of matter? How does the universe work?

When I was growing up in the Bronx, I used to watch my older brother playing with chemicals for hours. He was a whiz. I'd do all the chores in the house so he'd let me watch his experiments. Today he's in the novelty business. He sells things like whoopee cushions, booster license plates, and T-shirts with catchy sayings. These allow people to sum up their world view in a statement no wider than their chest. Science should have no less lofty a goal. My ambition is to live to see all of physics reduced to a formula so elegant and simple that it will fit easily on the front of a T-shirt.

Significant progress has been made through the centuries in the search for the ultimate T-shirt. Newton, for example, came up with gravity, a force that explains an amazing range of disparate phenomena: the tides, the fall of an apple, the orbits of the planets, and the clustering of galaxies. The Newton T-shirt reads $F = ma$. Later, Michael Faraday and James Clerk Maxwell unraveled the mystery of the electromagnetic spectrum. Electricity, magnetism, sunlight, radio waves, and x-rays, they found, are all manifestations of the same force. Any good campus bookstore will sell you a T-shirt with Maxwell's equations on it.

Today, many particles later, we have the standard model, which reduces all of reality to a dozen or so particles and four forces. The standard model represents all the data that have come out of all the accelerators since the Leaning Tower of Pisa. It organizes particles called quarks and leptons — six of each — into an elegant tabular array. One can diagram the entire standard model on a T-shirt, albeit a busy one. It's a hard-won simplicity, generated by an army of physicists who have traveled the same road. However, the standard-model T-shirt cheats. With its twelve particles and four forces, it is remarkably accurate. But it is also incomplete and, in fact, internally inconsistent. To have room on the T-shirt to make succinct excuses for the inconsistencies would require an X-tra large, and we'd still run out of shirt.

What, or who, is standing in our way, obstructing our search for the perfect T-shirt? This brings us back to our second mystery. Before we can complete the task begun by the ancient Greeks, we must consider the possibility that our quarry is laying false clues to confuse us. Sometimes, like a spy in a John le Carré novel, the experimenter must set a trap. He must force the culprit to expose himself.

THE MYSTERIOUS MR. HIGGS

Particle physicists are currently setting just such a trap. We're building a tunnel fifty-four miles in circumference that will contain the twin beam tubes of the Superconducting Super Collider, in which we hope to trap our villain.

And what a villain! The biggest of all time! There is, we believe, a wraithlike presence throughout the universe that is keeping us from understanding the true nature of matter. It's as if something, or someone, wants to prevent us from attaining the ultimate knowledge.

This invisible barrier that keeps us from knowing the truth is called the Higgs field. Its icy tentacles reach into every corner of the universe, and its scientific and philosophical implications raise large goose bumps on the skin of a physicist. The Higgs field works its black magic through — what else? — a particle. This particle goes by the name of the Higgs boson. The Higgs boson is a primary reason for building the Super Collider. Only the SSC will have the energy necessary to produce and detect the Higgs boson, or so we believe. This boson is so central to the state of physics today, so crucial to our final understanding of the structure of matter, yet so elusive, that I have given it a nickname: the God Particle. Why God Particle? Two reasons. One, the publisher wouldn't let us call it the Goddamn Particle, though that might be a more appropriate title, given its villainous nature and the expense it is causing. And two, there is a connection, of sorts, to another book, a *much* older one . . .

THE TOWER AND THE ACCELERATOR

> And the whole earth was of one language, and of one speech.
>
> And it came to pass, as they journeyed from the east, that they found a plain in the land of Shinar; and they dwelt there. And they said one to another, Go to, let us make brick, and burn them thoroughly. And they had brick for stone, and slime had they for mortar. And they said, Go to, let us build us a city and a tower, whose top *may reach* unto heaven; and let us make us a name, lest we be scattered abroad upon the face of the whole earth.
>
> And the Lord came down to see the city and the tower, which the children of men builded. And the Lord said, Behold, the people *is* one, and they have all one language; and this they begin to do: and now nothing will be restrained from them, which they have imagined to do. Go to, let us go down, and there confound their language, that they may not understand one another's speech.
>
> So the Lord scattered them abroad from thence upon the face of all the earth: and they left off to build the city. Therefore is the name of it called Babel.
>
> — Genesis 11:1–9

At one time, many millennia ago, long before those words were written, nature spoke but one language. Everywhere matter was the same — beautiful in its elegant, incandescent symmetry. But through the eons, it has been transformed, scattered throughout the universe in many forms, confounding those of us who live on this ordinary planet orbiting a mediocre star.

There have been times in mankind's quest for a rational understanding of the world when progress was rapid, breakthroughs abounded, and scientists were full of optimism. At other times utter confusion reigned. Frequently the most confused periods, times of intellectual crisis and total incomprehension, were themselves harbingers of the illuminating breakthroughs to come.

In the past few decades in particle physics, we have been in a period of such curious intellectual stress that the parable of the Tower of Babel seems appropriate. Particle physicists have been using their giant accelerators to dissect the parts and processes of the universe. The quest has, in recent years, been aided by astronomers and astrophysicists, who figuratively peer into giant telescopes to scan the heavens for residue sparks and ashes of a cataclysmic explosion that they are convinced took place 15 billion years ago, which they call the Big Bang.

Both groups have been progressing toward a simple, coherent, all-encompassing model that will explain everything: the structure of matter and energy, the behavior of forces in environments that range from the earliest moments of the infant universe with its exorbitant temperature and density to the relatively cold and empty world we know today. We were proceeding nicely, perhaps too nicely, when we stumbled upon an oddity, a seemingly adversarial force afoot in the universe. Something that seems to pop out of the all-pervading space in which our planets, stars, and galaxies are embedded. Something we cannot yet detect and which, one might say, has been put there to test and confuse us. Were we getting too close? Is there a nervous Grand Wizard of Oz who sloppily modifies the archaeological record?

The issue is whether physicists will be confounded by this puzzle or whether, in contrast to the unhappy Babylonians, we will continue to build the tower and, as Einstein put it, "know the mind of God."

> And the whole universe was of many languages, and of many speeches.
>
> And it came to pass, as they journeyed from the east, that they found a plain in the land of Waxahachie, and they dwelt there. And they said to one another, Go to, let us build a Giant Collider, whose collisions may reach back to the beginning of time. And they had superconducting magnets for bending, and protons had they for smashing.
>
> And the Lord came down to see the accelerator, which the children of men builded. And the Lord said, Behold the people are unconfounding my confounding. And the Lord sighed and said, Go to, let us go down, and there give them the God Particle so that they may see how beautiful is the universe I have made.
>
> — The Very New Testament, 11:1

· 2 ·

THE FIRST PARTICLE PHYSICIST

> He seemed surprised. "You found a knife that can cut off an atom?" he said. "In *this* town?"
> I nodded. "We're sitting on the main nerve right now," I said.
>
> — With apologies to Hunter S. Thompson

ANYONE CAN DRIVE (or walk or bicycle) into Fermilab, even though it is the most sophisticated scientific laboratory in the world. Most federal facilities are militant about preserving their privacy. But Fermilab is in the business of uncovering secrets, not keeping them. During the radical 1960s the Atomic Energy Commission told Robert R. Wilson, my predecessor and the lab's founding director, to devise a plan for handling student activists should they arrive at the gates of Fermilab. Wilson's plan was simple. He told the AEC he would greet the protesters alone, armed with a single weapon: a physics lecture. This was lethal enough, he assured the commission, to disperse even the bravest rabble-rousers. To this day, lab directors keep a lecture handy in case of emergencies. Let us pray we never have to use it.

Fermilab sits on 7,000 acres of converted corn fields five miles east of Batavia, Illinois, about an hour's drive west of Chicago. At the Pine Street entrance to the grounds stands a giant steel sculpture created by Robert Wilson, who besides being the first director was pretty much responsible for the building of Fermilab, an artistic, architectural, and scientific triumph. The sculpture, entitled *Broken Symmetry*, consists of three arches curving upward, as if to intersect at a point fifty feet above the ground. They don't make it, at least not cleanly. The three arms meet, but in an almost haphazard fashion, as if they had been built by different contractors who weren't talking to each other. The sculpture has an "oops" feel to it — not unlike our present universe. If you walk around the sculpture, the giant steel work appears jarringly asymmetrical from every angle. But if you lie on your back directly beneath it and look straight up, you will enjoy the one vantage point from which the sculpture is symmetrical. Wilson's work of art suits Fermilab perfectly, since the job description of the physicists here is to search for clues to what they suspect is a hidden symmetry in what appears to be a very asymmetrical universe.

As you drive into the grounds, you come across the most prominent structure on the site. Wilson Hall, Fermilab's sixteen-story central laboratory building, sweeps upward from the flat, flat land, somewhat like a Dürer drawing of hands held in prayer. The building was inspired by a cathedral Wilson visited in Beauvais, France, built in A.D. 1225. The Beauvais cathedral featured twin towers spanned by a chancel. Wilson Hall, completed in A.D. 1972, consists of twin towers (the two hands in prayer), joined by crossovers at several floors and one of the world's largest atriums. At the entrance to the high-rise is a reflecting pool with a tall obelisk at one end. The obelisk, Wilson's final artistic tribute to the lab, is known to all the researchers as Wilson's Last Construction.

Tangential to Wilson Hall is the raison d'être for the laboratory: the particle accelerator. Buried thirty feet beneath the prairie and describing a circle four miles around lies a stainless steel tube just a few inches in diameter. It weaves through a thousand superconducting magnets that guide protons around their circular track. The accelerator is filled with collisions and heat. Through this ring, protons race at near-light-speed velocities to their annihilation in head-to-head confrontations with their brethren antiprotons. These collisions momentarily generate temperatures of about 10,000 trillion (10^{16}) degrees above absolute zero, vastly higher than those found at the core of the sun or in the furious explosion of a supernova. Scientists here are time travelers more legitimate than those you'll find in science fiction movies. The last time such temperatures were "natural" was a tiny fraction of a second after the Big Bang, the birth of the universe.

Though underground, the accelerator ring can easily be discerned from above because of a twenty-foot-high berm of earth on the ground above the ring. (Imagine a very skinny, four-mile-around bagel.) Many people assume the berm's purpose is to absorb radiation from the machine, but it's really there because Wilson was an aesthetic sort of guy. After all the work of building the accelerator, he was disappointed that he couldn't tell where it was. So when the workmen dug out holes for cooling ponds around the accelerator, he had them pile up the dirt in this immense circle. To accent the circle, Wilson created a ten-foot-wide canal around it and installed circulating pumps that fire fountains of water into the air. The canal is functional as well as visual; it carries the cooling water for the accelerator. The whole thing is strangely beautiful. In satellite photos taken from three hundred miles above the earth, the berm-and-waterway — looking like a perfect circle from that height — is the sharpest feature on the northern Illinois landscape.

The 660 acres of land enclosed by the accelerator ring are a curious throwback. The laboratory is restoring the prairie inside the ring. Much of the original tall prairie grass, nearly choked out by European grasses over the past two centuries, has been replanted, thanks to several hundred volunteers who have harvested seeds from prairie remnants in the Chicago area. Trumpeter swans, Canada geese, and sandhill cranes make their home in surface-water collection lakes that dot the ring's interior.

Across the road, north of the main ring, is another restoration project — a pasture where a herd of about a hundred buffalo roam. The herd is made up of animals brought from Colorado and South Dakota, along with a few indigenous to Illinois, although buffalo have not flourished in the Batavia area for eight hundred years. Before then, herds were commonplace over the prairie where physicists now roam. Archaeologists tell us that buffalo hunting over the present Fermilab grounds goes back nine thousand years, as evidenced by all the arrowheads found in the region. It appears that for centuries a tribe of Native Americans from the nearby Fox River sent their hunters up to what is now Fermilab, where they camped out, hunted down the animals, and carried them back to the riverside settlement.

Some people find the present-day buffalo a trifle unsettling. Once, when I was promoting the lab on the Phil Donahue show, a lady who lived near the facility phoned in. "Dr. Lederman makes the accelerator seem relatively harmless," she complained. "If it is, why do they have all those buffalo? We all know they're extremely sensitive to radioactive material." She thought the buffalo were like canaries in a mine shaft, only trained to detect radiation instead of coal gas. I guess she figured that I kept one eye on the herd from my office in the high-rise, ready to run for the parking lot should one of them keel over. In truth, the buffalo are just buffalo. A Geiger counter works much better as a radiation detector and eats much less hay.

Drive east on Pine Street, away from Wilson Hall, and you come to several other important facilities, including the collider detector facility (CDF), designed to make most of our discoveries about matter, and the newly constructed Richard P. Feynman Computer Center, named after the great Cal Tech theorist who died just a few years ago. Keep driving and eventually you come to Eola Road. Take a right and drive straight for a mile or so, and you'll see a 150-year-old farmhouse on the left. That's where I lived as director: 137 Eola Road. That's not an official address. It's just the number I chose to put on the house.

It was Richard Feynman, in fact, who suggested that all physicists put a sign up in their offices or homes to remind them of how much we don't know. The sign would say simply this: 137. One hundred thirty-seven is the inverse of something called the fine-structure constant. This number is related to the probability that an electron will emit or absorb a photon. The fine-structure constant also answers to the name alpha, and it can be arrived at by taking the square of the charge of the electron divided by the speed of light times Planck's constant. What all that verbiage means is that this one number, 137, contains the crux of electromagnetism (the electron), relativity (the velocity of light), and quantum theory (Planck's constant). It would be less unsettling if the relationship between all these important concepts turned out to be one or three or maybe a multiple of pi. But 137?

The most remarkable thing about this remarkable number is that it is dimension-free. The speed of light is about 300,000 kilometers per second. Abraham Lincoln was 6 feet 6 inches tall. Most numbers come with dimensions. But it turns out that when you combine the quantities that make up alpha, all the units cancel! One hundred thirty-seven comes by itself; it shows up naked all over the place. This means that scientists on Mars, or on the fourteenth planet of the star Sirius, using whatever god-awful units they have for charge, speed, and their version of Planck's constant, will also get 137. It is a pure number.

Physicists have agonized over 137 for the past fifty years. Werner Heisenberg once proclaimed that all the quandaries of quantum mechanics would shrivel up when 137 was finally explained. I tell my undergraduate students that if they are ever in trouble in a major city anywhere in the world they should write "137" on a sign and hold it up at a busy street corner. Eventually a physicist will see that they're distressed and come to their assistance. (No one to my knowledge has ever tried this, but it should work.)

One of the wonderful (but unverified) stories in physics emphasizes the importance of 137 as well as illustrating the arrogance of theorists. According to this tale, a notable Austrian mathematical physicist of Swiss persuasion, Wolfgang Pauli, went to heaven, we are assured, and, because of his eminence in physics, was given an audience with God.

"Pauli, you're allowed one question. What do you want to know?"

Pauli immediately asked the one question that he had labored in vain to answer for the last decade of his life. "Why is alpha equal to one over one hundred thirty-seven?"

God smiled, picked up the chalk, and began writing equations on the blackboard. After a few minutes, She turned to Pauli, who waved his hand. "Das ist falsch!" [That's baloney!]

There's a true story also — a verifiable story — that takes place here on earth. Pauli was in fact obsessed with 137, and spent countless hours pondering its significance. The number plagued him to the very end. When Pauli's assistant visited the theorist in the hospital room in which he was placed prior to his fatal operation, Pauli instructed the assistant to note the number on the door as he left. The room number was 137.

That's where I lived: 137 Eola Road.

LATE NIGHT WITH LEDERMAN

Returning home one weekend night after a late supper in Batavia, I drove through the lab grounds. From several points on Eola Road, one can see the central lab building lit up against the prairie sky. Wilson Hall at 11:30 on a Sunday night is testimony to how strongly physicists feel about solving the remaining mysteries of the universe. Lights were blazing up and down the sixteen floors of the twin towers, each containing its quota of bleary-eyed researchers trying to work out the kinks in our opaque theories about matter and energy. Fortunately, I could drive home and go to bed. As director of the lab, my night-shift obligations were drastically reduced. I was able to sleep on problems rather than work on them. I was grateful that night to lie on a real bed rather than having to bunk down on the accelerator floor waiting for the data to come in. Nevertheless, I tossed and turned, worrying about quarks, Gina, leptons, Sophia . . . Finally, I resorted to counting sheep to get my mind off physics: ". . . 134, 135, 136, *137* . . ."

Suddenly I rose from between the sheets, a sense of urgency driving me from the house. I pulled my bicycle out of the barn, and — still clad in pajamas, my medals falling from my lapels as I pedaled — I rode in painfully slow motion toward the collider detector facility. It was frustrating. I knew I had some very important business to attend to, but I just couldn't get the bike to move any faster. Then I remembered what a psychologist had told me recently: that there is a kind of dream, called a lucid dream, in which the dreamer knows he is in a dream. Once you know this, said the psychologist, you can do anything you want inside the dream. The first step is to find some clue that you're dreaming and are not in real life. That was easy. I knew damn well this was a dream because of the italics. I hate italics. Too hard to read. I took control of my dream. "No more italics!" I screamed.

There. That's better. I put the bike into high gear and pedaled at light speed (hey, you can do anything in a dream) toward the CDF. Oops, too fast: I had circled the earth eight times and ended up back home. I geared down and pedaled at a gentle 120 miles per hour to the facility. Even at three in the morning the parking lot was fairly full; at accelerator labs the protons don't stop at nightfall.

Whistling a ghostly little tune, I entered the detector facility. The CDF is an industrial hangar–like building, with everything painted bright orange and blue. The various offices, computer rooms, and control rooms are all along one wall; the rest of the building is open space, designed to accommodate the detector, a three-story-tall, 5,000-ton instrument. It took some two hundred physicists and an equal number of engineers more than eight years to assemble this particular 10-million-pound Swiss watch. The detector is multicolored, radial in design, its components extending out symmetrically from a small hole in the center. The detector is the crown jewel of the lab. Without it, we cannot "see" what goes on in the accelerator tube, which passes through the center of the detector's core. What goes on, dead center in the detector, are the head-on collisions of protons and antiprotons. The radial spokes of the detector elements roughly match the radial spray of hundreds of particles produced in the collision.

The detector moves on rails that allow the enormous device to be moved out of the accelerator tunnel to the assembly floor for periodic maintenance. We usually schedule maintenance for the summer months, when electric rates are highest (when your electric bill runs more than $10 million a year, you do what you can to cut costs). On this night the detector was on-line. It had been moved back into the tunnel, and the passageway to the maintenance room had been plugged with a 10-foot-thick steel door that blocks the radiation. The accelerator is so designed that the protons and antiprotons collide (mostly) in the section of pipe that runs through the detector — the "collision region." The job of the detector, obviously, is to detect and catalogue the products of the head-on collisions between protons and p-bars (antiprotons).

Still in my pajamas, I made my way up to the second-floor control room, where the findings of the detector are continuously monitored. The room was quiet, as one would expect at this hour. No welders or other workmen roamed the facility making repairs or performing other maintenance tasks, as is common during the day shift. As usual, the lights in the control room were dim, to better see and read the distinctive bluish glow of dozens of computer monitors. The computers in the CDF control room are Macintoshes, just like the microcomputers you might buy to keep track of your finances or to play Cosmic Ozmo. They are fed information from a humongous "home-built" computer that works in tandem with the detector to sort through the debris created by the collisions between protons and antiprotons. The home-built thing is actually a sophisticated data acquisition system, or DAQ, designed by some of the brightest scientists in the fifteen or so universities around the world that collaborated to build the CDF monster. The DAQ is programmed to decide which of the hundreds of thousands of collisions each second are interesting or important enough to analyze and record on magnetic tape. The Macintoshes monitor the great variety of subsystems that collect data.

I surveyed the room, scanning the numerous empty coffee cups and the small band of young physicists, simultaneously hyper and exhausted, the result of too much caffeine and too many hours on shift. At this hour you find graduate students and young postdocs (new Ph.D.'s), who don't have enough seniority to draw decent shifts. Notable was the number of young women, a rare commodity in most physics labs. CDF's aggressive recruiting has paid off to the pleasure and profit of the group.

Over in the corner sat a man who didn't quite fit in. He was thin with a scruffy beard. He didn't look that different from the other researchers, but somehow I knew he wasn't a member of the staff. Maybe it was the toga. He sat staring into the Macintosh, giggling nervously. Imagine, laughing in the CDF control room! At one of the greatest experiments science has ever devised! I thought I'd better put my foot down.

LEDERMAN: Excuse me. Are you the new mathematician they were supposed to send over from the University of Chicago?

GUY IN TOGA: Right profession, wrong town. Name's Democritus. I hail from Abdera, not Chicago. They call me the Laughing Philosopher.

LEDERMAN: Abdera?

DEMOCRITUS: Town in Thrace, on the Greek mainland.

LEDERMAN: I don't remember requisitioning anyone from Thrace. We don't need a Laughing Philosopher. At Fermilab I tell all the jokes.

DEMOCRITUS: Yes, I've heard of the Laughing Director. Don't worry about it. I doubt if I'll be here long. Not given what I've seen so far.

LEDERMAN: So why are you taking up space in the control room?

DEMOCRITUS: I'm looking for something. Something very small.

LEDERMAN: You've come to the right place. Small is our specialty.

DEMOCRITUS: So I'm told. I've been looking for this thing for twenty-four hundred years.

LEDERMAN: Oh, you're *that* Democritus.

DEMOCRITUS: You know another one?

LEDERMAN: I get it. You're like the angel Clarence in *It's a Wonderful Life,* sent here to talk me out of suicide. Actually, I *was* thinking about slicing my wrists. We can't find the top quark.

DEMOCRITUS: Suicide! You remind me of Socrates. No, I'm no angel. That immortality concept came after my time, popularized by that softhead Plato.

LEDERMAN: But if you're not immortal, how can you be here? You died over two millennia ago.

DEMOCRITUS: There are more things in heaven and earth, Horatio, than are dreamt of in your philosophy.

LEDERMAN: Sounds familiar.

DEMOCRITUS: Borrowed it from a guy I met in the sixteenth century. But to answer your question, I'm doing what you call time traveling.

LEDERMAN: Time traveling? You figured out time travel in fifth-century-B.C. Greece?

DEMOCRITUS: Time is a piece of cake. It goes forward, it goes backward. You ride it in and out, like your California surfers. It's matter that's hard to figure. Why, we even sent some of our graduate students to your era. One, Stephenius Hawking, made quite a stir, I've heard. He specialized in "time." We taught him everything he knows.

LEDERMAN: Why didn't you publish this discovery?

DEMOCRITUS: Publish? I wrote sixty-seven books and would have sold a bunch, but the publisher just refused to advertise. Most of what you know about me you know through Aristotle's writings. But let me fill you in a little. I traveled — boy, did I travel! I covered more territory than any man in my time, making the most extensive investigations, and saw more climes and countries, and listened to more famous men . . .

LEDERMAN: But Plato hated your guts. Is it true he disliked your ideas so much that he wanted all your books burned?

DEMOCRITUS: Yes, and that superstitious old goat nearly succeeded. And then that fire in Alexandria really cooked my reputation. That's why you so-called moderns are so ignorant of time manipulation. Now all I hear about is Newton, Einstein . . .

LEDERMAN: So why this visit to Batavia in the 1990s?

DEMOCRITUS: Just checking up on one of my ideas, an idea that was unfortunately abandoned by my countrymen.

LEDERMAN: I bet you're speaking of the atom, the *atomos*.

DEMOCRITUS: Yes, the a-tom, the ultimate, indivisible, and invisible particle. The building block of all matter. I've been jumping ahead through time, to see how far man has come with refining my theory.

LEDERMAN: And your theory was . . .

DEMOCRITUS: You're baiting me, young man! You know very well what I believed. Don't forget, I've been time-hopping century by century, decade by decade. I'm well aware that the nineteenth-century chemists and the twentieth-century physicists have been playing around with my ideas. Don't get me wrong — you were right to do so. If only Plato had been as wise.

LEDERMAN: I just wanted to hear it in your own words. We know of your work primarily through the writings of others.

DEMOCRITUS: Very well. Here we go for the umpteenth time. If I sound bored, it's because I recently went through this with that fellow Oppenheimer. Just don't interrupt me with tedious musings about the parallels between physics and Hinduism.

LEDERMAN: Would you like to hear my theory about the role of Chinese food in mirror-symmetry violation? It's as valid as saying the world is made of air, earth, fire, and water.

DEMOCRITUS: Why don't you just keep quiet and let me start from the beginning. Here, take a seat next to this Macintosh thing and pay attention. Now, if you're going to understand my work, and the work of all of us atomists, we have to go back twenty-six hundred years. We have to start about two hundred years before I was born, with Thales, who flourished around 600 B.C. in Miletus, a hick town in Ionia, which you now call Turkey.

LEDERMAN: Thales was a philosopher, too?

DEMOCRITUS: And how! He was the *first* Greek philosopher. But philosophers in pre-Socratic Greece really knew a lot of things. Thales was an accomplished mathematician and astronomer. He sharpened his training in Egypt and Mesopotamia. Did you know he predicted an eclipse of the sun that occurred at the close of the war between the Lydians and Medes? He constructed one of the first almanacs — I understand you leave this task to farmers today — and he taught our sailors how to steer a ship at night by using the Little Bear constellation. He was also a political adviser, a shrewd businessman, and a fine engineer. Early Greek philosophers were respected not only for the aesthetic workings of their minds but also for their practical arts, or applied science, as you would put it. Is it any different today with physicists?

LEDERMAN: We have been known to do something useful now and then. But I'm sorry to say that our achievements are usually very narrowly focused, and very few of us know Greek.

DEMOCRITUS: Lucky for you I speak English then, yes? Anyhow, Thales, like me, kept asking himself a primary question: "What is the world made of, and how does it work?" Around us we see apparent chaos. Flowers bloom, then die. Floods destroy the land. Lakes become deserts. Meteors fall out of the sky. Whirlwinds appear apparently out of nowhere. From time to time a mountain explodes. Men grow old and turn to dust. Is there something permanent, an underlying identity, that persists through this constant change? Can all of this be reduced to rules so simple that our small minds can understand?

LEDERMAN: Did Thales come up with an answer?

DEMOCRITUS: Water. Thales said water was the primary and ultimate element.

LEDERMAN: How did he figure?

DEMOCRITUS: It's not such a crazy idea. I'm not totally sure what Thales was thinking. But consider: water is essential to growth, at least among plants. Seeds have a moist nature. Almost anything gives off water when heated. And water is the only substance known that can exist as solid, liquid, or gas — as water vapor or steam. Maybe he figured water could be transformed into earth if this process were carried further. I don't know. But Thales made a very great beginning for what you call science.

LEDERMAN: Not bad for a first try.

DEMOCRITUS: The impression around the Aegean is that Thales and his group were given a bad rap by the historians, especially Aristotle. Aristotle was obsessed by forces, by causation. You can hardly talk to him about anything else, and he picked on Thales and his friends in Miletus. Why water? And what force causes the change from rigid water to aethereal water? Why so many different forms of water?

LEDERMAN: In modern physics, er, in the physics of these times, forces are required in addition to —

DEMOCRITUS: Thales and his crowd may well have enmeshed the notion of cause into the very nature of his water-based matter. Force and matter unified! Let's save that for later. Then you can tell me about things you call gluons and supersymmetry and —

LEDERMAN [*frantically scratching his goose bumps*]: Uh, what else did this genius do?

DEMOCRITUS: He had some conventionally mystical ideas. He believed the earth floated on water. He believed that magnets have souls because they can move iron. But he believed in simplicity, that there is a unity to the universe, even though there are many varied material "things" around us. Thales combined a set of rational arguments with whatever mythological hangovers he had in order to give water a special role.

LEDERMAN: I suppose Thales believed the world was being carried by Atlas standing on a turtle.

DEMOCRITUS: Au contraire. Thales and his pals had this very important meeting, probably in the back room of a restaurant in downtown Miletus. After a certain quantity of Egyptian wine, they threw out Atlas and made a solemn agreement: "From this day forth, explanations and theories of how the world works will be based strictly upon logical arguments. No more superstition. No more appeals to Athena, Zeus, Hercules, Ra, Buddha, Lao-tzu. Let's see if we can find out for ourselves." This may have been the most important agreement ever made by humans. It was 650 B.C., probably a Thursday night, and it was the birth of science.

LEDERMAN: Do you think we've gotten rid of superstition now? Have you met our creationists? Our animal rights extremists?

DEMOCRITUS: Here at Fermilab?

LEDERMAN: No, but not far away. But tell me, when did this earth, air, fire, and water idea come in?

DEMOCRITUS: Hold your horses. There were a couple of other guys before we get to that theory. Anaximander, for one. He was a young associate of Thales' in Miletus. Anaximander also earned his spurs doing practical things, such as constructing a map of the Black Sea for Milesian sailors. Like Thales, he sought a primary building block of matter, but he decided it couldn't be water.

LEDERMAN: Another great advance in Greek thinking, no doubt. What was *his* candidate, baklava?

DEMOCRITUS: Have your laugh. We'll get to *your* theories soon enough. Anaximander was another practical genius and, like his mentor Thales, he used his spare time to join in the philosophical debate. Anaximander's logic was fairly subtle. He saw the world as being composed of warring opposites — hot and cold, wet and dry. Water puts out fire; the sun dries up water, et cetera. Therefore the primary substance of the universe cannot be water or fire or anything characterized by one of these opposites. No symmetry there. And you know how we Greeks loved symmetry. For example, if all matter was originally water, as Thales said, then heat or fire could never come into being, since water does not generate fire but obliterates it.

LEDERMAN: Then what *did* he propose as the primary substance?

DEMOCRITUS: He called it the *apeiron*, meaning "without boundaries." This first state of matter was an undifferentiated mass of enormous, possibly infinite, proportions. It was the primitive "stuff," neutral between opposites. This idea had a deep influence on my own thinking.

LEDERMAN: So this apeiron was something like your a-tom — except that it was an infinite substance as opposed to an infinitesimal particle? Didn't this just confuse things?

DEMOCRITUS: No, Anaximander was on to something. The apeiron was infinite, both in space and time, but it was also structureless; it had no component parts. It was nothing but apeiron through and through. And if you're going to decide on a primary substance, it had better have this quality. In fact, my point is to embarrass you by noting that after two thousand years, you are finally coming around to appreciating the prescience of my crowd. What Anaximander did was to invent the vacuum. I think your P. A. M. Dirac finally began to give the vacuum the properties it deserved in the 1920s. Anaxi's apeiron was the prototype of my own "void," a nothingness in which particles move. Isaac Newton and James Clerk Maxwell called it aether.

LEDERMAN: But what about the stuff, matter?

DEMOCRITUS: Listen to this [*pulls a parchment roll out of his toga, perches a pair of discount MagnaVision reading glasses on his nose*]: Anaximander says, "It is neither water nor any other of the so-called elements, but a different substance which is boundless, from which they come into being all the heavens and the worlds within them. Things perish into those things out of which they have their being . . . opposites are in the one and separated out." Now, I know you twentieth-century types are always talking about matter and antimatter created in the vacuum, also annihilating . . .

LEDERMAN: Sure, but . . .

DEMOCRITUS: When Anaximander says opposites were in the apeiron — call it a vacuum, or call it the aether — and were separated out, isn't that something like what you think?

LEDERMAN: Sort of, but I'm much more interested in what made Anaximander think these things.

DEMOCRITUS: Of course he didn't anticipate antimatter. But in a properly endowed vacuum, he thought that opposites could separate: hot and cold, wet and dry, sweet and sour. Today you add positive and negative, north and south. When they combine, they cancel their properties into the neutral apeiron. Isn't that neat?

LEDERMAN: How about democrat and republican? Was there a Greek named Republicas?

DEMOCRITUS: Very amusing. At least Anaximander attempted to explain the mechanism that creates diversity out of a primary element. And his theory led to a number of sub-beliefs, some of which you might even agree with. Anaximander believed, for example, that man evolved from lower animals, which in turn were descended from creatures in the sea. His greatest cosmological idea was to get rid of not only Atlas but even Thales' ocean that held up the earth. He knew you didn't need to hold up the earth. Picture the thing (not yet given spherical shape) suspended in infinite space. There is no place to go. Totally in accord with Newton's laws if, as these Greeks thought, there was nothing else. Anaximander also figured there had to be more than one world, or universe. In fact, he said there were an unlimited number of universes, all perishable, following one another in succession.

LEDERMAN: Like alternate universes on "Star Trek"?

DEMOCRITUS: Hold your commercials. The idea of innumerable worlds became very important to us atomists.

LEDERMAN: Wait a minute. I'm remembering something you wrote that gave me shivers in light of modern cosmology. I even memorized it. Let's see: "There are innumerable worlds of different sizes. In some there is neither sun nor moon, in others they are larger than in ours, and other worlds have more than one sun and more than one moon."

DEMOCRITUS: Yes, we Greeks held some ideas in common with your Captain Kirk. But we dressed a lot better. I'd rather compare my idea to the bubble universes that your inflationary cosmologists are publishing papers on these days.

LEDERMAN: That's really why I got spooked. Didn't one of your predecessors believe that air was the ultimate element?

DEMOCRITUS: You're thinking of Anaximenes, a younger associate of Anaximander's and the last of the Thales gang. He actually took a step backward from Anaximander and said there was a common primordial element, as Thales did — except Anaximenes said this element was air, not water.

LEDERMAN: He should have listened to his mentor; then he would have ruled out anything as mundane as air.

DEMOCRITUS: Yes, but Anaximenes did come up with a clever mechanism for explaining how various forms of matter are transformed from this primary substance. I understand from my readings that you're one of those experimentalists.

LEDERMAN: Yeah. You got a problem with that?

DEMOCRITUS: I've noticed your sarcasm toward so much of Greek theory. I suspect your prejudice comes from the fact that many of these ideas, while plausibly suggested by the world around us, do not lend themselves to incisive experimental verification.

LEDERMAN: True. Experimenters dearly love ideas that can be verified. It's how we make a living.

DEMOCRITUS: Then you may have more respect for Anaximenes, since his beliefs were based on observation. He theorized that the various elements of matter were separated out of air via condensation and rarefaction. Air can be reduced to moisture and vice versa. Heat and cold transform air into different substances. To demonstrate how heat is connected to rarefaction and cold to condensation, Anaximenes advised people to conduct this experiment: breathe out with your lips nearly closed, and the air will emerge cold. But if you open your mouth wide, your breath will be warmer.

LEDERMAN: Congress would love Anaximenes. His experiments are cheaper than mine. And all that hot air . . .

DEMOCRITUS: I get it, but I wanted to dispel your idea that we ancient Greeks never did any experiments. The main problem with thinkers such as Thales and Anaximenes was their belief that substances can be transformed: water can become earth; air can become fire. Can't happen. This snag in our early philosophy wasn't really addressed until two of my contemporaries came along—Parmenides and Empedocles.

LEDERMAN: Empedocles is the earth, air, et cetera guy, right? Remind me about Parmenides.

DEMOCRITUS: He is often called the father of idealism, since much of his thought was picked up by that idiot Plato, but in fact he was a hard core materialist. He talked a lot about Being, but this Being was material. Essentially, Parmenides held that Being can neither come to be nor pass away. Matter doesn't just pop in and out of existence. It's there and we can't destroy it.

LEDERMAN: Let's go down to the accelerator and I'll show you how wrong he is. We pop matter in and out of existence all the time.

DEMOCRITUS: Okay, okay. But this is an important concept. Parmenides was embracing an idea that was dear to us Greeks: oneness. Wholeness. What exists, exists. It is complete and enduring. I suspect you and your colleagues also embrace unity.

LEDERMAN: Yes, it's an enduring and endearing concept. We strive for unity in our beliefs whenever we can. Grand Unification is one of our current obsessions.

DEMOCRITUS: And, in fact, you don't just pop new matter into existence by will alone. I believe you have to add energy to the process.

LEDERMAN: True, and I have the electric bill to prove it.

DEMOCRITUS: So, in a way, Parmenides wasn't that far off. If you include both matter and energy in what he calls Being, then he's right. It can neither come to be nor pass away, at least not in a total sort of way. And yet our senses tell another story. We see trees burn to the ground. The fire can then be destroyed by water. The hot air of summer can evaporate the water. Flowers appear, then die. It was Empedocles who saw a way around this apparent contradiction. He agreed with Parmenides that matter must be conserved, that it cannot appear or disappear willy-nilly. But he disagreed with Thales and Anaximenes that one kind of matter can become another. How, then, does one account for the constant change one sees around us? There are only four kinds of matter, said Empedocles. His famous earth, air, fire, and water. They do not change into other types of matter, but are unchangeable and ultimate *particles*, which form the concrete objects of the world.

LEDERMAN: Now you're talking.

DEMOCRITUS: Thought you'd like that. Objects come into being through the mingling of these elements, and they cease to be through the separation of elements. But the elements themselves — earth, air, fire, water — neither come into being nor pass away but remain unchanged. Obviously I disagree with him as to the identity of these particles, but in principle he made an important intellectual leap. There are only a few basic ingredients in the world, and you construct objects by mixing them together in a multitude of ways. For example, Empedocles said that bone is composed of two parts earth, two parts water, and four parts fire. How he came up with this recipe escapes me at the moment.

LEDERMAN: We tried the air-earth-fire-water mixture and all we got was hot, bubbling mud.

DEMOCRITUS: Leave it to a "modern" to bring the discussion down a notch.

LEDERMAN: What about forces? None of you Greeks seem to realize you need forces as well as particles.

DEMOCRITUS: I have my doubts, but Empedocles would agree. He saw that you needed forces to fuse these elements into other objects. He came up with two: love and strife — love to draw things together, strife to separate them. Not very scientific, perhaps, but don't the scientists in your age have a similar system of beliefs for the universe? A number of particles and a set of forces? Often given whimsical names?

LEDERMAN: In a way, yes. We have what we call the "standard

model." It holds that everything we know about the universe can be explained by the interactions of a dozen particles and four forces.

DEMOCRITUS: There you go. Empedocles' world view doesn't sound all that different, does it? He said the universe could be explained with four particles and two forces. You've just added a couple more, but the structure of both models is similar, no?

LEDERMAN: Sure, but we don't go along with the content: fire, earth, strife . . .

DEMOCRITUS: Well, I suppose you have to show something for two thousand years of hard work. But, no, I don't hold with the content of Empedocles' theory either.

LEDERMAN: Then what do you believe in?

DEMOCRITUS: Ah, now we get down to business. The work of Parmenides and Empedocles set the stage for my own work. I believe in the a-tom, or atom, that which cannot be cut. The atom is the building block of the universe. All of matter is composed of various arrangements of atoms. It is the smallest thing in the universe.

LEDERMAN: You had the instruments necessary to find invisible objects in fifth-century-B.C. Greece?

DEMOCRITUS: Not exactly "find."

LEDERMAN: Then what?

DEMOCRITUS: Perhaps "discover" is a better word. I discovered the atom through Pure Reason.

LEDERMAN: What you're saying is that you just thought about it. You didn't bother to do any experiments.

DEMOCRITUS [*gesturing to indicate the far reaches of the laboratory*]: There are some experiments that the mind can do better than even the largest, most precise instrument.

LEDERMAN: What gave you the idea of atoms? It was, I must admit, a brilliant hypothesis. But it goes way beyond what went before.

DEMOCRITUS: Bread.

LEDERMAN: Bread? Someone paid you to come up with the idea?

DEMOCRITUS: Not that kind of bread. This was in the era before federal grants. I mean real bread. One day, during a prolonged fast, someone walked into my study carrying a loaf of bread just out of the oven. I knew it was bread before I saw it. I thought: some invisible essence of bread traveled ahead and reached my Grecian nose. I made a note about odors and thought about other "traveling essences." A small pool of water shrinks and eventually dries up. Why? How? Can invisible essences of water leap out of the pool and travel long distances like my warm bread? Lots of little things like that — you see, you think, you talk about it. My friend Leucippus and I argued for days and days, sometimes until the sun rose and our wives came after us with clubs. We finally decided that if each substance was made of atoms, invisible because they were too small for our human eyes, we would have too many different types: water atoms, iron atoms, daisy petal atoms, bee foreleg atoms — a system so ugly as to be un-Greek.

Then we got a better idea. Have only a few different styles of atoms, like smooth, rough, round, angular, and have a selected number of different shapes, but have an infinite supply of each kind. Then put them in empty space. (Boy, you should have seen all the beer we drank to understand empty space! How do you define "nothing at all"?) Let these atoms move about at random. Let them move incessantly, occasionally colliding, sometimes sticking and collecting together. Then one collection of atoms makes wine, another makes the glass in which it is served, ditto feta cheese, baklava, and olives.

LEDERMAN: Didn't Aristotle argue that these atoms should naturally fall?

DEMOCRITUS: That's his problem. Ever watch motes of dust dancing in a beam of sunlight that enters a darkened room? The dust moves in any and all directions, just like atoms.

LEDERMAN: How did you imagine the *indivisibility* of atoms?

DEMOCRITUS: It took place in the mind. Imagine a knife of polished bronze. We ask our servant to spend his entire day honing the edge until it can sever a blade of grass held at its distant end. Finally satisfied, I begin to act. I take a piece of cheese . . .

LEDERMAN: Feta?

DEMOCRITUS: Of course. Then I cut the cheese in two with the knife. Then again and again, until I have a speck of cheese too small to hold. Now I think that if I myself were much smaller, the speck would appear large to me, and I could hold it, and with my knife honed even sharper, cut it again and again. Now I must again, in my mind, reduce myself to the size of a pimple on an ant's nose. I continue cutting the cheese. If I repeat the process enough, do you know what the result will be?

LEDERMAN: Sure, a feta-compli.

DEMOCRITUS [*groans*]: Even the Laughing Philosopher chokes on a lousy pun. If I may continue . . . Eventually I will come to a piece of stuff so hard that it can never be cut, even given enough servants to sharpen the knife for a hundred years. I believe the smallest object cannot be cut as a matter of necessity. It is unthinkable that we can continue to cut forever, as some so-called learned philosophers say. Now I have the ultimate uncuttable object, the atomos.

LEDERMAN: And you came up with this idea in fifth-century-B.C. Greece?

DEMOCRITUS: Yes, why? Your ideas today are so much different?

LEDERMAN: Well, actually, they're pretty much the same. It's just that we hate the fact that you published first.

DEMOCRITUS: However, what you scientists call the atom is not what I had in mind.

LEDERMAN: Oh, that's the fault of some nineteenth-century chemists. No, nobody today believes the atoms on the periodic table of the elements — hydrogen, oxygen, carbon, et cetera — are indivisible objects. Those guys jumped the gun. They thought they had found your atoms. But they were still many cuts away from the ultimate cheese.

DEMOCRITUS: And today you have found it?

LEDERMAN: Found *them*. There's more than one.

DEMOCRITUS: Well, of course. Leucippus and I believed there were many.

LEDERMAN: I thought Leucippus didn't really exist.

DEMOCRITUS: Tell that to Mrs. Leucippus. Oh, I know some scholars think he was a fictitious figure. But he was as real as this Macintosh thing [*thumps top of computer*], whatever it is. Leucippus was from Miletus, like Thales and the others. And we worked out our atomic theory together, so it's hard to remember who came up with what. Just because he was a few years older, people say he was my teacher.

LEDERMAN: But it was *you* who insisted there were many atoms.

DEMOCRITUS: Yes, that I remember. There are an infinite number of indivisible units. They differ in size and shape, but beyond that they have no real quality other than solidity, impenetrability.

LEDERMAN: They have shape but are otherwise structureless.

DEMOCRITUS: Yes, that's a good way of putting it.

LEDERMAN: So, in your standard model, as it were, how did you relate the qualities of atoms to the stuff they made?

DEMOCRITUS: Well, it's not quite so specific. We figured out that sweet things, for example, are made of smooth atoms, while bitter things are made of sharp atoms. We know that because they hurt the tongue. Liquids are made up of round atoms, while metal atoms have little locks to hold them together. That's why metals are so hard. Fire is composed of small, spherical atoms, as is the soul of man. As Parmenides and Empedocles theorized, nothing real can be born or destroyed. The objects we see around us change constantly, but that's because they are made of atoms, which can assemble and disassemble.

LEDERMAN: How does this assembling and disassembling happen?

DEMOCRITUS: The atoms are in constant motion. Sometimes they combine when they happen to have shapes that are capable of interlocking. And this creates objects large enough to see: trees, water, dolmades. This constant motion can also lead to atoms detaching themselves and to the apparent change in matter we see around us.

LEDERMAN: But new matter, in terms of atoms, is neither created nor destroyed?

DEMOCRITUS: No. That is an illusion.

LEDERMAN: If all substance is created of these essentially featureless atoms, why are objects so different? Why are rocks hard, for instance, and sheep soft?

DEMOCRITUS: Easy. Hard things have less empty space in them. The atoms are packed tighter. Soft things have more space.

LEDERMAN: So you Greeks accepted the concept of space. The void.

DEMOCRITUS: Sure. My partner Leucippus and I invented the atom. Then we needed someplace to put it. Leucippus got himself all tied up in knots (and a little drunk) trying to define the empty space in which we could put our atoms. If it is empty, it is nothing, and how can you define nothing? Parmenides had ironclad proof that empty space cannot exist. We finally decided his proof didn't exist. [*Chuckle.*] Heck of a problem. Took a lot of retsina. During the time of air-earth-fire-water, the void was considered the fifth essence — quintessential is your word. It gave us quite a problem. You moderns accept nothingness unflinchingly?

LEDERMAN: One has to. Nothing works without, well, nothing. But even today it's a difficult and complex concept. However, as you reminded us, our "nothing," the vacuum, is constantly filling up with theoretical concepts: aether, radiation, a negative energy sea, Higgs. Like attic storage space. I don't know what we'd do without it.

DEMOCRITUS: You can imagine how difficult it was in 420 B.C. to explain the void. Parmenides had denied the reality of empty space. Leucippus was the first to say there could be no motion without a void, therefore a void had to exist. But Empedocles had a clever retort that fooled people for a time. He said that motion could take place without empty space. Look at a fish swimming through the ocean, he said. The water parts for the fish's head, then instantaneously moves into the space left by the moving fish at the tail. The two, fish and water, are always in contact. Forget about empty space.

LEDERMAN: And people bought this argument?

DEMOCRITUS: Empedocles was a bright man, and he had effectively demolished void arguments before. The Pythagoreans, for example — contemporaries of Empedocles — accepted the void for the obvious reason that units had to be kept apart.

LEDERMAN: Weren't they the philosophers who refused to eat beans?

DEMOCRITUS: Yes, and that's not such a bad idea in any era. They had some other trivial beliefs, like you shouldn't sit on a bushel or stand on your own toenail clippings. But they also did some interesting things with math and geometry, as you well know. On this void business, though, Empedocles had them because they said the void is filled with air. Empedocles destroyed this argument simply by showing that air was corporeal.

LEDERMAN: So how did you come to accept the void? You had respect for the thinking of Empedocles, no?

DEMOCRITUS: Indeed, and this point defeated me for a long time. I have trouble with emptiness. How do I describe it? If it is truly nothing, then how can it exist? My hands touch your desk here. On the way to the desk top, my palm feels the gentle rush of air that fills the void between me and the desk's surface. Yet air cannot be the void itself, as Empedocles so ably pointed out. How can I imagine my atoms if I cannot feel the void in which they must move? And yet, if I want to somehow account for the world by atoms, I must first define something that seems to be undefinable because it is devoid of properties.

LEDERMAN: So what did you do?

DEMOCRITUS [*laughing*]: I decided not to worry. I a-voided the issue.

LEDERMAN: Oi Vay!

DEMOCRITUS: Σορρψ. [Sorry.] Seriously, I solved the problem with my knife.

LEDERMAN: Your imaginary knife that cuts cheese into atoms?

DEMOCRITUS: No, a real knife, cutting, say, a real apple. The blade must find empty places where it can penetrate.

LEDERMAN: What if the apple is composed of solid atoms, packed together with no space?

DEMOCRITUS: Then it would be impenetrable, because atoms are impenetrable. No, all matter that we can see and feel is cuttable if you have a sharp enough blade. Therefore the void exists. But mostly I said to myself back then, and I believe it still, that one must not forever be stalled by logical impasses. We go on, we continue as if nothingness can be accepted. This will be an important exercise if we are to continue to search for a key to how everything works. We must be prepared to risk falls as we pick our way along the knife edge of logic. I suppose you modern experimentalists would be shocked by this attitude. You need to prove each and every point in order to progress.

LEDERMAN: No, your approach is very modern. We do the same thing. We make assumptions, or we'd never get anywhere. Sometimes we even pay attention to what theorists say. And we have been known to bypass puzzles, leaving them for future physicists to solve.

DEMOCRITUS: You're starting to make some sense.

LEDERMAN: So, to sum up, your universe is quite simple.

DEMOCRITUS: Nothing exists except atoms and empty space; everything else is opinion.

LEDERMAN: If you've figured it all out, why are you here, at the tail end of the twentieth century?

DEMOCRITUS: As I said, I've been time-hopping to see when and if the opinions of man finally coincide with reality. I know that my countrymen rejected the a-tom, the ultimate particle. I understand that people in 1993 not only accept it but believe they have found it.

LEDERMAN: Yes and no. We believe there is an ultimate particle, but not quite the way you said.

DEMOCRITUS: How so?

LEDERMAN: First of all, while you believe in the a-tom as the essential building block, you actually believe there are many kinds of a-toms: liquids have round a-toms; a-toms for metals have locks; smooth a-toms form sugar and other sweet things; sharp a-toms make up lemons, sour stuff. Et cetera.

DEMOCRITUS: And your point is?

LEDERMAN: Too complicated. Our a-tom is much simpler. In your model there would be too large a variety of a-toms. You might as well have one for each type of substance. We hope to find but one single "a-tom."

DEMOCRITUS: I admire such a quest for simplicity, but how could such a model work? How do you get variety from one a-tom, and just what is this a-tom?

LEDERMAN: At this stage we have a small number of a-toms. We call one type of a-tom "quark" and another type "lepton," and we recognize six forms of each type.

DEMOCRITUS: How are they like my a-tom?

LEDERMAN: They are indivisible, solid, structureless. They are invisible. They are . . . small.

DEMOCRITUS: How small?

LEDERMAN: We think the quark is pointlike. It has no dimension, and, unlike your a-tom, it therefore has no shape.

DEMOCRITUS: No dimension? Yet it exists, it is solid?

LEDERMAN: We believe it to be a mathematical point, and then the issue of its solidity is moot. The apparent solidity of matter depends on the details of how quarks combine with one another and with leptons.

DEMOCRITUS: This is hard to think about. But give me time. I do understand your theoretical problem here. I believe I can accept this quark, this substance with no dimension. However, how can you explain the variety of the world around us — trees and geese and Macintoshes — with so few particles?

LEDERMAN: The quarks and leptons combine to make everything else in the universe. And we have six of each. We can make billions of different things with just two quarks and one lepton. For a while we thought that was all one needed. But nature wants more.

DEMOCRITUS: I agree that twelve particles is a lot simpler than my numerous a-toms, but twelve is still a large number.

LEDERMAN: The six kinds of quarks are perhaps different manifestations of the same thing. We say there are six "flavors" of quarks. What this allows us to do is to combine the various quarks to make up all sorts of matter. But one doesn't have to have a separate flavor of quark for each type of object in the universe — one for fire, one for oxygen, one for lead — as is necessary in your model.

DEMOCRITUS: How do these quarks combine?

LEDERMAN: There is a strong force between quarks, a very curious kind of force that behaves very differently from the electrical forces, which are also involved.

DEMOCRITUS: Yes, I know about this electricity business. I had a brief talk with that Faraday fellow back in the nineteenth century.

LEDERMAN: A brilliant scientist.

DEMOCRITUS: Perhaps so, but his math was terrible. He would never have made it in Egypt, where I studied. But I digress. You say a strong force. Are you referring to this gravitational force I've heard about?

LEDERMAN: Gravity? Much too weak. The quarks are actually held together by particles we call gluons.

DEMOCRITUS: Ah, your gluons. Now we're talking about a whole new kind of particle. I thought the quarks were it, that they made matter.

LEDERMAN: They do. But don't forget about forces. There are also particles we call gauge bosons. These bosons have a mission. Their job is to carry information about the force from particle A to particle B and back again to A. Otherwise, how would B know that A is exerting a force on it?

DEMOCRITUS: Wow! Eureka! What a Grecian idea! Thales would love it.

LEDERMAN: The gauge bosons or force carriers or, as we call them, mediators of the force have properties — mass, spin, charge — which in fact determine the behavior of the force. So, for example, the photons, which carry the electromagnetic force, have zero mass, enabling them to travel very fast. This indicates that the force has a very long reach. The strong force, carried by zero-mass gluons, also reaches out to infinity, but the force is so strong that quarks can never get very far from one another. The heavy W and Z particles, which carry what we call the weak force, have a short reach. They work only over very tiny distances. We have a particle for gravity, which we have named the "graviton," even though we have yet to see one or even write down a good theory for one.

DEMOCRITUS: And this is what you call "simpler" than my model?

LEDERMAN: How did you atomists account for the various forces?

DEMOCRITUS: We didn't. Leucippus and I knew that the atoms had to be in constant motion, and we simply accepted this idea. We gave no reason why the world should originally have this restless atomic motion, except perhaps in the Milesian sense that the cause of motion is part of the attribute of the atom. The world is what it is, and one has to accept certain basic characteristics. With all your theories about the four different forces, can you disagree with this idea?

LEDERMAN: Not really. But does this mean that the atomists believed strongly in fate, or chance?

DEMOCRITUS: Everything existing in the universe is the fruit of chance and necessity.

LEDERMAN: Chance and necessity — two opposing concepts.

DEMOCRITUS: Nevertheless, nature obeys them both. It is true that a poppy seed always gives rise to a poppy, never a thistle. That's necessity at work. But the number of poppy seeds formed by the collisions of atoms may well have strong elements of chance.

LEDERMAN: What you're saying is that nature deals us a particular poker hand, which is a matter of chance. But that hand has necessary consequences.

DEMOCRITUS: A vulgar simile, but yes, that's the way it works. This is so alien to you?

LEDERMAN: No, what you've just described is something like one of the fundamental beliefs of modern physics. We call it quantum theory.

DEMOCRITUS: Oh yes, those young Turks in the nineteen-twenties and thirties. I didn't tarry in that era for long. All those fights with that Einstein fellow — never did make much sense to me.

LEDERMAN: You didn't enjoy those wonderful debates between the quantum cabal — Niels Bohr, Werner Heisenberg, Max Born, and their crowd — and such physicists as Erwin Schrödinger and Albert Einstein, who argued against the idea of chance determining nature's way?

DEMOCRITUS: Don't get me wrong. Brilliant men, all of them. But their arguments always concluded with one party or the other bringing up the name of God and Her supposed motivations.

LEDERMAN: Einstein said he couldn't accept that God plays dice with the universe.

DEMOCRITUS: Yes, they always pull the God trump card when the debate goes poorly. Believe me, I had enough of that in ancient Greece. Even my defender Aristotle raked me over the coals for my beliefs in chance and for accepting motion as a given.

LEDERMAN: How did you like quantum theory?

DEMOCRITUS: Definitely I liked it, I think. Later I met Richard Feynman, and he confided that he had never understood quantum theory either. I always had trouble with . . . Wait a minute! You've changed the subject. Let's get back to those "simple" particles you were prattling about. You were explaining how the quarks stick together to make up . . . to make what?

LEDERMAN: Quarks are building blocks of a large class of objects that we call hadrons. This is a Greek word meaning "heavy."

DEMOCRITUS: Really!

LEDERMAN: It's the least we can do. The most famous object made of quarks is the proton. It takes three quarks to make a proton. In fact, it takes three quarks to make the many cousins of the proton, but with six different quarks, there are plenty of combinations of three quarks — I think it's two hundred sixteen. Most of these hadrons have been discovered and given Greek-letter names like lambda (Λ), sigma (Σ), et cetera.

DEMOCRITUS: The proton is one of these hadrons?

LEDERMAN: And the most popular in our present universe. You can stick three quarks together to get a proton or a neutron, for instance. Then you can make an atom by adding an electron, which belongs to the class of particles called leptons, to one proton. That particular atom is called hydrogen. With eight protons and an equal number of neutrons and eight electrons you can build an oxygen atom. The neutrons and protons huddle together in a tiny clump that we call the nucleus. Stick two hydrogen atoms and one oxygen atom together and you get water. A little water, a little carbon, some oxygen, a few nitrogens, and sooner or later you have gnats, horses, and Greeks.

DEMOCRITUS: And it all starts with quarks.

LEDERMAN: Yup.

DEMOCRITUS: And that's all you need.

LEDERMAN: Not exactly. You need something that allows atoms to stay together and then to stick to other atoms.

DEMOCRITUS: The gluons again.

LEDERMAN: No, they only stick quarks together.

DEMOCRITUS: Γοοδ γριεφ! [Good grief!]

LEDERMAN: That's where Faraday and the other electricians, such as Chuck Coulomb, come in. They studied the electrical forces that hold electrons to the nucleus. Atoms attract each other by a complicated dance of nuclei and electrons.

DEMOCRITUS: These electrons, they are also behind electricity?

LEDERMAN: It's one of their main bags.

DEMOCRITUS: So these are gauge bosons, too, like photons and W's and Z's?

LEDERMAN: No, electrons are particles of matter. They belong to the lepton family. Quarks and leptons make up matter. Photons, gluons, W's, Z's, and gravitons make up forces. One of the most intriguing developments today is that the very distinction between force and matter is blurring. It's all particles. A new simplicity.

DEMOCRITUS: I like my system better. My complexity seems simpler than your simplicity. So what are the other five leptons?

LEDERMAN: There are three varieties of neutrinos, plus two leptons called the muon and the tau. But let's not get into that now. The electron is by far the most important lepton in today's global economy.

DEMOCRITUS: So I should worry only about the electron and the six quarks. These explain the birds, the sea, the clouds . . .

LEDERMAN: In truth, almost everything in the universe today is composed of only two of the quarks — the up and the down — and the electron. The neutrino zings around the universe freely and pops out of our radioactive nuclei, but most of the other quarks and leptons must be manufactured in our laboratories.

DEMOCRITUS: Then why do we need them?

LEDERMAN: That's a good question. We believe this: there are twelve basic particles of matter. Six quarks, six leptons. Only a few exist in abundance today. But they were all here on an equal footing during the Big Bang, the birth of the universe.

DEMOCRITUS: And who believes all this, the six quarks and six leptons? A handful of you? A few renegades? All of you?

LEDERMAN: All of us. At least, all the intelligent particle physicists. But this concept is pretty much accepted by all scientists. They trust us on this one.

DEMOCRITUS: So where do we disagree? I said there was an uncuttable atom. But there were many, many of them. And they combined because they had complementary shape characteristics. You say there are only six or twelve such "a-toms." And they do not have shapes, but they combine because they have complementary electrical charges. Your quarks and leptons are also uncuttable. Now, are you sure there are only twelve?

LEDERMAN: Well . . . depends on how you count. There are also six antiquarks and six antileptons and —

DEMOCRITUS: Γρεατ Ζευσэσ υνδερπαντσ! [Great Zeus's underpants!]

LEDERMAN: It's not as bad as it sounds. We agree much more than we disagree. But in spite of what you told me, I am still amazed that such a primitive, ignorant heathen could come up with the atom, which we call the quark. What kind of experiments did you do to verify the idea? Here we spend billions of drachmas to test each concept. How did you work so cheaply?

DEMOCRITUS: We did it the old-fashioned way. Not having a Department of Energy or a National Science Foundation, we had to use Pure Reason.

LEDERMAN: So you spun your theories out of whole cloth.

DEMOCRITUS: No, even we ancient Greeks had clues from which we molded our ideas. As I said, we saw that poppy seeds always grow into poppies. The spring always comes after the winter. The sun rises and sets. Empedocles studied water clocks and whirling buckets. One can form conclusions by keeping one's eyes open.

LEDERMAN: "You can observe a lot just by looking," as one of my contemporaries once said.

DEMOCRITUS: Exactly! Who is this sage, so Grecian in his perspective?

LEDERMAN: Yogi Berra.

DEMOCRITUS: One of your greatest philosophers, no doubt.

LEDERMAN: You could say that. But why do you distrust experiment?

DEMOCRITUS: The mind is better than the senses. It contains *trueborn* knowledge. The second kind of knowledge is bastard knowledge, which comes from the senses — sight, hearing, smell, taste, touch. Think about it. The drink that tastes sweet to you may taste sour to me. A woman who appears beautiful to you is nothing to me. An ugly child appears beautiful to its mother. How can we trust such information?

LEDERMAN: Then you do not think we can measure the object world? Our senses simply manufacture sensory information?

DEMOCRITUS: No, our senses do not create knowledge from the void. Objects shed their atoms. That is how we can see them or smell them — like that loaf of bread I told you about. These atoms/images enter through our organs of sense, which are passages to the soul. But the images are distorted as they pass through the air, which is why objects very far off may not be seen at all. The senses give no reliable information about reality. Everything is subjective.

LEDERMAN: To you there is no objective reality?

DEMOCRITUS: Oh, there's an objective reality. But we are not able to perceive it accurately. When you are sick, foods taste different. Water might seem warm to one hand and not the other. It is all a matter of the temporary arrangement of the atoms in our bodies and their reaction to the equally temporary combination in the object being sensed. The truth must be deeper than the senses.

LEDERMAN: The object being measured and the measuring instrument — in this case, the body — interact with each other and change the nature of the object, thus obscuring the measurement.

DEMOCRITUS: An awkward way of thinking about it, but yes. What are you getting at?

LEDERMAN: Well, instead of thinking of this as bastard knowledge, one could see it as a matter of *uncertainty* of measurement, or sensation.

DEMOCRITUS: I can live with that. Or, to quote Heraclitus, "The senses are bad witnesses."

LEDERMAN: Is the mind any better, even though you call it the source of "trueborn" knowledge? The mind, in your world view, is a property of what you call the soul, which in turn is also composed of atoms. Are not these atoms also in constant motion, and interacting with distorted atoms from the exterior? Can one make an absolute separation between sense and thought?

DEMOCRITUS: You make a good point. As I have said in the past, "Poor Mind, it is from us." From the senses. Still, Pure Reason is less misleading than the senses. I remain skeptical of your experiments. I find these huge buildings with all their wires and machines almost laughable.

LEDERMAN: Perhaps they are. But they stand as monuments to the difficulty of trusting what we can see and touch and hear. Your comments about the subjectivity of measurement were, for us, learned slowly in the sixteenth to eighteenth centuries. Little by little we learned to reduce observation and measurement to objective acts like writing numbers in notebooks. We learned to examine a hypothesis, an idea, a process of nature from many angles, in many laboratories by many scientists, until the best approximations to objective reality emerged — by consensus. We made wonderful instruments to help us observe, but we learned to be skeptical about what they revealed until it was repeated in many places by many techniques. Finally, we subjected the conclusions to the test of time. If some young SOB a hundred years later and juicing for a reputation shakes it up, so be it. We rewarded him with praises and prizes. We learned to suppress our envy and fear and to love the bastard.

DEMOCRITUS: But what about authority? Most of what the world learned about my work came from Aristotle. Talk about authority. People were exiled, imprisoned, and buried if they disagreed with old Aristotle. The atom idea barely made it to the Renaissance.

LEDERMAN: It's much better now. Not perfect, but better. Today we can almost define a good scientist by how skeptical he is of the establishment.

DEMOCRITUS: By Zeus, this is good news. What do you pay mature scientists who don't do windows or experiments?

LEDERMAN: Obviously, you're applying for a job as a theorist. I don't hire many of those, though the hours are good. Theorists never schedule meetings on Wednesday because it kills two weekends. Besides, you're not as anti-experiment as you make yourself out to be. Whether you like the idea or not, you did conduct experiments.

DEMOCRITUS: I did?

LEDERMAN: Sure. Your knife. It was a mind experiment, but an experiment nonetheless. By cutting that piece of cheese in your mind over and over again, you reached your theory of the atom.

DEMOCRITUS: Yes, but that was all in the mind. Pure Reason.

LEDERMAN: What if I could show you that knife?

DEMOCRITUS: What are you talking about?

LEDERMAN: What if I could show you a knife that could cut matter forever, until it finally cut off an a-tom.

DEMOCRITUS: You found a knife that can cut off an atom? In *this* town?

LEDERMAN [*nodding*]: We're sitting on the main nerve right now.

DEMOCRITUS: This laboratory, it is your knife?

LEDERMAN: The particle accelerator. Beneath our feet particles are spiraling through a four-mile-around tube and crashing into each other.

DEMOCRITUS: And this is how you cut away at matter to get down to the a-tom?

LEDERMAN: Quarks and leptons, yes.

DEMOCRITUS: I'm impressed. And you're sure there's nothing smaller?

LEDERMAN: Well, yes; absolutely sure, I think, maybe.

DEMOCRITUS: But not positive. Otherwise you would have stopped cutting.

LEDERMAN: "Cutting" teaches us something about the properties of quarks and leptons even if there aren't little people running around inside them.

DEMOCRITUS: There's one thing I forgot to ask. The quarks — they're all pointlike, dimensionless; they have no real size. So, outside of their electrical charges, how do you tell them apart?

LEDERMAN: They have different masses.

DEMOCRITUS: Some are heavy, some are light?

LEDERMAN: Da.

DEMOCRITUS: I find that puzzling.

LEDERMAN: That they have different masses?

DEMOCRITUS: That they weigh anything at all. *My* atoms have no weight. Doesn't it bother you that your quarks have mass? Can you explain it?

LEDERMAN: Yes, it bothers us a lot, and no, we can't explain it. But that's what our experiments indicate. It's even worse with the gauge bosons. The sensible theories say that their masses should be zero, nothing, zilch! But . . .

DEMOCRITUS: Any ignorant Thracian tinker would find himself in the same predicament. You pick up a rock. It feels heavy. You pick up a tuft of wool. It feels light. It follows from living in this world that atoms — quarks, if you will — have different weights. But again, the senses are bad witnesses. Using Pure Reason, I don't see why matter should have any mass at all. Can you explain it? What gives particles their mass?

LEDERMAN: It's a mystery. We're still struggling with this idea. If you stick around the control room until we are into Chapter 8 of this book, we'll clear it all up. We suspect that mass comes from a field.

DEMOCRITUS: A field?

LEDERMAN: Our theoretical physicists call it the Higgs field. It pervades all of space, the apeiron, cluttering up your void, tugging on matter, making it heavy.

DEMOCRITUS: Higgs? Who is Higgs? Why don't you people name something after me — the democriton! By its sound you *know* it interacts with all other particles.

LEDERMAN: Sorry. Theorists always name things after one another.

DEMOCRITUS: What is this field?

LEDERMAN: The field is represented by a particle we call the Higgs boson.

DEMOCRITUS: A particle! I like this idea already. And you have found this Higgs particle in your accelerators?

LEDERMAN: Well, no.

DEMOCRITUS: So you found it where?

LEDERMAN: We haven't found it yet. It exists only in the collective physicist mind. Kind of like Impure Reason.

DEMOCRITUS: Why do you believe in it?

LEDERMAN: Because it has to exist. The quarks, the leptons, the four known forces — none of these make complete sense unless there is a massive field distorting what we see, skewing our experimental results. By deduction, the Higgs is out there.

DEMOCRITUS: Spoken like a Greek. I like this Higgs field. Well, look, I must go. I've heard that the twenty-first century has a special on sandals. Before I continue on to the future, do you have any ideas about when and where I should go to see some greater progress in the search for my atom?

LEDERMAN: Two times, two different places. First, I suggest you come back here to Batavia in 1995. After that, try Waxahachie, Texas, around, say, 2005.

DEMOCRITUS [*snorting*]: Oh, come on. You physicists are all alike. You think everything's going to be cleared up in a couple of years. I visited Lord Kelvin in 1900 and Murray Gell-Mann in 1972, and they both assured me that physics had ended; everything was completely understood. They said to come back in six months and all the kinks would be worked out.

LEDERMAN: I'm not saying that.

DEMOCRITUS: I hope not. I've been following this road for twenty-four hundred years. It's not so easy.

LEDERMAN: I know. I say to come back in '95 and 2005 because I think you'll find some *interesting* events then.

DEMOCRITUS: Such as?

LEDERMAN: There are six quarks, remember? We've found only five of them, the last one here at Fermilab in 1977. We need to find the sixth and final quark — the heaviest quark; we call it the top quark.

DEMOCRITUS: You'll start looking in 1995?

LEDERMAN: We're looking now, as I speak. The whirling particles beneath our feet are being cut apart and examined meticulously in search of this quark. We haven't found it yet. But by 1995 we will have found it . . . or proved it doesn't exist.

DEMOCRITUS: You can do that?

LEDERMAN: Yes, our machine is that powerful, that precise. If we find it, then everything is in order. We will have further solidified the idea that the six quarks and six leptons are your a-toms.

DEMOCRITUS: And if you don't . . .

LEDERMAN: Then everything crumbles. Our theories, our standard model, will be next to worthless. Theorists will be leaping out of second-story windows. They'll be sawing at their wrists with butter knives.

DEMOCRITUS [*laughing*]: Won't that be fun! You're right. I need to come back to Batavia in 1995.

LEDERMAN: It might spell the end of your theory, too, I might add.

DEMOCRITUS: My ideas have survived a long time, young man. If the a-tom isn't a quark or a lepton, it will turn up as something else. Always has. But tell me. Why 2005? And where is this Waxahachie?

LEDERMAN: In Texas, in the desert, where we're building the largest particle accelerator in history. In fact, it will be the largest scientific tool of any kind built since the great pyramids. (I don't know who designed the pyramids, but my ancestors did all the work!) The Superconducting Super Collider, our new machine, should be in full swing by 2005 — give or take a few years, depending on when Congress approves the funding.

DEMOCRITUS: What will your new accelerator find that this one here cannot?

LEDERMAN: The Higgs boson. It will go after the Higgs field. Try to capture the Higgs particle. We hope it will find out for the first time why things are heavy and why the world looks so complicated when you and I know that, deep down, the world is simple.

DEMOCRITUS: Like a Greek temple.

LEDERMAN: Or a shul in the Bronx.

DEMOCRITUS: I must see this new machine. And this particle. The Higgs boson — not a very poetic name.

LEDERMAN: I call it the God Particle.

DEMOCRITUS: Better. Though I prefer a lowercase "g." But tell me: you're an experimenter. What physical evidence have you amassed so far for this Higgs particle?

LEDERMAN: None. Zero. In fact, outside of Pure Reason, the evidence would convince most sensible physicists that the Higgs does not exist.

DEMOCRITUS: Yet you persist.

LEDERMAN: The negative evidence is only preliminary. Besides, we have an expression in this country . . .

DEMOCRITUS: Yes?

LEDERMAN: "It ain't over till it's over."

DEMOCRITUS: Yogi Berra?

LEDERMAN: Yup.

DEMOCRITUS: A genius.

On the northern rim of the Aegean, in the Greek province of Thrace, the town of Abdera sits at the mouth of the river Nestos. As in many other cities in this part of the world, history is written into the very stones of the hills that overlook the supermarkets, parking lots, and cinemas. Some 2,400 years ago, the town was on the busy land route from the motherland of ancient Greece to the important possessions in Ionia, now the western part of Turkey. Abdera was in fact settled by Ionian refugees fleeing from the armies of Cyrus the Great.

Imagine living in Abdera in the fifth century before Christ. In this land of goatherds, natural events weren't necessarily assigned scientific causes. Lightning strikes were thunderbolts hurled from atop Mount Olympus by an angry Zeus. Whether one enjoyed a calm sea or suffered a tidal wave depended on the mercurial moods of Poseidon. Feasts or famines came at the whim of Ceres, the goddess of agriculture, rather than atmospheric conditions. Imagine, then, the focus and integrity of a mind that could ignore the popular beliefs of the age and come up with concepts harmonious with quark and quantum theory. In ancient Greece, as now, progress was an accident of genius — of individuals with vision and creativity. But even for a genius, Democritus was far ahead of his time.

He is probably best known for two of the most scientifically intuitive quotes ever uttered by an ancient: "Nothing exists except atoms and space; everything else is opinion" and "Everything existing in the universe is the fruit of chance and necessity." Of course, we must credit Democritus's heritage — the colossal achievements of his predecessors in Miletus. These men defined the mission: a single order underlies the chaos of our perceptions; furthermore, we are capable of comprehending that order.

It probably helped Democritus that he traveled. "I covered more territory than any man in my time, making the most extensive investigations, and saw more climes and countries and listened to more famous men." He learned astronomy in Egypt and mathematics in Babylonia. He visited Persia. But the stimulation to his atomistic theory came from Greece, as did his predecessors Thales, Empedocles, and perhaps, of course, Leucippus.

And he published! The Alexandrian catalogue listed more than sixty works: physics, cosmology, astronomy, geography, physiology, medicine, sensation, epistemology, mathematics, magnetism, botany, poetic and musical theory, linguistics, agriculture, painting, and other topics. Almost none of his published work survived intact; we know about Democritus primarily from fragments and the testimony of later Greek historians. Like Newton, he also wrote on magic and alchemical discoveries. What kind of man was this?

Historians refer to him as the Laughing Philosopher, moved to mirth by the follies of mankind. He was probably rich; most of the Greek philosophers were. We know he disapproved of sex. Sex is so pleasurable, Democritus said, that it overwhelms one's consciousness. Maybe that was his secret, and perhaps we should ban sex among our theorists so they can think better. (Experimenters don't need to think and would be exempt from the rule.) Democritus valued friendship but thought ill of women. He didn't want children, because educating them would have interfered with his philosophy. He purported to dislike everything violent and passionate.

It is hard to accept this as true. He was no stranger to violence; his atoms were in constant violent motion. And it took passion to believe what Democritus believed. He remained true to his beliefs, though they brought him no fame. Aristotle respected him, but Plato, as mentioned, wanted all of his books burned. In his hometown Democritus was outshone by another philosopher, Protagoras, the most eminent of the Sophists, a school of philosophers who hired themselves out as teachers of rhetoric to wealthy young men. When Protagoras left Abdera and went to Athens, he was received enthusiastically. Democritus, on the other hand, said, "I went to Athens and no one knew me."

Democritus believed in a lot of other things that we didn't cover in our mythical dream conversation, which was pieced together with a smattering of quotes from Democritus's writings and seasoned with some imagination. I took liberties, but not with Democritus's basic beliefs, though I allowed myself the luxury of changing his mind about the value of experiments. I'm confident there's no way he could resist the appeal of seeing his mythical "knife" come alive in the bowels of Fermilab.

Democritus's work on the void was revolutionary. He knew, for instance, that there is no top, bottom, or middle in space. Although this idea was first suggested by Anaximander, it was still quite an accomplishment for a human born on this planet with its geocentric populace. The concept that there is no up or down is still difficult for most people, in spite of TV scenes from space capsules. One of Democritus's further-out beliefs was that there are innumerable worlds of different sizes. These worlds are at irregular distances, more in one direction and fewer in another. Some are flourishing, others declining. Here they come into being. There they die, destroyed by collisions with one another. Some of the worlds have no animal or vegetable life nor any water. Odd stuff, yet this perception can be related to modern cosmological ideas associated with what is called the "inflationary universe," out of which can spring numerous "bubble universes." This from a laughing philosopher who trekked around the Greek empire more than two millennia ago.

As for his famous quote about everything being "the fruit of chance or necessity," we find the same paradox most dramatically in quantum mechanics, one of the great theories of the twentieth century. Individual collisions of atoms, said Democritus, have necessary consequences. There are strict rules. However, which collisions are more frequent, which atoms preponderate in a particular location — these are elements of chance. Carried to its logical conclusion, this notion means that the creation of an almost ideal earth-sun system is a matter of luck. In the modern quantum-theory resolution of this conundrum, certainty and regularity emerge as events that are averages over a distribution of reactions of varying probability. As the number of random processes contributing to the average increases, one can predict with increasing certainty what will happen. Democritus's notion is compatible with our present belief. One cannot say with certainty what fate will befall a given atom, but one can foretell accurately the consequences of the motions of zillions of atoms colliding randomly in space.

Even his distrust of the senses provides remarkable insight. He points out that our sense organs are made of atoms, which collide with the atoms of the object being sensed, thereby constraining our perceptions. As we shall see in Chapter 5, his way of expressing this problem is resonant with another of the great discoveries of this century, the Heisenberg uncertainty principle. The act of measuring affects the particle being measured. Yes, there is some poetry here.

What is Democritus's place in the history of philosophy? Not very high by conventional standards — certainly not high compared with that of virtual contemporaries such as Socrates, Aristotle, and Plato. Some historians treat his atomic theory as a kind of curious footnote to Greek philosophy. Yet there is at least one potent minority opinion. The British philosopher Bertrand Russell said that philosophy went downhill after Democritus and did not recover until the Renaissance. Democritus and his predecessors were "engaged in a disinterested effort to understand the world," wrote Russell. Their attitude was "imaginative and vigorous and filled with the delight of adventure. They were interested in everything — meteors and eclipses, fishes and whirlwinds, religion and morality; with a penetrating intellect they combined the zest of children." They were not superstitious but genuinely scientific, and they were not greatly influenced by the prejudices of their age.

Of course Russell, like Democritus, was a serious mathematician, and these guys stick together. It's only natural that a mathematician would have a bias toward such rigorous thinkers as Democritus, Leucippus, and Empedocles. Russell pointed out that although Aristotle and others reproached the atomists for not accounting for the original motion of the atoms, Leucippus and Democritus were far more scientific than their critics by not bothering to ascribe purpose to the universe. The atomists knew that causation must start from something, and that no cause can be assigned to this original something. Motion was simply a given. The atomists asked mechanistic questions and gave mechanistic answers. When they asked "Why?" they meant: what was the *cause* of an event? When their successors — Plato, Aristotle, and so on — asked "Why?" they were searching for the *purpose* of an event. Unfortunately, this latter course of inquiry, said

Russell, "usually arrives, before long, at a Creator, or at least an Artificer." This Creator must then be left unaccounted for, unless one wishes to posit a super-Creator, and so on. This kind of thinking, said Russell, led science up a blind alley, where it remained trapped for centuries.

Where do we stand today compared to Greece circa 400 B.C.? Today's experiment-driven "standard model" is not all that dissimilar to Democritus's speculative atomic theory. We can make anything in the past or present universe, from chicken soup to neutron stars, with just twelve particles of matter. Our a-toms come in two families: six quarks and six leptons. The six quarks are named the up, the down, the charm, the strange, the top (or truth), and the bottom (or beauty). The leptons include the familiar electron, the electron neutrino, the muon, the muon neutrino, the tau, and the tau neutrino.

But note that we said "past or present" universe. If we're talking about our present environment only, from the South Side of Chicago to the edge of the universe, we can get by nicely with even fewer particles. For quarks, all we really need are the up and the down, which can be used in different combinations to assemble the nucleus of the atom (the kind in the periodic table). Among the leptons, we can't live without the good old electron, which "orbits" the nucleus, and the neutrino, which is essential in many kinds of reactions. But why do we need the muon and the tau particles? Or the charm, the strange, and the heavier quarks? Yes, we can make them in our accelerators or observe them in cosmic ray collisions. But why are they here? More about these "extra" a-toms later.

LOOKING THROUGH THE KALEIDOSCOPE

The fortunes of atomism went through a lot of ups and downs, fits and starts, before we arrived at our standard model. It started with Thales saying all is water (atom count: 1). Empedocles came up with air-earth-fire-water (count: 4). Democritus had an uncomfortable number of shapes but only one concept (count: ?). Then there was a long historical pause, although atoms remained a philosophical concept discussed as such by Lucretius, Newton, Roger Joseph Boscovich, and many others. Finally atoms were reduced to experimental necessity by John Dalton in 1803. Then, firmly in the hands of chemists, the number of atoms increased — 20, 48, and by the early years of this century, 92. Soon nuclear chemists began making new ones (count: 112 and rising). Lord Rutherford took a giant step back to simplicity when he discovered (circa 1910) that Dalton's atom wasn't indivisible but contained a nucleus plus electrons (count: 2). Oh yes, there was also the photon (count: 3). In 1930, the nucleus was found to house neutrons as well as protons (count: 4). Today, we have 6 quarks, 6 leptons, 12 gauge bosons and, if you want to be mean, you can count the antiparticles and the colors, because quarks come in three shades (count: 60). But who's counting?

History suggests that we may find things, call them "prequarks," thus reducing the total number of basic building blocks. But history isn't always right. The newer concept is that we are looking through a glass, darkly — that the proliferation of "a-toms" in our standard model is a consequence of how we look. A children's toy, the kaleidoscope, shows lovely patterns by using mirrors to add complexity to a simple pattern. A star pattern is seen to be an artifact of a gravitational lens. As now conceived, the Higgs boson — the God Particle — may well provide the mechanism that reveals a simple world of pristine symmetry behind our increasingly complex standard model.

This brings us back to an old philosophical debate. Is this universe real? If so, can we know it? Theorists don't often grapple with this problem. They simply accept objective reality at face value, like Democritus, and go about their calculations. (A smart choice if you're going to get anywhere with a pencil and pad.) But an experimenter, tormented by the frailty of his instruments and his senses, can break out in a cold sweat over the task of measuring this reality, which can be a slippery thing when you lay a ruler down on it. Sometimes the numbers that come out of an experiment are so strange and unexpected that they raise the hairs on a physicist's neck.

Take this problem of mass. The data we have gathered on the masses of the quarks and the W and Z particles are absolutely baffling. The leptons — the electron, muon, and tau — present us with particles that appear identical in every way except for their mass. Is mass real? Or is it an illusion, an artifact of the cosmic environment? One opinion bubbling up in the literature of the 1980s and '90s is that something pervades this empty space and provides atoms with an illusory weight. That "something" will one day manifest itself in our instruments as a particle.

In the meantime, nothing exists except atoms and empty space; everything else is opinion.

I can hear old Democritus giggling.

· Interlude A ·
A TALE OF TWO CITIES

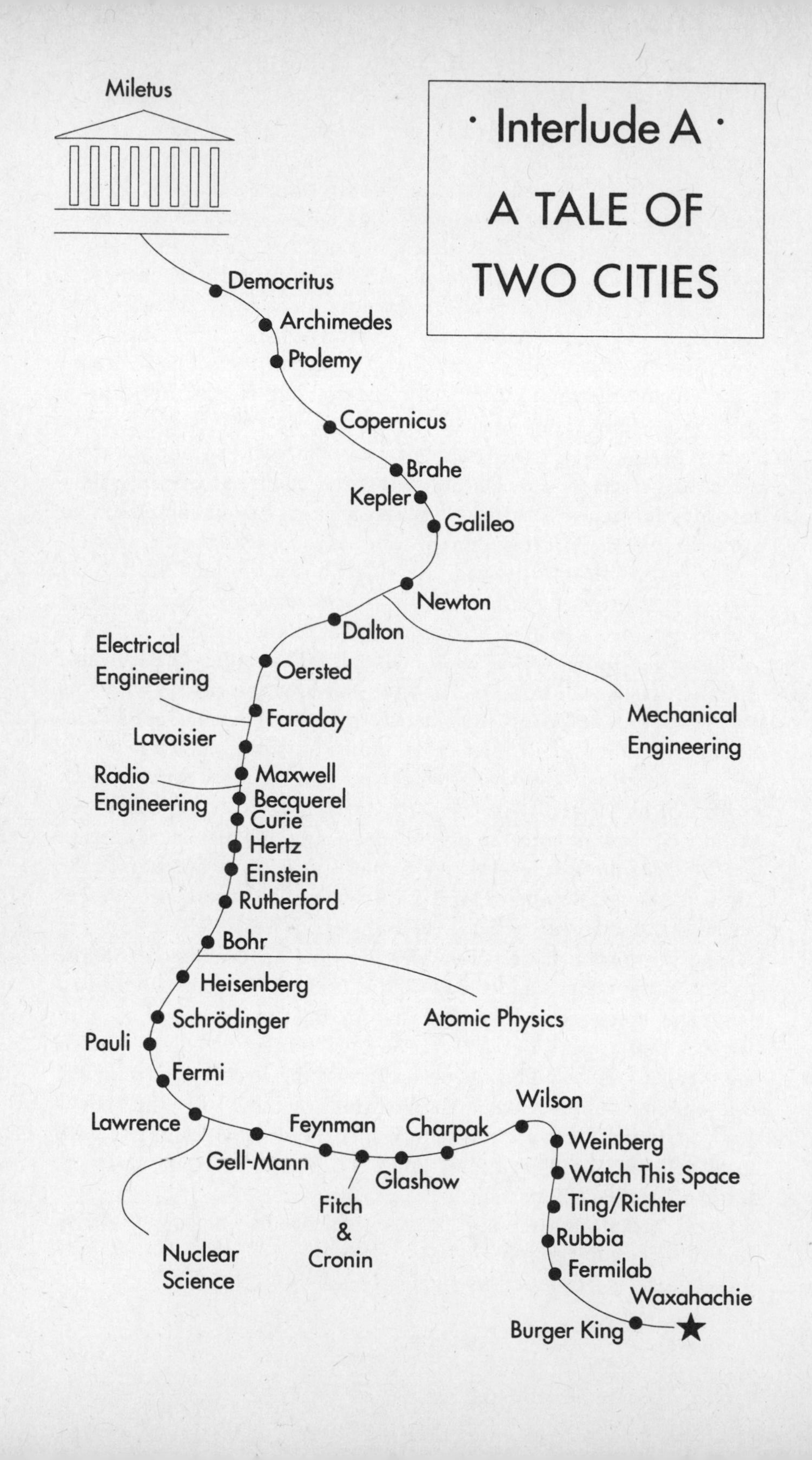

· 3 ·

LOOKING FOR THE ATOM: THE MECHANICS

> To you who are preparing to mark the 350th anniversary of the publication of Galileo Galilei's great work, *Dialoghi sui due massimi sistemi del mondo,* I would like to say that the Church's experience, during the Galileo affair and after it, has led to a more mature attitude and to a more accurate grasp of the authority proper to her. I repeat before you what I stated before the Pontifical Academy of Sciences on 10 November 1979: "I hope that theologians, scholars and historians, animated by a spirit of sincere collaboration, will study the Galileo case more deeply and, in frank recognition of wrongs, from whichever side they come, will dispel the mistrust that still forms an obstacle, in the minds of many, to a fruitful concord between science and faith."
>
> — His Holiness Pope John Paul II, 1986

VINCENZO GALILEI hated mathematicians. This might seem odd, since he was a highly skilled mathematician himself. First and foremost, however, he was a musician, a lutenist of great repute in sixteenth-century Florence. In the 1580s he turned his skills to musical theory and found it lacking. The blame, said Vincenzo, lay with a mathematician who had been dead for two thousand years, Pythagoras.

Pythagoras, a mystic, was born on the Greek island of Samos about a century before Democritus. He spent most of his life in Italy, where he organized the Pythagoreans, a kind of secret society of men who held a religious regard for numbers and whose lives were governed by a set of obsessive taboos. They refused to eat beans or to pick up objects they had dropped. When they awakened in the morning, they took care to smooth out the sheets to eradicate the impressions of their bodies. They believed in reincarnation, refusing to eat or beat dogs in case they might be long-lost friends.

They were obsessed with numbers. They believed that things were numbers. Not just that objects could be enumerated, but that they *were* numbers, such as 1, 2, 7, or 32. Pythagoras thought of numbers as shapes and came up with the idea of squares and cubes of numbers, terms that stay with us today. (He also talked about "oblong" and "triangular" numbers, terms we no longer think about.)

Pythagoras was the first to divine a great truth about right triangles. He pointed out that *the sum of the squares of the sides is equal to the square of the hypotenuse,* a formula that is hammered into every teenage brain that wanders into a geometry classroom from Des Moines to Ulan Bator. This reminds me of the time one of my students was conscripted into the army and, with a group of fellow buck privates, was being lectured about the metric system by his sergeant.

SERGEANT: In the metric system water boils at ninety degrees.
PRIVATE: Begging your pardon, sir, it boils at one hundred degrees.
SERGEANT: Of course. How stupid of me. It's a right angle that boils at ninety degrees.

The Pythagoreans loved to study ratios, proportions. They came up with the "golden rectangle," the perfect shape, whose proportions are evident in the Parthenon and many other Greek structures and found in Renaissance paintings.

Pythagoras was the first cosmic guy. It was he (and not Carl Sagan) who coined the word *kosmos* to refer to everything in our universe, from human beings to the earth to the whirling stars overhead. *Kosmos* is an untranslatable Greek word that denotes the qualities of order and beauty. The universe is a *kosmos,* he said, an ordered whole, and each of us humans is also a *kosmos* (some more than others).

If Pythagoras were alive today, he would live in the Malibu hills or perhaps Marin County. He'd hang out at health-food restaurants accompanied by an avid following of bean-hating young women with names like Sundance Acacia or Princess Gaia. Or maybe he'd be an adjunct professor of mathematics at the University of California at Santa Cruz.

But I digress. The crucial fact for our story is that the Pythagoreans were lovers of music, to which they brought their obsession with numbers. Pythagoras believed consonance in music depended on "sonorous numbers." He claimed that the perfect consonances were intervals of the musical scale that can be expressed as ratios between the numbers 1, 2, 3, and 4. These numbers add up to 10, the perfect number in the Pythagorean world view. The Pythagoreans brought their musical instruments to their gatherings, which turned into jam sessions. We don't know how good they were, there being no compact disk recorders at the time. But one later critic made an educated guess.

Vincenzo Galilei figured that the Pythagoreans must have had a collective tin ear, given their ideas about consonance. His ear told him that Pythagoras was dead wrong. Other practicing musicians of the sixteenth century also paid no attention to these ancient Greeks. Yet the Pythagoreans' ideas endured even into Vincenzo's day, and the sonorous numbers were still a respected component of musical theory, if not practice. The greatest defender of Pythagoras in sixteenth-century Italy was Gioseffo Zarlino, the foremost music theorist of his day and also Vincenzo's teacher.

Vincenzo and Zarlino entered into a bitter debate over the matter, and Vincenzo came up with a method of proving his point that was revolutionary for the time: *he experimented.* By setting up experiments with strings of different lengths or strings of equal length but different tensions, he found new, non-Pythagorean mathematical relationships in the musical scale. Some claim that Vincenzo was the first person to dislodge a universally accepted mathematical law through experimentation. At the very least, he was in the forefront of a movement that replaced the old polyphony with modern harmony.

We know there was at least one interested spectator at these musical experiments. Vincenzo's eldest son watched as he measured and calculated. Exasperated by the dogma of musical theory, Vincenzo railed at his son about the stupidity of mathematics. We don't know his exact words, but in my mind I can hear Vincenzo screaming something like, "Forget about these theories with dumb numbers. Listen to what your ear tells you. Don't let me ever hear you talking about becoming a mathematician!" He taught the boy well, turning him into a competent performer on the lute and other instruments. He trained his son's senses, teaching him to detect errors in timing, an essential ability for a musician. But he wanted his eldest son to forsake both music and mathematics. A typical father, Vincenzo wanted him to become a doctor, wanted him to have a decent income.

Watching those experiments had a greater impact on the young man than Vincenzo could have imagined. The boy was especially fascinated by an experiment in which his father applied various tensions to his strings by hanging different weights from their ends. When plucked, these weighted strings acted as pendulums, and this may have started the young Galilei thinking about the distinctive ways objects move in this universe.

The son's name, of course, was Galileo. To modern eyes his achievements are so luminous it is difficult to see anyone else in that period of history. Galileo ignored Vincenzo's diatribes against the spuriousness of pure mathematics and became a math professor. But as much as he loved mathematical reasoning, he made it subservient to observation and measurement. In fact, his adroit blending of the two is frequently cited as the true beginning of the "scientific method."

GALILEO, ZSA ZSA, AND ME

Galileo was a new beginning. In this chapter and the one that follows, we will see the creation of classical physics. We'll meet an awesome set of heroes: Galileo, Newton, Lavoisier, Mendeleev, Faraday, Maxwell, and Hertz, among others. Each attacked the problem of finding the ultimate building block of matter from a new angle. For me this is an intimidating chapter. All of these people have been written about time and again. The physics is well-covered territory. I feel like Zsa Zsa Gabor's seventh husband. I know what to do, but how do you make it interesting?

Thanks to the post-Democritan thinkers, there was little action in science from the time of the atomists until the dawn of the Renaissance. That's one reason the Dark Ages were so dark. The nice thing about particle physics is that we can ignore almost two thousand years of intellectual thought. Aristotelian logic — geocentric, human-centered, religious — dominated Western culture during this period, creating a sterile environment for physics. Of course, Galileo didn't spring full-grown from a complete desert. He gave much credit to Archimedes, Democritus, and the Roman poet-philosopher Lucretius. No doubt he studied and built on other predecessors who are now known well only to scholars. Galileo accepted Copernicus's world view (after careful checking), and that determined his personal and political future.

We'll see a departure from the Greek method in this period. No longer is Pure Reason good enough. We enter an era of experimentation. As Vincenzo told his son, between the real world and pure reason (that is, mathematics) there lie the senses and, most important, measurement. We'll meet several generations of measurers as well as theorists. We'll see how the interplay between these camps helped forge a magnificent intellectual edifice, known now as classical physics. Their work did not benefit just scholars and philosophers. From their discoveries emerged technologies that changed the way humans live on this planet.

Of course, measurers are nothing without their measuring sticks, their instruments. It was a period of wonderful scientists, but also of wonderful instruments.

BALLS AND INCLINATIONS

Galileo gave particular attention to the study of motion. He may or may not have dropped rocks from the Leaning Tower, but his logical analysis of how distance, time, and speed are related probably predated the experiments he did carry out. Galileo studied how things move, not by allowing objects to fall free, but by using a trick, a substitute, the inclined plane. Galileo reasoned that the motion of a ball rolling down a smooth, slanting board would bear a close relationship to that of a ball in free fall, but the plane would have the enormous advantage of slowing the motion enough that it could be measured.

In principle he could check this reasoning by starting with very gentle inclinations — raising one end of his six-foot-long board by a few inches to create a gentle slide — and by repeating his measurements with increasing inclinations until the speed became too great to measure precisely. This would give him confidence in extending his conclusions to the ultimate inclination, a vertical free fall.

Now, he needed something to time his rolling balls. Galileo's visit to the local shopping mall to buy a stopwatch failed; the invention was still three hundred years away. Here is where his father's training came in. Remember that Vincenzo refined Galileo's ear for musical beats. A march, for example, might have a beat every half second. At that beat a competent musician, as Galileo was, can detect an error of about one sixty-fourth of a second.

Galileo, lost in a land without timepieces, decided to make a sort of musical instrument out of his inclined plane. He strung a series of lute strings at intervals across the plane. Now when he rolled a ball down, it made a click as it passed over each string. Galileo then slid the strings up and down until the beat of each interval was exactly the same to his ear. He sang a march tune, releasing the ball on one beat, and when the strings were finally set just right, the ball struck each lute string precisely on successive beats, each a half second apart. When Galileo measured the spaces between the strings — *mirabile dictu!* — he found that they increased geometrically down the plane. In other words, the distance from start to the second string was four times the distance from start to the first string. The distance from start to the third string was nine times the first interval; the fourth string was sixteen times farther down the plane than the first; and so on, even though each gap between strings always represented a half second. (The ratios of the intervals, 1 to 4 to 9 to 16, can also be expressed as squares: 1^2, 2^2, 3^2, 4^2, and so on.)

But what happens if one raises the plane a bit, making the inclination steeper? Galileo worked many angles and found this same relationship, this sequence of squares, at each inclination, from gentle to less gentle, until the motion proceeded too swiftly for his "clock" to record distances accurately enough. The crucial thing is that Galileo demonstrated that a falling object doesn't just drop, but drops faster and faster and faster over time. It *accelerates*, and the acceleration is constant.

Being a mathematician, he came up with a formula to describe this motion. The distance s that a falling body covers is equal to a number A times the square of the time t it takes to cover the distance. In the ancient language of algebra, we abbreviate this by: $s = At^2$. The constant A changed with each inclination of the plane. A represents the crucial concept of acceleration, that is, the increase of speed as the object continues to fall. Galileo was able to deduce that speed changes with time in a simpler way than distance, increasing simply with the time (rather than with the square of the time).

The inclined plane, the trained ear's ability to measure times to a sixty-fourth of a second, and the ability to measure distances to somewhat better than a tenth of an inch gave Galileo the precision he needed to make his measurements. Galileo later invented a clock based upon the regular period of the pendulum. Today the Bureau of Standards' atomic cesium clocks keep time to a precision better than one millionth of a second per year! These clocks are rivaled by nature's own timepieces: astronomical pulsars, which are whirling neutron stars that sweep beams of radio waves across the cosmos with a regularity you can set your watch to. They may in fact be more precise than the atomic pulse in the cesium atom. Galileo would have been entranced by this deep connection between astronomy and atomism.

Well, what is so important about $s = At^2$?

It was the first time, as far as we know, that motion was correctly described mathematically. The crucial concepts of acceleration and velocity were sharply defined. Physics is a study of matter and motion. The movement of projectiles, the motion of atoms, the whirl of planets and comets must all be described quantitatively. Galileo's mathematics, confirmed by experiment, provided the starting point.

Lest all of this sound too easy, we should note that Galileo's obsession with the law of free fall lasted for decades. He even got the law wrong in one publication. Most of us, being basically Aristotelians (did you know that you, dear reader, are a basic Aristotelian?), would guess that the speed of the fall would depend on the weight of the ball. Galileo, because he was smart, reasoned otherwise. But is it so crazy to think that heavy things should fall faster than light things? We do so because nature misleads us. Smart as Galileo was, he had to do careful experiments to show that the *apparent* dependence of a body's time of fall on its weight comes from the friction of the ball on the plane. So he polished and polished to decrease the effect of friction.

THE FEATHER AND THE PENNY

Extracting a simple law of physics from a set of measurements is not so simple. Nature hides the simplicity in a thicket of complicating circumstances, and the experimenter's job is to prune away these complications. The law of free fall is a splendid example. In freshman physics we hold a feather and a penny at the top of a tall glass tube and drop them simultaneously. The penny falls rapidly and clinks to the bottom in less than a second. The feather floats gently down, arriving in five or six seconds. Such observations led Aristotle to postulate his law that heavier objects fall faster than light ones. Now we pump the air out of the tube and repeat the experiment. Feather and penny drop with equal times. Air resistance obscures the law of free fall. To make progress, we must remove this complicating feature to get the simple law. Later, if it is important, we can learn how to add this effect back in to arrive at a more complex but more applicable law.

The Aristotelians believed that an object's "natural" state was to be at rest. Push a ball along a plane and it comes to rest, no? Galileo knew all about imperfect conditions, and that understanding led to one of the great discoveries. He read physics in inclined planes as Michelangelo saw magnificent bodies in slabs of marble. He realized, however, that because of friction, air pressure, and other imperfect conditions, his inclined plane was not ideal for studying the forces on various objects. What happens, he pondered, if you have an ideal plane? Like Democritus mentally sharpening his knife, you mentally polish the plane until it attains the ultimate smoothness, completely free of friction. Then you stick it in an evacuated chamber to get rid of air resistance. And you extend the plane to infinity. You make sure the plane is absolutely horizontal. Now when you give a tiny nudge to the perfectly polished ball sitting on your smooth, smooth plane, how far will it roll? For how long will it roll? (As long as all of this is in the mind, the experiment is possible and cheap.)

The answer is forever. Galileo reasoned thus: when a plane, even an earthly imperfect plane, is tilted up, a ball, started by a push from the bottom, will go slower and slower. If the plane is tilted down, a ball released at the top will go faster and faster. Therefore, using the intuitive sense of continuity of action, he concluded that a ball on a flat plane will neither slow down nor speed up but continue forever. Galileo had made an intuitive jump to what we now call Newton's first law of motion: a body in motion tends to remain in motion. Forces are not necessary for motion, only for changes in motion. In contrast to the Aristotelian view, a body's natural state is motion with constant velocity. Rest is the special case of zero velocity, but in the new view that is no more natural than any other constant velocity. For anyone who has driven a car or a chariot, this is a counterintuitive idea. Unless you keep your foot on the pedal or keep whipping the horse, the vehicle will halt. Galileo saw that to find the truth you must mentally attribute ideal conditions to your instrument. (Or drive your car on an ice-slicked road.) It was Galileo's genius to see how to remove natural obfuscations such as friction and air resistance to establish a set of fundamental relations about the world.

As we shall see, the God Particle itself is a complication imposed upon a simple, beautiful universe, perhaps in order to hide this dazzling symmetry from an as yet undeserving humanity.

THE TRUTH OF THE TOWER

The most famous example of Galileo's ability to strip complications away from simplicity is the Leaning Tower of Pisa story. Many experts doubt that this fabled event ever took place. Stephen Hawking, for one, writes that the story is "almost certainly untrue." Why, Hawking asks, would Galileo bother dropping weights from a tower with no accurate way of timing their descent when he already had his inclined plane to work with? Shades of the Greeks! Hawking, the theorist, is using Pure Reason here. That doesn't cut it with a guy like Galileo, an experimenter's experimenter.

Stillman Drake, the biographer of choice of Galileo, believes the Leaning Tower story is true for a number of sound historical reasons. But it also fits Galileo's personality. The Tower experiment was not really an experiment at all but a demonstration, a media happening, the first great scientific publicity stunt. Galileo was showing off, and showing up his critics.

Galileo was an irascible sort of guy — not really contentious, but quick of temper and a fierce competitor when challenged. He could be a pain in the ass when annoyed, and he was annoyed by foolishness in all its forms. An informal man, he ridiculed the doctoral gowns that were required attire at the University of Pisa and wrote a humorous poem called "Against the Toga" that was appreciated most by the younger and poorer lecturers, who could ill afford the robes. (Democritus, who loves togas, didn't enjoy the poem at all.) The older professors were less than amused. Galileo also wrote attacks on his rivals using various pseudonyms. His style was distinct, and not too many people were fooled. No wonder he had enemies.

His worst intellectual rivals were the Aristotelians, who believed that a body moves only if driven by some force and that a heavy body falls faster than a light one because it has a greater pull toward the earth. The thought of testing these ideas never occurred to them. Aristotelian scholars pretty much ruled the University of Pisa and, for that matter, most universities in Italy. As you can imagine, Galileo wasn't a big favorite of theirs.

The stunt at the Leaning Tower of Pisa was directed at this group. Hawking is right that it wouldn't have been an ideal experiment. But it was an event. And as in any staged event, Galileo knew in advance how it was going to come out. I can see him climbing the tower in total darkness at three in the morning and tossing a couple of lead balls down at his postdoc assistants. "You should feel both balls hitting you in the head simultaneously," he yells at his assistant. "Holler if the big one hits you first." But he didn't really have to do that, because he had already reasoned that both balls should strike the ground at the same instant.

Here's how his mind worked: let us suppose, he said, that Aristotle was right. The heavy ball will land first, meaning that it will accelerate faster. Let us then tie the heavy ball to the light ball. If the light ball is indeed slower, it will hold back the heavy ball, making it fall more slowly. However, by tying them both together, we have created an even heavier object, and this combination object should fall faster than each ball individually. How do we solve this dilemma? Only one solution satisfies all conditions: both balls must fall at the same rate of speed. That is the only conclusion that gets around the slower/faster conundrum.

According to the story, Galileo spent a good part of the morning dropping lead balls from the tower, proving his point to interested observers and scaring the heck out of everybody else. He was wise enough to not use a penny and a feather but instead unequal weights of very similar shapes (such as a wooden ball and a hollow lead sphere of the same radius) to roughly equalize the air resistance. The rest is history, or it should be. Galileo had demonstrated that free fall is utterly independent of mass (though he didn't know why, and it would take Einstein, in 1915, to really understand it). The Aristotelians were taught a lesson they never forgot — or forgave.

Is this science or show biz? A little of both. It's not only experimenters who are so inclined. Richard Feynman, the great theorist (but one who always showed a passionate interest in experiment), thrust himself into the public eye when he was on the commission investigating the *Challenger* space shuttle disaster. As you may recall, there was a controversy over the ability of the shuttle's O-rings to withstand low temperatures. Feynman ended the controversy with one simple act: when the TV cameras were on him, he tossed a bit of O-ring into a glass of ice water and let the audience view its loss of elasticity. Now, don't you suspect that Feynman, like Galileo, knew in advance what was going to happen?

In fact, in the 1990s, Galileo's Tower experiment has emerged with a brand-new intensity. The issue involves the possibility that there is a "fifth force," a hypothetical addition to Newton's law of gravitation that would produce an extremely small difference when a copper ball and, say, a lead ball are dropped. The difference in time of fall through, say, one hundred feet might be less than a billionth of a second, unthinkable in Galileo's time but merely a respectable challenge with today's technology. So far, evidence for the fifth force, which appeared in the late 1980s, has all but vanished, but keep watching your newspaper for updates.

GALILEO'S ATOMS

What did Galileo think about atoms? Influenced by Archimedes, Democritus, and Lucretius, Galileo was intuitively an atomist. He taught and wrote about the nature of matter and light over many decades, especially in his book *The Assayer* of 1622 and his last work, the great *Dialogues Concerning the Two New Sciences.* He seemed to believe that light consisted of pointlike corpuscles and that matter was similarly constructed.

Galileo called atoms "the smallest quanta." Later he pictured an "infinite number of atoms separated by an infinite number of voids." The mechanistic view is closely tied to the mathematics of infinitesimals, a precursor to the calculus that would be invented sixty years later by Newton. Here we have a rich lode of paradox. Take a simple circular cone — a dunce cap? — and think of slicing it horizontally, parallel to its base. Let's examine two contiguous slices. The top of the lower piece is a circle, the bottom of the upper piece is a circle. Since they were previously in direct contact, point to point, they have the same radius. Yet the cone is continuously getting smaller, so how can the circles be the same? However, if each circle is composed of an infinite number of atoms and voids, one can imagine that the upper circle contains a smaller though still infinite number of atoms. No? Let's remember that we are in 1630 or so and dealing with exceedingly abstract ideas — ideas that were almost two hundred years from experimental test. (One way around this paradox is to ask how thick the knife is that slices the cone. I think I hear Democritus giggling again.)

In *Dialogues Concerning the Two New Sciences,* Galileo presents his last thoughts on atom structure. In this hypothesis, according to recent historical scholars, atoms are reduced to the mathematical abstraction of points, lacking any dimension, clearly indivisible and uncuttable, but devoid of the shapes that Democritus had envisioned.

Here Galileo moves the idea closer to its most modern version, the pointlike quarks and leptons.

ACCELERATORS AND TELESCOPES

Quarks are even more abstract and difficult to visualize than atoms. No one has ever "seen" one, so how can they exist? Our proof is indirect. Particles collide in an accelerator. Sophisticated electronics receive and process electrical pulses generated by particles in a variety of sensors in the detector. A computer interprets the electronic impulses from the detector, reducing them to a bunch of zeroes and ones. It sends these results to a monitor in our control room. We look at the representation of ones and zeroes and say, "Holy cow, a quark!" It seems so far-fetched to the layman. How can we be so sure? Couldn't the accelerator or the detector or the computer or the wire from the computer to the monitor have "manufactured" the quark? After all, we never see the quark with our own God-given eyes. Oh, for a time when science was simpler! Wouldn't it be great to be back in the sixteenth century? Or would it? Ask Galileo.

Galileo built, according to his records, a huge number of telescopes. He tested his telescope, in his own words, "a hundred thousand times on a hundred thousand stars and other objects." He trusted the thing. Now I have this little mental picture. Here's Galileo with all his graduate students. He's looking out the window with his telescope and describing what he sees, and they're all scribbling it down: "Here's a tree. It's got a branch this way and a leaf that way." After he tells them what he sees through the telescope, they all get on their horses or wagons — maybe a bus — and go across the field to look at the tree close up. They compare it to Galileo's description. That's how you calibrate an instrument. You do that ten thousand times. So a critic of Galileo describes the meticulous nature of the testing and says, "If I follow these experiments on terrestrial objects, the telescope is superb. I trust it, even though it interposes something between the God-given eye and the God-given object. Nevertheless, it does not fool you. On the other hand, if you look up at the sky, there's a star. And if you look through the telescope, there are two stars. It's totally cracked!"

Okay, those weren't his exact words. But one critic did use words to this effect to dispute Galileo's claim that Jupiter has four moons. Since the telescope allowed him to see more than could be seen with the naked eye, it must be lying. A math professor also dismissed Galileo, saying he, too, could find four moons of Jupiter if given enough time "to build them into some glasses."

Anyone who uses an instrument runs into this problem. Is the instrument "manufacturing" the results? Galileo's critics seem foolish today, but were they off the wall or just scientific conservatives? Some of both, no doubt. In 1600 people believed that the eye had an active role in vision; the eyeball, given to us by God, interpreted the visual world for us. Today we know the eye is no more than a lens with a bunch of receptors in it that passes visual information along to our brain's visual cortex, where we actually "see." The eye is in fact a mediator between the object and the brain, just as the telescope is. Do you wear eyeglasses? You're already modifying. In fact, among devout Christians and philosophers in sixteenth-century Europe, wearing spectacles was considered almost sacrilegious, even though they had been around for three hundred years. One notable exception was Johannes Kepler, who was very religious but who nonetheless wore specs because they helped him see; this was fortunate, given that he became the greatest astronomer of his time.

Let's accept that a well-calibrated instrument can provide a good approximation of reality. As good perhaps as the ultimate instrument, our brain. Even the brain must be calibrated at times, and safeguards and fudge factors applied to compensate for distortion. For example, even if you have 20/20 vision, a few glasses of wine can double the number of friends around you.

THE CARL SAGAN OF 1600

Galileo helped pioneer the acceptance of instruments, an accomplishment whose importance to science and experimentation cannot be overemphasized. What sort of person was he? He comes across as a deep thinker with a subtle mind, capable of intuitive insights that would be the envy of any theoretical physicist today, but with energy and technical skills that included lens polishing and the construction of many instruments, including telescopes, the compound microscope, and the pendulum clock. Politically he alternated from docile conservatism to bold, slashing attacks on his opponents. He must have been a dynamo of activity, constantly engaged, for he left behind an enormous correspondence and monumental volumes of published works. He was a popularizer, giving public lectures to huge audiences after the supernova of 1604, writing in a lucid, vulgarized Latin. No one comes as close as he does to being the Carl Sagan of his day. Not too many faculties would have granted him tenure, so vigorous was his style and so stinging his criticism, at least before his condemnation.

Was Galileo the complete physicist? As complete as one can find in history, in that he combined consummate skills of both the experimenter and the theorist. If he had faults, they fell on the theoretical side. Although this combination was relatively common in the eighteenth and nineteenth centuries, in today's age of specialization it is rare. In the seventeenth century, much of what would be called "theory" was in such close support of experiment as to defy separation. We shall soon see the advantage of having a great experimenter followed by a great theorist. In fact, by Galileo's time there had already been one such pivotal succession.

THE MAN WITH NO NOSE

Let me backtrack for a minute, because no book about instrument and thought, experiment and theory, is complete without two names that go together like Marx and Engels, Emerson and Thoreau, or Siegfried and Roy. I'm speaking of Brahe and Kepler. They were strictly astronomers, not physicists, but they warrant a brief digression.

Tycho Brahe was one of the more bizarre characters in the history of science. This Danish nobleman, born in 1546, was a measurer's measurer. Unlike atomistic physicists, who look downward, he looked up at the heavens, and he did it with unprecedented precision. Brahe constructed all manner of instruments for measuring the positions of the stars, planets, comets, the moon. Brahe missed the telescope's invention by a couple of decades, so he built elaborate sighting devices — azimuthal semicircles, Ptolemaic rulers, brass sextants, azimuthal quadrants, parallactic rulers — that he and his assistants used with the naked eye to nail down coordinates of stars and other heavenly bodies. Most of these variations on today's sextants consisted of crossarms with arcs between them. The astronomers used the quadrants like rifles, lining up stars by looking through metal sights attached to the ends of the arms. The arcs connecting the crossarms functioned like the protractors you used in school, enabling the astronomers to measure the angle of the sightline to the star, planet, or comet being observed.

There was nothing particularly new about the basic concept of Brahe's instruments, but he defined the state of the art. He experimented with different materials. He figured out how to make these cumbersome gadgets easily rotatable in the vertical or horizontal plane, and at the same time fixed them in place so that he could track celestial objects from the same point night after night. Most of all, Brahe's measuring devices were *big*. As we shall see when we get to the modern era, big is not always, but usually, better. Tycho's most famous instrument was the mural quadrant, which had a radius of six meters, or about eighteen feet! It took forty strong men to wrestle it into place — a veritable Super Collider of its day. The degrees marked off on its arc were so far apart that Brahe was able to divide each of the sixty minutes of arc in each degree into six subdivisions of ten seconds each. In simpler terms, Brahe's margin of error was the width of a needle held at arm's length. All this done with the naked eye! To give you some idea of the man's ego, inside the quadrant's arc was a life-size portrait of Brahe himself.

You'd think such fastidiousness would indicate a nerdy kind of man. Tycho Brahe was anything but. His most unusual feature was his nose — or lack of one. When Brahe was a twenty-year-old student, he got into a furious argument with a student named Manderup Parsbjerg over a mathematical point. The quarrel, which took place at a celebration at a professor's house, got so heated that friends had to separate the two. (Okay, maybe he was a little nerdy, fighting over formulas rather than girls.) A week later Brahe and his rival met again at a Christmas party, had a few drinks, and began the math argument anew. This time they couldn't be cooled down. They adjourned to a dark spot beside a graveyard and went at each other with swords. Parsbjerg ended the duel quickly by slicing off a good chunk of Brahe's nose.

This nose episode would haunt Brahe all his life. There are two stories concerning what he did in the way of cosmetic surgery. The first, most likely apocryphal, is that he commissioned a whole set of artificial noses made of different materials for different occasions. But the story accepted by most historians is almost as good. This version has Brahe ordering a permanent nose made of gold and silver, skillfully painted and shaped to look like a real nose. Reportedly he carried a little box of glue with him, which he applied whenever the nose became wobbly. The nose was the butt of jokes. One scientific rival claimed that Brahe made his astronomical observations through his nose, using it as a sight vane.

Despite these difficulties, Brahe did have an advantage over many scientists today — his noble birth. He was friends with King Frederick II, and after he became famous because of his observations of a supernova in the constellation Cassiopeia, the king gave him the island of Hven in The Sound to use as an observatory. Brahe was also given rule over all the tenants of the island, the rents derived therefrom, and extra funds from the king. In this fashion, Tycho Brahe became the world's first laboratory director. And what a director he was! With his rents, a grant from the king, and his own fortune, he led a regal existence. He missed only the benefits of dealing with funding agencies in twentieth-century America.

The two-thousand-acre island became an astronomer's paradise, replete with workshops for the artisans who made the instruments, a windmill, a paper mill, and nearly sixty fish ponds. For himself, Brahe built a magnificent home and observatory on the island's highest point. He called it Uraniborg, or "heavenly castle," and enclosed it within a walled square that contained a printing office, servants' quarters, and kennels for Brahe's watchdogs, plus flower gardens, herbaries, and some three hundred trees.

Brahe eventually left the island under less than pleasant circumstances after his benefactor, King Frederick, died of an excess of Carlsberg or whatever mead was popular in Denmark in 1600. The fief of Hven reverted to the crown, and the new king subsequently gave the island to one Karen Andersdatter, a mistress he had picked up at a wedding party. Let this be a lesson to all lab directors, as to their status in the world and their replaceability in the eyes of the powers that be. Fortunately, Brahe landed on his feet, moving his data and instruments to a castle near Prague where he was welcomed to continue his work.

It was the regularity of the universe that prompted Brahe's interest in nature. As a fourteen-year-old he had been fascinated by the total eclipse of the sun predicted for August 21, 1560. How could men understand the motions of the stars and planets so finely that they could foretell their positions years in advance? Brahe left an enormous legacy: a catalogue of the positions of exactly one thousand fixed stars. It surpassed Ptolemy's classic catalogue and destroyed many of the old theories.

A great virtue of Brahe's experimental technique was his attention to possible errors in his measurements. He insisted, and this was unprecedented in 1580, that measurements be repeated many times and that each measurement be accompanied by an estimate of its accuracy. He was far ahead of his time in his dedication to presenting data together with the limits of their trustworthiness.

As a measurer and observer, Brahe had no peer. As a theorist, he left much to be desired. Born just three years after the death of Copernicus, he never fully accepted the Copernican system, which held that the earth orbited the sun rather than vice versa, as Ptolemy had stated many centuries earlier. Brahe's observations proved to him that the Ptolemaic system didn't work but, educated as an Aristotelian, he could never bring himself to believe that the earth rotated, nor could he give up the belief that the earth was at the center of the universe. After all, he reasoned, if the earth really moved and you fired a cannonball in the direction of the earth's rotation, it should go farther than if you fired it in the opposite direction, but that is not the case. So Brahe came up with a compromise: the earth stayed immobile at the center of the universe, but contrary to the Ptolemaic system, the planets revolved about the sun, which in turn circled the earth.

THE MYSTIC DELIVERS

Through his career, Brahe had many superb assistants. The most brilliant of all was a strange, mystical mathematician-astronomer named Johannes Kepler. A devout German-born Lutheran, Kepler would have preferred to be a clergyman, had not mathematics offered him a way of making a living. In truth, he failed the ministerial qualifying exams and stumbled into astronomy with a strong minor in astrology. Even so, he was destined to become the theorist who would discern simple and profound truths in Brahe's mountain of observational data.

Kepler, a Protestant at an unfortunate time (the Counter Reformation was sweeping Europe), was a frail, neurotic, nearsighted man, with none of the self-assurance of a Brahe or a Galileo. The entire Kepler family was a trifle offbeat. Kepler's father was a mercenary, his mother was tried as a witch, and Johannes himself was occupied much of the time with astrology. Fortunately, he was good at it, and it paid some bills. In 1595 he constructed a calendar for the city of Graz that predicted bitter cold weather, peasant uprisings, and invasions by the Turks — all events that came to pass. In fairness to Kepler, he was not alone in moonlighting as an astrologer. Galileo cast horoscopes for the Medicis, and Brahe also dabbled in the art, although he wasn't so good at it: from the lunar eclipse of October 28,

1566, Brahe predicted the death of Sultan Suleiman the Magnificent. Unfortunately the sultan was already dead at the time.

Brahe treated his assistant rather shabbily — more like a postdoc, which Kepler was, than as a peer, which he certainly deserved to be. The sensitive Kepler bristled under the insult, and the two had many fallings-out and an equal number of reconciliations, for Brahe did come to appreciate Kepler's brilliance.

In October 1601, Brahe attended a dinner party and, as was his wont, drank far too much. According to the strict etiquette of the day, it was improper to leave the table during a meal, and when he finally made a mad dash for the bathroom, it was too late. "Something of importance" had burst inside him. Eleven days later he was dead. Having already appointed Kepler as his chief assistant, on his deathbed Brahe bequeathed to him all of the data he had acquired over his illustrious, well-funded career, and beseeched Kepler to use his analytical mind to create a grand synthesis that would further an understanding of the heavens. Of course, Brahe added that he expected Kepler to follow the Tychonian hypothesis of a geocentric universe.

Kepler agreed to the dying man's wish, no doubt with fingers crossed, because he thought Brahe's system was nuts. But the data! The data were nonpareil. Kepler pored over the information, looking for patterns in the motions of the planets. Kepler rejected the Tychonian and Ptolemaic systems out of hand for their clumsiness. But he had to start somewhere. So he began with the Copernican system as a model because, with its system of spherical orbits, it was the most elegant thing around.

The mystic in Kepler also embraced the idea of a centrally positioned sun, which not only illuminated all the planets but provided a force, or motive as it was then called, for the movements of the planets. He didn't quite know how the sun did this — he guessed it was something like magnetism — but he paved the way for Newton. He was among the first to promote the idea that a force is needed to make sense of the solar system.

Just as important, he found that the Copernican system didn't quite jibe with Brahe's data. The surly old Dane had taught Kepler well, instilling in him the inductive method: lay down a foundation of observations, and only then ascend to the causes of things. Despite his mysticism and his awe of, and obsession with, geometric forms, Kepler stuck faithfully to the data. He emerged from his study of Brahe's observations — especially the data on Mars — with three laws of planetary motion, which, almost four hundred years later, still serve as the basis of modern planetary astronomy. I won't go into the details of these laws here, except to say that his first law destroyed the lovely Copernican notion of circular orbits, a concept that had remained unquestioned since the days of Plato. Kepler established that the planets trace out ellipses in their orbital paths with the sun at one focus. The eccentric Lutheran had saved Copernicanism and freed it from the cumbersome epicycles of the Greeks; he did so by making sure his theories followed Brahe's observations to the precise minute of arc.

Ellipses! Pure mathematics! Or is it pure nature? If, as Kepler discovered, planets move in perfect ellipses with the sun at one focus, then nature must love mathematics. Something — maybe God — looks down on the earth and says, "I like mathematical form." It is easy to demonstrate nature's love of mathematical forms. Pick up a rock and throw it. It traces out a very good parabola. If there were no air, it would be a perfect parabola. In addition to being a mathematician, God is kind. She hides complexity when the mind isn't ready for it. We now know that the orbits are not perfect ellipses (because of the pull of the planets on one another) but the deviations were far too small to see with Brahe's apparatus.

Kepler's genius was often obscured in his books by massive amounts of spiritual clutter. He believed that comets were evil omens, that the universe was divided into three regions corresponding to the Holy Trinity, and that the tides were the breathing of the earth, which he likened to an enormous living animal. (This idea of earth-as-organism has been resurrected today in the form of the Gaia hypothesis.)

Even so, Kepler had a great mind. The stiff-upper-lipped Sir Arthur Eddington, one of the most eminent physicists of his time, in 1931 called Kepler "the forerunner of the modern theoretical physicist." Eddington lauded Kepler for demonstrating an outlook similar to that of the theorists of the quantum age. Kepler didn't look for a concrete mechanism to explain the solar system, according to Eddington, but "was guided by a sense of mathematical form, an aesthetic instinct for the fitness of things."

POPE TO GALILEO: DROP DEAD

In 1597, long before he had worked out the troublesome details, Kepler wrote to Galileo urging him to support the Copernican system. With typical religious fervor, he told Galileo to "believe and step forth." Galileo refused to come out of the Ptolemaic closet. He needed proof. That proof came from an instrument, the telescope.

The nights of January 4 to 15, 1610, must be recorded as among the most important in the history of astronomy. On those dates, using a new and improved telescope that he had constructed, Galileo saw, measured, and tracked four tiny "stars" moving near the planet Jupiter. He was forced to conclude that these bodies were moving in circular orbits around Jupiter. This conclusion converted Galileo to the Copernican view. If bodies could orbit Jupiter, the notion that all planets and stars orbit the earth is wrong. Like most late converts, whether to a scientific notion or to a religious or political conviction, he became a fierce and unwavering advocate of Copernican astronomy. History credits Galileo, but we must here also honor the telescope, which in his capable hands opened the heavens.

The long and complex story of his conflict with the reigning authority has often been told. The Church sentenced him to life imprisonment for his astronomical beliefs. (The sentence was later commuted to permanent house arrest.) It wasn't until 1822 that a reigning pope officially declared that the sun could be at the center of the solar system. And it took until 1985 for the Vatican to acknowledge that Galileo was a great scientist and that he had been wronged by the Church.

THE SOLAR SPONGE

Galileo was guilty of a less celebrated heresy, one that is closer to the heart of our mystery than the orbits of Mars and Jupiter. In his first scientific expedition to Rome to report on his work with physical optics, he brought with him a little box containing rock fragments discovered by alchemists in Bologna. The rocks glowed in the dark. Today this luminescent mineral is known as barium sulfide. But in 1611 alchemists called it by the much more poetic name "*solar sponge.*"

Galileo brought chunks of solar sponge to Rome to aid him in his favorite pastime: annoying the hell out of his Aristotelian colleagues. As the Aristotelians sat in the dark watching the glow from the barium sulfide, their rogue colleague's point did not escape them. Light was a *thing*. Galileo had held the rock in the sun, then brought the rock into the darkness and the light had been carried inside with it. This belied the Aristotelian notion that light was simply a quality of an illuminated medium, that it was incorporeal. Galileo had separated the light from its medium, had moved it around at will. To an Aristotelian Catholic, this was like saying you could take the sweetness of the Holy Virgin and place it in a mule or a stone. And what exactly did light consist of? Invisible corpuscles, Galileo reasoned. Particles! Light possessed a mechanical action. It could be transmitted, strike objects, reflect off them, penetrate them. Galileo's realization that light was corpuscular led him to accept the idea of *indivisible* atoms. He wasn't sure how the solar sponge worked, but perhaps a special rock could attract luminous corpuscles as a magnet attracts iron shavings, though he didn't subscribe to this theory literally. In any case, ideas such as these deepened Galileo's already precarious position with Catholic orthodoxy.

Galileo's historical legacy seems to be inextricably tied to the Church and religion, but he wouldn't have viewed himself as a professional heretic or, for that matter, a wrongly accused saint. For our purposes, he was a physicist, and a great one, far beyond his advocacy of Copernicanism. He broke new ground in many fields. He blended experiments and mathematical thinking. When an object moves, he said, it's important to quantify its motion with a mathematical equation. He always asked "How do things move? How? How?" He didn't ask "Why? Why is this sphere falling?" He was aware that he was just describing motion, a difficult enough task for his time. Democritus might have wisecracked that Galileo wanted to leave something for Newton to do.

THE MASTER OF THE MINT

> Most Merciful Sir:
>
> I am going to be murdered, although perhaps you may think not but 'tis true. I shall be murdered the worst of all murders. That is in the face of Justice unless I am rescued by your merciful hands.

Thus wrote the convicted counterfeiter William Chaloner — the most colorful and resourceful outlaw of his time — in 1698 to the official who had finally succeeded in capturing, prosecuting, and convicting him. Chaloner had threatened the integrity of English currency, then largely in the form of gold and silver coins.

The object of this desperate appeal was one Isaac Newton, warden (soon to become master) of the Mint. Newton was doing his job, which was to supervise the Mint, oversee a vast recoinage, and protect the currency against counterfeiters and clippers, those who shaved some of the precious metal off the coins and passed them as whole. This position, something like the Secretary of the Treasury, mixed the high politics of parliamentary infighting with the prosecution of thugs, crooks, thieves, launderers, and other riffraff who preyed on the currency of the realm. The crown awarded Newton, the preeminent scientist of his day, the job as a sinecure while he worked on more important things. But Newton took the job seriously. He invented the technique of fluting the edges of coins to defeat clippers. He personally attended the hangings of counterfeiters. The position was a far cry from the serene majesty of Newton's earlier life, when his obsession with science and mathematics generated the most profound advance in the history of natural philosophy, one that would not be clearly surpassed until, possibly, the theory of relativity in the 1900s.

In one of the quirks of chronology, Isaac Newton was born in England the same year (1642) that Galileo died. You can't talk about physics without talking about Newton. He was a scientist of transcendent importance. The influence of his achievements on human society rivals that of Jesus, Mohammed, Moses, and Gandhi, as well as Alexander the Great, Napoleon, and their ilk. Newton's universal law of gravitation and the methodology he created occupy the first half dozen chapters of every textbook on physics; understanding them is essential to anyone pursuing a scientific or engineering career. Newton has been called modest because of his famous statement "If I have seen further than most it is because I have stood on the shoulders of giants," which most assume refers to men such as Copernicus, Brahe, Kepler, and Galileo. Another interpretation, however, is that he was simply twitting his primary scientific rival and nemesis, the very short Robert Hooke, who claimed, not without some justice, to have discovered gravity first.

I have counted more than twenty serious biographies of Newton. And the literature that analyzes, interprets, extends, comments on Newton's life and science is enormous. Richard Westfall's 1980 biography includes ten dense pages of sources. Westfall's admiration for his subject is boundless:

> It has been my privilege at various times to know a number of brilliant men, men whom I acknowledge without hesitation to be my intellectual superiors. I have never, however, met one against

> whom I was unwilling to measure myself, so that it seemed reasonable to say that I was half as able, or a third, or a fourth, but in every case, a finite fraction. The end result of my study of Newton has served to convince me that with him there is no measure. He has become for me wholly other, one of the tiny handful of supreme geniuses who have shaped the categories of the human intellect.

The history of atomism is one of reductionism — the effort to reduce all the operations of nature to a small number of laws governing a small number of primordial objects. The most successful reductionist of all was Isaac Newton. It would be another 250 years before his possible equal would emerge from the masses of *Homo sapiens* in the town of Ulm, Germany, in 1879.

THE FORCE BE WITH US

To have a sense of how science works, one must study Newton. Yet the Newtonian drill for the students in Physics 101 all too often obscures the power and sweep of his synthesis. Newton developed a quantitative and yet comprehensive description of the physical world that accorded with factual descriptions of how things behave. His legendary connection of the falling apple to the periodic moon captures the awesome power of mathematical reasoning. How the apple falls to the earth and precisely how the moon orbits the earth are included in one all-encompassing idea. Newton wrote: "I wish we could derive the rest of the phenomena of nature by the same level of reasoning from mechanical principles, for I am inclined by many reasons to suspect that they may all depend on certain forces."

By Newton's day how objects moved was known: the trajectory of the thrown stone, the regular swing of the pendulum, the motion down the inclined plane, the free fall of disparate objects, the stability of structures, the shape of a drop of water. What Newton did was organize these and many other phenomena in a single system. He concluded that any change of motion is caused by force and that the response of an object to the force is related to a property of the object he called "mass." Every schoolchild knows that Newton came up with three laws of motion. His first law is a restatement of Galileo's discovery that no force is required for steady, unchanging motion. What we're concerned with here is the second law. It centers around force but is inextricably entwined with one of the mysteries of our story: mass. And it prescribes *how* force changes motion.

Generations of textbooks have struggled with definitions and logical consistencies of Newton's second law, which is written like this: $F = ma$. Eff equals emm ay, or the *force* is equal to the *mass* multiplied by the *acceleration*. In this equation Newton defines neither the force nor the mass, and thus it is never clear whether this represents a definition or a law of physics. Nevertheless, one struggles through it somehow to arrive at the most useful physical law ever devised. This simple equation is awesome in its power and, despite its innocent appearance, can be a frightening thing to solve. Awrrk! Ma-a-a-ath! Don't worry, we'll just talk about it, not really do it. Besides, this handy prescription is the key to the mechanical universe, so there is motivation to stay with it. (We shall be dealing with two Newtonian formulas. For our purposes, let's call this formula I.)

What is a? This is the very same quantity, acceleration, that Galileo defined and measured in Pisa and Padua. It can be the acceleration of any object, be it a stone, a pendulum bob, a projectile of soaring beauty and menace, or the *Apollo* spacecraft. If we put no limit on the domain of our little equation, then a represents the motion of planets, stars, or electrons. Acceleration is the rate at which a speed changes. Your car's accelerator pedal is truly named. If you go from 10 mph to 40 mph in 5 minutes, you have achieved some value of a. If you go from 0 to 60 mph in 10 seconds, you have achieved a much greater acceleration.

What is m? Glibly, it is a property of matter. It is measured by the response of an object to a force. The larger the m, the smaller the response (a) to the imposed force. This property is often called inertia, and the full name given to m is "inertial mass." Galileo invoked inertia in understanding why a body in motion "tends to preserve that motion." We can certainly use the equation to distinguish among masses. Apply the same force — we'll get to what force is later — to a series of objects and use a stopwatch and ruler to measure the resulting motion, the quantity a. Objects having different m's will have different a's. Set up a long series of such experiments comparing the m's of a large number of objects. Once we do this successfully, we can arbitrarily fabricate a standard object, exquisitely wrought of some durable metal. Print on this object "1.000 kilogram" (that's our unit of mass) and place it in a vault at the Bureau of Standards in major capitals of the world (world peace helps). Now we have a way of attributing a value, a number m, to any object. It's simply a multiple or a fraction of our one-kilogram standard.

Okay, enough about mass, what is *F*? The force. What's that? Newton called it the "crowding of one body on another" — the causative agent for change of motion. Isn't our reasoning somewhat circular? Probably, but not to worry; we can use the law to compare forces acting on a standard body. Now comes the interesting part. Forces are provided to us by a bountiful nature. Newton supplies the equation. Nature supplies the force. Keep in mind that the equation works for any force. At the moment we know of four forces in nature. In Newton's day scientists were just beginning to learn about one of them, gravity. Gravity causes objects to fall, projectiles to soar, pendulums to swing. The entire earth, pulling on all objects on or near its surface, generates the force that accounts for the large variety of possible motions and even the lack of motion.

Among other things, we can use $F = ma$ to explain the structure of stationary objects like the reader sitting in her chair or, a more instructive example, standing on her bathroom scale. The earth pulls down on the reader with a force. The chair or scale pushes up on the reader with an equal and opposite force. The sum of the two forces on the reader is zero, and there is no motion. (All of this happens after she goes out and buys this book.) The bathroom scale tells what it cost to cancel the pull of gravity — 60 kilograms or, in the nations of low culture, not yet in the metric system, 132 pounds. "Ohmygod, the diet starts tomorrow." That's the force of gravity acting on the reader. This is what we call "weight," simply the pull of gravity. Newton knew that your weight would change, slightly if you were in a deep valley or on a high mountain, greatly if on the moon. But the mass, the stuff in you that resists a force, doesn't change.

Newton did not know that the pushes and pulls of floors, chairs, strings, springs, wind, and water are fundamentally electrical. It didn't matter. The origin of the force was irrelevant to the validity of his famous equation. He could analyze springs, cricket bats, mechanical structures, the shape of a drop of water or of the planet earth itself. Given the force, we can calculate the motion. If the force is zero, the *change* in speed is zero; that is, the body continues its motion at constant speed. If you throw a ball up, its speed decreases until, at the apex of its path, it stops and then descends with increasing speed. The force of gravity does this, being directed down. Throw a ball into the outfield. How do we understand the graceful arc? We decompose the motion into two parts, an up-and-down part and a horizontal part (indicated by the shadow of the ball on the ground).

The horizontal part has no force (like Galileo, we must neglect the resistance of air, which is a small complicating factor). So the horizontal motion is at constant speed. Vertically, we have the ascent and the descent into the glove of the fielder. The composed motion? A parabola! Yeow! There She goes again, showing off her command of geometry.

Assuming we know the mass of the ball and can measure its acceleration, its precise motion can be calculated by $F = ma$. Its path is determined: it will describe a parabola. But there are many parabolas. A weakly batted ball barely reaches the pitcher; a powerful smash causes the center fielder to race backward. What is the difference? Newton called such variables the starting or initial conditions. What is the initial speed? What is the initial direction? It can range from straight up (in which case the batter gets bopped on his head) to almost horizontal, where the ball falls quickly to the ground. In all cases the trajectory is determined by the speed and direction at the start of the motion — that is, the initial conditions.

WAIT!!!

Now comes a deeply philosophical point. Given a set of initial conditions for a certain number of objects, and given a knowledge of the forces acting on these objects, their motions can be predicted . . . forever. Newton's world view is predictable and determined. For example, suppose that everything in the world is made of atoms — a bizarre thought to raise on page 90 of this book. Suppose we know the initial motion of each of the billions and billions of atoms, and suppose we know the force on each atom. Suppose some cosmic, mother-of-all-computers could grind out the future location of all these atoms. Where will they all be at some future time, for example on Coronation Day? The outcome would be predictable. Among these billions of atoms would be a small subset called "reader" or "Leon Lederman" or "the pope." Predicted, determined . . . with free choice merely an illusion created by a mind with self-interest. Newtonian science was apparently deterministic. The role of the Creator was reduced by post-Newtonian philosophers to winding up the world spring and setting it into operation. Thereafter, the universe could run very well without Her. (Cooler heads dealing with these problems in the 1990s would demur.)

Newton's impact on philosophy and religion was as profound as his influence on physics. All out of that key equation $\vec{F} = m\vec{a}$. The arrows remind the student that forces and their consequent accelerations point in some direction. Lots of quantities — mass, temperature, volume, for example — don't point in any direction in space. But "vectors," quantities such as force, velocity, and acceleration, all get little arrows.

Before we leave "Eff equals emm ay," lets dwell a bit on its power. It is the basis of our mechanical, civil, hydraulic, acoustic, and other types of engineering; it is used to understand surface tension, the flow of fluids in pipes, capillary action, the drift of continents, the propagation of sound in air and in steel, the stability of structures like the Sears Tower or one of the most wonderful of all bridges, the Bronx-Whitestone Bridge, arching gracefully over the waters of Pelham Bay. As a boy, I would ride my bike from my home on Manor Avenue to the shores of Pelham Bay, where I watched the construction of this lovely structure. The engineers who designed that bridge had an intimate understanding of Newton's equation; now, as our computers become faster and faster, our ability to solve problems using $F = ma$ ever increases. Ya did good, Isaac Newton!

I promised three laws and have delivered only two. The third law is stated as "action equals reaction." More precisely it asserts that whenever an object A exerts a force on an object B, B exerts an equal and opposite force on A. The power of this law is that it is a requirement for all forces, no matter how generated — gravitational, electrical, magnetic, and so on.

ISAAC'S FAVORITE *F*

The next most profound discovery of Isaac N. had to do with the one specific force he found in nature, the $\vec{F}$ of gravity. Remember that the F in Newton's second law merely means force, any force. When one chooses a force to plug into the equation, one must first define, quantify that force so the equation will work. That means, God help us, another equation.

Newton wrote down an expression for F (gravity) — that is, for when the relevant force is gravity — called the universal law of gravitation. The idea is that all objects exert gravitational forces on one another. These forces depend on how far apart the objects are and how much stuff each object has. Stuff? Wait a minute. Here Newton's partiality for the atomic nature of matter came in. He reasoned that the force of gravity acts on all atoms of the object, not only, for example, those on the surface. The force is exerted by the earth on the apple as a whole. Every atom of the earth pulls on every atom of the apple. And also, we must add, the force is exerted by the apple on the earth; there is a fearful symmetry here, as the earth must move up an infinitesimal amount to meet the falling apple. The "universal" attribute of the law is that this force is everywhere. It's also the force of the earth on the moon, of the sun on Mars, of the sun on Proxima Centauri, its nearest star neighbor at a distance of 25,000,000,000,000 miles. In short, Newton's law of gravity applies to all objects anywhere. The force reaches out, diminishing with the amount of separation between the objects. Students learn that it is an "inverse-square law," which means that the force weakens as the square of the distance. If the separation of two objects doubles, the force weakens to one fourth of what it was; if the distance triples, the force diminishes to one ninth, and so on.

WHAT'S PUSHING UP?

As I've mentioned, force also points — down for gravity on the surface of the earth, for example. What is the nature of the counterforce, the "up" force, the action of the chair on the backside of the sitter, the impact of wooden bat on baseball, or hammer on the nail, the push of helium gas that expands the balloon, the "pressure" of water that propels a piece of wood up if it is forced beneath the surface, the "boing" that holds you up when you lie on bedsprings, the depressing inability of most of us to walk through a wall? The surprising, almost shocking, answer is that all of these "up" forces are different manifestations of the electrical force.

This idea may seem alien at first. After all, we don't feel electric charges pushing us upward when we stand on the scale or sit on the sofa. The force is indirect. As we have learned from Democritus (and experiments in the twentieth century), most of matter is empty space and everything is made of atoms. What keeps the atoms together, and accounts for the rigidity of matter, is the electric force. (The resistance of solids to penetration has to do with the quantum theory, too.) This force is very powerful. There is enough of it in a small metal bathroom scale to offset the pull of the entire earth's gravity. On the other hand, you wouldn't want to stand in the middle of a lake or step off your tenth-floor balcony. In water and especially in air, the atoms are too far apart to offer the kind of rigidity that will offset your weight.

Compared to the electrical force that holds matter together and gives it its rigidity, the gravitational force is extremely weak. How weak? I do the following experiment in a physics class I teach. I take a length of wood, say a one-foot-long piece of two-by-four, and draw a line around it at the six-inch mark. I hold up the two-by-four vertically and label the top half "top" and the bottom half "bot." Holding top, I ask, "Why does bot stay up when the entire earth is pulling down on it?" Answer: "It is firmly attached to top by the cohesive electrical forces of the atoms in the wood. Lederman is holding top." Right.

To estimate how much stronger the electrical force of top pulling up on bot is than the gravitational force (earth pulling down on bot), I use a saw to cut the wood in half along the dividing line. (I've always wanted to be a shop teacher.) At this point I've reduced the electrical forces of top on bot to essentially zero with my saw. Now, about to fall to the floor, two-by-four bot is conflicted. Two-by-four top, its electrical power thwarted by the saw, is still pulling up on bot with its gravity force. The earth is pulling down on bot with gravity. Guess which wins. The bottom half of the two-by-four drops to the floor.

Using the equation for the law of gravity, we can calculate the difference between the two gravitational forces. It turns out that the earth's gravity on bot wins out by being more than one billion times stronger than top's gravity on bot. (Trust me on this one.) Conclusion: The electrical force of top on bot before the saw cut was *at least* one billion times stronger than the gravitational force of top on bot. That's the best we can do in a lecture hall. The actual number is 10^{41}, or a one followed by forty-one zeroes!! Let's write that out:

100,000,000,000,000,000,000,000,000,000,000,000,000,000,000

The number 10^{41} can't be appreciated, no way, but perhaps this will help. Consider an electron and a positron one hundredth of an inch apart. Calculate their gravitational attraction. Now calculate how far apart they would have to be to reduce their electrical force to the value of their gravitational attraction. The answer is some thousand trillion miles (fifty light-years). This assumes that the electric force decreases as the square of the distance — just like the gravitational force. Does that help? Gravity dominates the many motions Galileo first studied because every bit of the planet earth pulls on the things near its surface. In the study of atoms and smaller objects, the gravitational effect is too small to be noticed. In many other phenomena, gravity becomes irrelevant. For example, in the collision of two billiard balls (physicists love collisions as a tool for understanding), the influence of the earth is removed by doing the experiment on a table. The vertical downward pull of gravity is countered by the upward push of the table. What then remains are the horizontal forces that come into play when ball strikes ball.

THE MYSTERY OF THE TWO MASSES

Newton's law of universal gravitation provided the *F* in all cases in which gravitation is relevant. I mentioned that he wrote his *F* so that the force of any object, say the earth, on any other object, say the moon, would depend on the "gravitational stuff" contained in the earth times the gravitational stuff contained in the moon. To quantify this profound truth, Newton came up with another formula, around which we have been dancing. In words, the force of gravity between any two objects, call them A and B, is equal to some numerical constant (usually denoted by the symbol *G*) times the stuff in A (let's denote it by M_A) times the stuff in B (M_B) all divided by the square of the distance between object A and object B. In symbols:

$$F = G\frac{M_A \times M_B}{R^2}$$

We'll call this Formula II. Even diehard innumerates will recognize the economy of our formula. For concreteness you can think of *A* as the earth and *B* as the moon, although in Newton's powerful synthesis the formula applies to all bodies. A specific equation for that two-body system would look like this:

$$F = G\frac{M_{earth} \times M_{moon}}{R^2}$$

The earth-moon distance, *R*, is about 250,000 miles. The constant *G*, if you want to know, is 6.67×10^{-11} in units that measure the *M*'s in kilograms and *R* in meters. This precisely known constant measures the strength of the gravitational force. You don't need to memorize this number or even care about it. Just note that the 10^{-11} means that it is very small. *F* becomes really significant only when at least one of the *M*'s is huge, like all the "stuff" in the earth. If a vengeful Creator could make *G* equal to zero, life would end pretty quickly. The earth would head off on a tangent to its elliptical orbit around the sun and global warming would be dramatically reversed.

The exciting thing is *M*, which we call gravitational mass. I said it measures the amount of stuff in the earth and the moon, the stuff that, by our formula, creates the gravity force. "Wait a second," I hear somebody in the back row groaning. "You've got two masses now. The mass (m) in $F = ma$ (formula I) and the mass (M) in our new formula II. What gives?" Very perceptive. Rather than being a disaster, this is a challenge.

Let's call these two different kinds of masses big *M* and little *m*. Big *M* is the gravitational stuff in an object that *pulls on another object*. Little *m* is inertial mass, the stuff in an object that *resists a force* and determines the resulting motion. These are two quite different attributes of matter. It was Newton's insight to understand that the experiments carried out by Galileo (remember Pisa!) and many others strongly suggested that $M = m$. The gravitational stuff is precisely equal to the inertial mass that appears in Newton's second law.

THE MAN WITH TWO UMLAUTS

Newton did not understand why the two quantities are equal; he just accepted it. He even did some clever experiments to study their equality. His experiments showed equality to about 1 percent. That is, $M/m = 1.00$; *M* divided by *m* results in a 1 to two decimal places. More than two hundred years later, this number was dramatically improved. Between 1888 and 1922, a Hungarian nobleman, Baron Roland Eötvös, in an incredibly clever series of experiments using pendulum bobs of aluminum, copper, wood, and various other materials, proved that the equality of these two very different properties of matter was accurate to better than five parts in a billion. In math this says: $M(\text{gravity})/m(\text{inertia}) = 1.000\ 000\ 000$ plus or minus .000 000 005. That is, it lies between 1.000 000 005 and .999 999 995.

Today we have confirmed this ratio to more than twelve zeroes past the decimal point. Galileo proved in Pisa that two unequal spheres fall at the same rate. Newton showed why. Since big *M* equals little *m*, the force of gravity is proportional to the mass of the object. The gravitational mass (M) of a cannonball might be a thousand times greater than that of a ball bearing. That means the gravitational force on it will be a thousand times greater. But that also means that its inertial mass (*m*) will muster a thousand times more resistance to the force than the inertial mass of the ball bearing. If the two objects are dropped from the tower, the two effects cancel. The cannonball and the ball bearing hit the ground at the same time.

The equality of *M* and *m* was an incredible coincidence, and it tormented scientists for centuries. It was the classical equivalent of 137. And in 1915 Einstein incorporated this "coincidence" into a profound theory known as the general theory of relativity.

Baron Eötvös's research on *M* and *m* was his most noteworthy scientific work but by no means his major contribution to science. Among other things, he was a pioneer in punctuation. Two umlauts! More important, Eötvös became interested in science education and in the training of high school teachers, a subject near and dear to me. Historians have noted how Baron Eötvös's educational efforts led to an explosion of genius — such luminaries as the physicists Edward Teller, Eugene Wigner, Leo Szilard, and the mathematician John von Neumann all came out of Budapest during the Eötvös era. The production of Hungarian scientists and mathematicians in the early twentieth century was so prolific that many otherwise calm observers believe Budapest was settled by Martians in a plan to infiltrate and take over the planet.

The work of Newton and Eötvös is dramatically illustrated by space flight. We have all seen the space capsule video: the astronaut releases his pen, which hovers near him in a delightful demonstration of "weightlessness." Of course, the man and his pen aren't really weightless. The force of gravity is still at work. The earth tugs on the gravitational mass of capsule, astronaut, and pen. Meanwhile, the motion in orbit is determined by the inertial masses, given by formula I. Since the two masses are equal, the motion is the same for all objects. Astronauts, pen, and capsule move together in a dance of weightlessness.

Another approach is to think of the astronaut and pen in free fall. As the capsule orbits the earth, it is actually falling toward the earth. That's what orbiting is. The moon, in a sense, is falling toward the earth; it just never gets there because the surface of the spherical earth falls away at the same rate. So if our astronaut is free falling and his pen is free falling, they're in the same position as the two weights dropped from the Leaning Tower. In the capsule or in free fall, if the astronaut could manage to stand on a scale, it would read zero. Hence the term "weightlessness." In fact, NASA uses the free-fall technique for training astronauts. In simulations of weightlessness, astronauts are taken to high altitude in a jet, which flies a series of forty or so parabolas (there's that form again). On the dive side of the parabola, the astronauts experience free fall . . . weightlessness. (Not without some discomfort, however. The plane is unofficially known as the "vomit comet.")

Space-age stuff. But Newton knew all about the astronaut and his pen. Back in the seventeenth century, he could have told you what would happen on the space shuttle.

THE GREAT SYNTHESIZER

Newton led a semireclusive life in Cambridge, with frequent visits to his family estate in Lincolnshire, at a time when most of the other great scientific minds of England were hanging out in London. From 1684 to 1687 he toiled over what was to be his major work, the *Philosophiae Naturalis Principia Mathematica*. This work synthesized all of Newton's previous studies in mathematics and mechanics, much of which had been incomplete, tentative, ambivalent. The *Principia* was a complete symphony, encompassing all of his past twenty years of effort.

To write the *Principia*, Newton had to recalculate, rethink, review, and collect new data — on the passage of comets, on the moons of Jupiter and Saturn, on the tides in the estuary of the Thames River, and more and more. It is here that he began to insist on absolute space and time and it is here that he expressed with rigor his three laws of motion. Here he developed the concept of mass as the quantity of stuff in a body: "The quantity of matter is that which rises conjointly from its density and its magnitude."

This frenzy of creative production had its side effects. According to the testimony of an assistant who lived with him:

> So intent, so serious upon his studies that he eats very sparingly, nay, oft times he forgets to eat at all. . . . At rare times when he decided to dine in the Hall, he would . . . go out into the street, stop, realize his mistake, would hastily turn back and, instead of going into the Hall, return to his Chambers. . . . He would occasionally begin to write at his desk standing, without giving himself the leisure to draw a chair.

Such is the obsession of the creative scientist.

The *Principia* hit England, indeed Europe, like a bombshell. Rumors of the publication spread rapidly, even before it emerged from the printers. Among physicists and mathematicians, Newton's reputation was already large. The *Principia* catapulted him to legend status and brought him to the attention of philosophers such as John Locke and Voltaire. It was a smash. Disciples, acolytes, and even such eminent critics as Christiaan Huygens and Gottfried Leibniz joined in praise of the awesome reach and depth of the work. His archrival, Robert "Shorty" Hooke, paid Newton's *Principia* the ultimate compliment, asserting that it was plagiarized from him.

When I last visited Cambridge University, I asked to see a copy of the *Principia*, expecting to find it in a glass case in a helium atmosphere. No, there it was, first edition, on the bookshelf in the physics library! This is a book that changed science.

Where did Newton get his inspiration? Again, there was a substantial literature on planetary motion, including some very suggestive work by Hooke. These sources probably had as much influence as the intuitive power suggested by the timeworn tale of the apple. As the story goes, Newton saw an apple fall one late afternoon with an early moon in the sky. That was his link. The earth exerts its gravitational pull on the apple, a terrestrial object, but the force continues and can reach the moon, a celestial object. The force causes the apple to drop to the ground. It causes the moon to circle the earth. Newton plugged in his equations, and it all made sense. By the mid-1680s Newton had combined celestial mechanics with terrestrial mechanics. The universal law of gravitation accounted for the intricate dance of the solar system, the tides, the gathering of stars in galaxies, the gathering of galaxies in clusters, the infrequent but predictable visits of Halley's comet, and more. In 1969, NASA sent three men to the moon in a rocket. They needed space-age technology for the equipment, but the key equations programmed into NASA's computers to chart the trajectory to the moon and back were three centuries old. All Newton's.

THE TROUBLE WITH GRAVITY

We've seen that on the atomic scale, say the force of an electron on a proton, the gravitational force is so small that we'd need a one followed by forty-one zeros to express its weakness. That's like . . . *weak!* On the macroscopic scale the inverse-square law is verified by the dynamics of our solar system. It can be checked in the laboratory only with great difficulty, using a sensitive torsion balance. But the trouble with gravity in the 1990s is that it is the only one of the four known forces that does not conform to the quantum theory. As mentioned earlier, we have identified force-carrying particles associated with the weak, strong, and electromagnetic forces. But a gravity-related particle still eludes us. We have given a name to the hypothetical carrier of the gravity force — it's called the graviton — but we haven't detected it yet. Large, sensitive devices have been built to detect gravity waves, which would emerge from some cataclysmic astronomical event out there — for example, a supernova, or a black hole that eats an unfortunate star, or the unlikely collision of two neutron stars. No such event has yet been detected. However, the search goes on.

Gravity is our number-one problem as we attempt to combine particle physics with cosmology. Here we are like the ancient Greeks, waiting and watching for something to happen, not able to experiment. If we could slam two stars together instead of two protons, then we'd see some real effects. If the cosmologists are right and the Big Bang is really a good theory — and I was assured recently at a meeting that it's still a good theory — then at some early phase all the particles in the universe were in a very small location. The energy per particle became *huge*. The gravitational force, strengthened by all that energy, which is equivalent to mass, becomes a respectable force in the domain of the atom. The atom is ruled by the quantum theory. Without bringing the gravitational force into the family of quantum forces, we'll never understand the details of the Big Bang or, in fact, the deep, deep structure of elementary particles.

ISAAC AND HIS ATOMS

Most Newtonian scholars agree that he believed in a particle-like structure of matter. Gravity was the one force Newton treated mathematically. He reasoned that the force between bodies, whether they be earth and moon or earth and apple, must be a consequence of the force between constituent particles. I would hazard a guess that Newton's invention of the calculus is not unrelated to his belief in atoms. To understand the earth-moon force, for example, one has to apply our formula II. But what do we use for R, the earth-moon distance? If earth and moon were very small, there would be no problem in assigning R. It would be the distance between the centers of the objects. However, to know how the force of a very small particle of earth influences the moon and to add up all the forces of all the particles requires the invention of integral calculus, which is a way of adding an infinite number of infinitesimals. In fact, Newton invented calculus in and around that famous year, 1666, when the physicist claimed his mind was "remarkably fit for invention."

In the seventeenth century there was precious little evidence for atomism. In the *Principia,* Newton said we must extrapolate from sensible experiences to understand the workings of the microscopic particles that make up bodies: "Because the hardness of the whole arises from the hardness of the parts, we . . . justly infer the hardness of the undivided particles not only of the bodies we feel but of all others."

His research in optics led him, like Galileo, to interpret light as a stream of corpuscles. At the end of his book *Opticks,* he reviewed current ideas on light and took this breathtaking plunge:

> Have not the Particles of Bodies certain Powers, Virtues or Forces by which they act at a distance, not only on the rays of Light for reflecting, refracting, and inflecting them, but also upon one another for producing a great part of the phenomena of nature? For it is well known that bodies act on one another by the Attractions of Gravity, Magnetism, and Electricity, and these instances show the tenor and course of nature and *make it not improbable that there may be more attractive powers than these . . . others which reach to small distances as hitherto escape observations; and perhaps electrical attractions may reach to small distances even without being excited by Friction.* [emphasis mine]

Here is prescience, insight, and even, if you like, hints of the grand unification that is the Holy Grail of physicists in the 1990s. Was not Newton calling here for a search for forces within the atom, known today as the strong and weak forces? Forces that work only at "small distances," unlike gravity? He went on to write:

> All these things being considered, it seems probable to me that God in the Beginning formed matter in solid, massy, hard, impenetrable, moveable particles . . . and these primitive particles being solids . . . so very hard as never to wear out or break in pieces, no ordinary power being able to divide what God Himself made one in the first creation.

The evidence was weak, but Newton set a course for physicists that would wind its way relentlessly toward the microworld of quarks and leptons. The quest for an extraordinary force to divide "what God himself made one" is the active frontier of particle physics today.

SPOOKY STUFF

In the second edition of *Opticks,* Newton hedged his conclusions with a series of Queries. These questions are so perceptive — and so open-ended — that one can find anything one wants in them. But it is not so far-fetched to believe that Newton may have anticipated, in some deeply intuitive way, the wave-particle duality of quantum theory. One of the most disturbing ramifications of Newton's theory is the problem of action at a distance. The earth pulls on an apple. It falls to the ground. The sun pulls on the planets; they orbit elliptically. How? How can two bodies, with nothing but space between them, transmit force to each other? One popular model of the time hypothesized an aether, some invisible and insubstantial medium pervading all space, through which object A could make contact with object B.

As we shall see, the aether idea was seized upon by James Clerk Maxwell to carry his electromagnetic waves. This idea was destroyed by Einstein in 1905. But like Pauline's, aether's perils come and go, and today we believe that some new version of aether (really the void of Democritus and Anaximander) is the hiding place of the God Particle.

Newton eventually rejected the notion of an aether. His atomistic view would have required a particulate aether, which he found objectionable. Also the aether would have to transmit forces without impeding the motion of, for example, the planets in their inviolate orbits.

Newton's attitude is illustrated by this paragraph of his *Principia:*

> There is a cause without which these motive forces would not be propagated through the spaces round about; whether that cause be of some central body (such as a magnet in the center of the magnetic force), or anything else that does not yet appear. For I have design only to give a mathematical notion of these forces, without considering their physical causes and feats.

At this, the audience, if they were physicists at a modern seminar, would stand up and cheer, because Newton hits the very modern theme that the test of a theory is its agreement with experiment and observation. So what if Newton (and his present-day admirers) didn't know *Why gravity?* What creates gravity? That is a philosophical question until someone shows that gravity is a consequence of some deeper concept, some symmetry perhaps of higher-dimensional space-time.

Enough of philosophy. Newton advanced our quest for the a-tom enormously by establishing a rigorous scheme of predicting, of synthesis that could be applied to a vast array of physical problems. As these principles caught on, they had, as we have seen, a profound influence on practical arts such as engineering and technology. Newtonian mechanics, and its new mathematics, is truly the base of a pyramid upon which all the layers of physical sciences and technology are built. Its revolution represented a change in the perspective of human thinking. Without this change, there would have been no industrial revolution and no continuing systematic search for new knowledge and new technology. This marks a transition from a static society waiting for something to happen to a dynamic society seeking understanding, knowing that understanding implies control. And the Newtonian imprint gave reductionism a powerful boost.

Newton's contributions to physics and mathematics and his commitment to an atomistic universe are clearly documented. What remains misty is the influence on his scientific work of his "other life," his extensive research in alchemy and his devotion to occult religious philosophy, especially Hermetic ideas that harked back to ancient Egyptian priestly magic. These activities were very largely hidden. As Lucasian professor at Cambridge (Stephen Hawking is the current incumbent) and later as a member of the London political establishment, Newton could not let his devotion to these subversive religious practices be known, for that would have brought him extreme embarrassment if not total disgrace.

We may leave the last comment on Newton's work to Einstein:

> Newton, forgive me; you found the only way which, in your age, was just about possible for a man of highest thought — and creative power. The concepts, which you created, are even today still guiding our thinking in physics, although we now know that they will have to be replaced by others farther removed from the sphere of immediate experience, if we aim at a profounder understanding of relationships.

THE DALMATIAN PROPHET

A final note on this first stage, the age of mechanics, the great era of classical physics. The phrase "ahead of his time" is overused. I'm going to use it anyway. I'm not referring to Galileo or Newton. Both were definitely right on time, neither late nor early. Gravity, experimentation, measurement, mathematical proofs . . . all these things were in the air. Galileo, Kepler, Brahe, and Newton were accepted — heralded! — in their own time, because they came up with ideas that the scientific community was ready to accept. Not everyone is so fortunate.

Roger Joseph Boscovich, a native of Dubrovnik who spent much of his career in Rome, was born in 1711, sixteen years before Newton's death. Boscovich was a great supporter of Newton's theories, but he had some problems with the law of gravitation. He called it a "classical limit," an adequate approximation where distances are large. He said that it was "very nearly correct but that differences from the law of inverse squares do exist even though they are very slight." He speculated that this classical law must break down altogether at the atomic scale, where the forces of attraction are replaced by an oscillation between attractive and repulsive forces. An amazing thought for a scientist in the eighteenth century.

Boscovich also struggled with the old action-at-a-distance problem. Being a geometer more than anything else, he came up with the idea of *fields of force* to explain how forces exert control over objects at a distance. But wait, there's more!

Boscovich had this other idea, one that was real crazy for the eighteenth century (or perhaps any century). Matter is composed of invisible, indivisible a-toms, he said. Nothing particularly new there. Leucippus, Democritus, Galileo, Newton, and others would have agreed with him. Here's the good part: Boscovich said these particles had no size; that is, they were geometrical points. Clearly, as with so many ideas in science, there were precursors to this — probably in ancient Greece, not to mention hints in Galileo's works. As you may recall from high school geometry, a point is just a place; it has no dimensions. And here's Boscovich putting forth the proposition that matter is composed of particles that have no dimensions! We found a particle just a couple of decades ago that fits such a description. It's called a quark.

We'll get back to Mr. Boscovich later.

· 4 ·

STILL LOOKING FOR THE ATOM: CHEMISTS AND ELECTRICIANS

> The scientist does not defy the universe. He accepts it. It is his dish to savor, his realm to explore; it is his adventure and never-ending delight. It is complaisant and elusive but never dull. It is wonderful both in the small and in the large. In short, its exploration is the highest occupation for a gentleman.
>
> — I. I. Rabi

AN ADMISSION: the physicists haven't been the only ones searching for Democritus's atom. Chemists have certainly made their mark, especially during the long era (circa 1600–1900) that saw the development of classical physics. The difference between chemists and physicists is not really insurmountable. I started out as a chemist but switched to physics partly because it was easier. Since then I have frequently noted that some of my best friends talk to chemists.

The chemists did something that the physicists before them hadn't done. They did experiments relevant to atoms. Galileo, Newton, et al., despite their considerable experimental accomplishments, dealt with atoms on a purely theoretical basis. They weren't lazy; they just didn't have the equipment. It was up to the chemists to conduct the first experiments that made atoms reveal their presence. In this chapter we'll dwell on the rich experimental evidence that supported the existence of Democritus's a-tom. We'll see many false starts, some red herrings, and misinterpreted results, always the bane of the experimenter.

THE MAN WHO DISCOVERED NINE INCHES OF NOTHING

Before we get to the hard-core chemists, we must mention one scientist, Evangelista Torricelli (1608–1647), who bridged the gap between the mechanics and the chemists in the attempt to restore atomism as a valid scientific concept. To repeat, Democritus said, "Nothing exists except atoms and empty space; everything else is opinion." Thus, to prove the validity of atomism, you need atoms, but you also need empty space between them. Aristotle opposed the very idea of a vacuum, and even during the Renaissance the Church continued to insist that "nature abhors a vacuum."

That's where Torricelli came in. He was one of Galileo's disciples in that scientist's latter days, and in 1642 Galileo set him to work on a problem. The Florentine well diggers had observed that in suction pumps water will not rise more than 10 meters. Why should this be? The initial hypothesis, advanced by Galileo and others, was that vacuum was a "force" and that the partial vacuum produced by the pumps propelled the water upward. Galileo obviously didn't want to be personally bothered with the well diggers' problem, so he delegated it to Torricelli.

Torricelli figured out that the water was not being pulled up by the vacuum at all, but rather *pushed up* by normal air pressure. When the pump lowers the air pressure above the column of water, the normal air outside the pump pushes down harder on the ground water, forcing water in the pipe upward. Torricelli checked out his theory the year after Galileo died. He reasoned that since mercury is 13.5 times denser than water, air should be able to lift mercury only 1/13.5 times as high as water — or about 30 inches. Torricelli obtained a thick glass tube about 1 meter (about 39 inches) long that was closed at the bottom, open at the top, and did a simple experiment. He filled the tube to the brim with mercury, covered the top with a stopper, then turned the tube upside down, placed it in a bowl of mercury and pulled out the stopper. Some of the mercury poured down out of the tube into the dish. But as Torricelli had predicted, 30 inches of the liquid metal remained in the tube.

This pivotal event in physics is often referred to as the invention of the first barometer, which of course it was. Torricelli noted that the height of the mercury varied from day to day, measuring fluctuations in the atmospheric pressure. For our purposes, however, there was a greater significance. Let's forget about the 30 inches of mercury filling up most of the tube. What's important to us is those 9 odd inches at the top. Those few inches at the top of the tube — the closed end — contained nothing. Really nothing. No mercury, no air, nothing. Well, hardly anything. It's a fair vacuum, but it contains some mercury vapor, the amount depending on the temperature. The vacuum is about 10^{-6} torr. (A torr, after Evangelista, is a measure of pressure; 10^{-6} torr is about one billionth of the normal pressure of the atmosphere.) Modern pumps can get to 10^{-11} torr and better. In any case, Torricelli had attained the first artificially created high-quality vacuum. There was no backing off from this conclusion. Nature may or may not abhor a vacuum, but she has to put up with it. Now that we had proved the existence of empty space, we needed some atoms to put there.

SQUEEZING GAS

Enter Robert Boyle. This Irish-born chemist (1627–1691) was criticized by his peers for being too much a physicist and too little a chemist in his way of thinking, but clearly his accomplishments belong primarily to the realm of chemistry. He was an experimentalist whose experiments often came to naught, yet he advanced the idea of atomism in England and on the continent. He was sometimes known as the Father of Chemistry and the Uncle of the Earl of Cork.

Influenced by Torricelli's work, Boyle became fascinated with vacuums. He hired Robert Hooke, the same Hooke who loved Newton so much, to build an improved air pump for him. The air pump inspired an interest in gases, which Boyle came to realize were a key to atomism. He may have had some help here from Hooke, who pointed out that the pressure a gas exerts on the walls of its container — such as air straining against the sides of a balloon — might result from a torrent of atoms. We don't see individual indentations from the atoms inside a balloon because there are billions and billions of them, which simulate a smooth outward push.

Like Torricelli's, Boyle's experiment involved mercury. Taking a seventeen-foot, J-shaped tube, he sealed the short end; then he poured mercury into the long open end to close off the bottom curve of the J. He then continued to add mercury to the open end. The more he poured, the smaller the space available for the air trapped in the short end. Correspondingly, the air pressure in the small volume increased, as he could easily measure by the extra height of mercury in the open end of the tube. Boyle discovered that the volume of the gas varied inversely with the pressure on it. The pressure on the gas trapped in the closed end results from the extra weight of the mercury plus the atmosphere pushing down on the open end. If he doubled the pressure by adding mercury, the volume of air decreased to one half.

Triple the pressure, and the volume shrank to a third, and so on. This effect became known as Boyle's law, a staple of chemistry to this day.

More important is a stunning implication of this experiment: air, or any gas, can be compressed. One way to understand this is to think of the gas as composed of particles separated by empty space. Under pressure, the particles are pushed closer together. Does this prove that atoms exist? Unfortunately, other explanations can be imagined, and Boyle's experiment only provided evidence consistent with the idea of atomism. The evidence was strong enough, however, to help convince Isaac Newton, among others, that an atomic theory of matter was the way to go. Boyle's compression experiment at the very least challenged the Aristotelian assumption that matter was continuous. There remained the problem of liquids and solids, which could not be squeezed with the same ease as gases. This didn't mean they aren't composed of atoms, just that they have less empty space.

Boyle was a champion of experimentation, which, despite the feats of Galileo and others, was still viewed with suspicion in the seventeenth century. Boyle carried on a long debate with Benedict Spinoza, the Dutch philosopher (and lens grinder), over the question of whether experiment could provide proof. To Spinoza only logical thought was proof; experiment was simply a tool for confirming or refuting an idea. Such great scientists as Huygens and Leibniz also doubted the value of experiment. Experimenters have always had an uphill battle.

Boyle's efforts to prove the existence of atoms (he preferred the term "corpuscles") advanced the science of chemistry, which was in a bit of a mess at the time. The prevailing belief of the day was still the old idea of elements, going back to the air, earth, fire, and water of Empedocles and modified through the years to include salt, sulfur, mercury, phlegm (phlegm?), oil, spirit, acid, and alkali. By the seventeenth century these were not just the simplest substances comprising matter according to the prevailing theory, they were believed to be the *essential ingredients of everything*. Acid, to take one example, was expected to be present in every compound. How confused chemists must have been! With these criteria even the simplest chemical reaction must have been impossible to analyze. Boyle's corpuscles led the way to a more reductionist, and simpler, method of analyzing compounds.

THE NAME GAME

One of the problems faced by chemists in the seventeenth and eighteenth centuries was that the names given to various chemicals made no sense. Antoine-Laurent Lavoisier (1743–1794) changed all that in 1787 with his classic work, *Méthode de Nomenclature Chimique.* Lavoisier could be called the Isaac Newton of chemistry. (Perhaps chemists call Newton the Lavoisier of physics.)

He was an amazing character. An accomplished geologist, Lavoisier was also a pioneer in scientific agriculture, an able financier, and a social reformer who had a hand in fomenting the French Revolution. He established a new system of weights and measures that led to the metric system, in use today in civilized nations. (In the 1990s the United States, not to be left too far behind, is inching toward the metric system.)

The previous century had produced a mountain of data, but they were hopelessly disorganized. The names of substances — pompholyx, colcothar, butter of arsenic, flowers of zinc, orpiment, martial ethiop — were colorful, but gave no clue to an underlying order. One of Lavoisier's mentors once told him, "The art of reasoning is nothing more than a language well arranged," and Lavoisier took this to heart. The Frenchman eventually shouldered the task of rearranging and renaming all of chemistry. He changed martial ethiop to iron oxide; orpiment became arsenic sulfide. The various prefixes, like "ox" and "sulf," and suffixes, like "ide" and "ous," helped organize and catalogue the countless numbers of compounds. What's in a name? Sometimes nomenclature is destiny. Would Archibald Leach have gotten all those movie roles if he hadn't changed his name to Cary Grant?

It wasn't quite that simple for Lavoisier. Before revising the nomenclature, he had to revise chemical theory itself. Lavoisier's major contributions to chemistry had to do with the nature of gases and the nature of combustion. Eighteenth-century chemists believed that if you heated water, you transmuted it to air, which they believed was the only true gas. Lavoisier's studies led to the first realization that any given element could exist in three states: solid, liquid, and "vapor." He also determined that the act of combustion was a chemical reaction in which substances such as carbon, sulfur, and phosphorus combined with oxygen. He displaced the theory of phlogiston, an Aristotelian-like obstacle to a true understanding of chemical reactions. More than this, Lavoisier's style of research — based on precision, exquisite experimental technique, and critical analysis of the assembled data — set chemistry on its modern course. Although Lavoisier's direct contribution to atomism was minor, without his groundwork scientists in the following century could not have discovered the first direct proof of the existence of atoms.

THE PELICAN AND THE BALLOON

Lavoisier was fascinated with water. At the time, many scientists were still convinced that water was a basic element, one that could not be split into smaller components. Some also believed in transmutation, thinking that water could be transmuted into earth, among other things. There were experiments to back this up. If you boil a pot of water long enough, eventually a solid residue will form on the surface. That's water being transmuted into another element, these scientists would say. Even the great Robert Boyle believed in transmutation. He had done experiments showing that plants grow by soaking up water. Ergo, water is transformed into stems, leaves, flowers, and so on. You can see why so many people distrusted experiment. Such conclusions are enough to make you start agreeing with Spinoza.

Lavoisier saw that the flaw in these experiments was one of measurement. He conducted his own experiment by boiling distilled water in a special vessel called a pelican. The pelican was so designed that the water vapor produced by boiling was trapped and condensed in a spherical cap, from which it returned to the boiling pot through two handlelike tubes. In this way no water was lost. Lavoisier carefully weighed the pelican and the distilled water, then boiled the water for 101 days. The long experiment produced an appreciable amount of solid residue. Lavoisier then weighed each element: the pelican, the water, and the residue. The water weighed *exactly the same* after 101 days of boiling, which tells us something about Lavoisier's meticulous technique. The pelican, however, weighed slightly less. The weight of the residue was equal to the weight lost by the vessel. Therefore the residue in the boiling water was not transmuted water but dissolved glass, silica, from the pelican. Lavoisier had shown that experimentation without precise measurement is worthless, even misleading. Lavoisier's chemical balance was his violin; he played it to revolutionize chemistry.

So much for transmutation. But many people, Lavoisier included, still believed water was a basic element. Lavoisier ended that illusion when he invented an apparatus with a double nozzle. He would shoot a different gas through each nozzle, hoping they would combine and form a third substance. One day he decided to work with oxygen and hydrogen, expecting them to mix together into some kind of acid. What he got was water. He described it as "pure as distilled water." Why not? He was making it from scratch. Obviously, water was not an element but a substance that could be manufactured from two parts hydrogen, one part oxygen.

In 1783 a historical event occurred that would indirectly further chemistry. The Montgolfier brothers demonstrated the first manned air flights with hot-air balloons. Soon thereafter J. A. C. Charles, a physics teacher no less, rose to a height of 10,000 feet in a balloon filled with hydrogen. Lavoisier was impressed; he saw in such balloons the possibility of rising above the clouds to study meteors. Soon thereafter he was named to a committee to explore methods of cheaply producing gas for the balloons. Lavoisier set up a large-scale operation to produce hydrogen by decomposing water into its constituent parts by percolating it through a gun barrel filled with hot iron rings.

At this point, no one with any sense still believed that water was an element. But there was a bigger surprise for Lavoisier. He was splitting apart water now in vast quantities, and the numbers always came out the same. Water yielded oxygen and hydrogen in a weight ratio of eight to one every time. Clearly, some sort of neat mechanism was at work here, a mechanism that might be explained by an argument based on atoms.

Lavoisier did not speculate much about atomism, except to say that simple indivisible particles are at work in chemistry and we don't know much about them. Of course, he never had the opportunity to sit back in retirement and write his memoirs, in which he might have elaborated further on atoms. An early supporter of the Revolution, Lavoisier fell out of favor during the Reign of Terror and was sent to the guillotine in 1794 at the age of fifty.

The day after Lavoisier's execution, the geometer Joseph Louis Lagrange summed up the tragedy: "It took them only an instant to cut off that head, and a hundred years may not produce another like it."

BACK TO THE ATOM

The implications of Lavoisier's work were investigated a generation later by a modest, middle-class English schoolteacher named John

Dalton (1766–1844). In Dalton we have at last our made-for-TV-movie image of a scientist. He appears to have led a totally uneventful private life and never married, saying that "my head is too full of triangles, chemical processes, and electrical experiments, etcetera, to think much of marriage." A big day for him was a walking tour and maybe attendance at a Quaker meeting.

Dalton started out as a humble teacher in a boarding school, where he filled his spare hours reading the works of Newton and Boyle. He put in over a decade at this job before landing a position as a professor of mathematics at a college in Manchester. When he arrived he was informed that he would also have to teach chemistry. He complained about twenty-one hours of teaching per week! In 1800 he resigned to open his own teaching academy, which gave him the time to pursue his chemical research. Until he unveiled his atomic theory of matter shortly after the turn of the century (between 1803 and 1808), Dalton was still considered little more than an amateur in the scientific community. As far as we know, Dalton was the first to formally resurrect Democritus's term *atom* to mean the tiny indivisible particles that make up matter. There was a difference, however. Recall that Democritus said atoms of different substances had different shapes. In Dalton's scheme, weight played the crucial role.

Dalton's atomic theory was his most important contribution. Whether it was "in the air" (it was) or whether history gives far too much credit to Dalton (as some historians say), no one questions the tremendous influence of the atomic theory on chemistry, a discipline that soon became one of the most pervasively influential sciences. That the first experimental "proof" of the reality of atoms came from chemistry is also most appropriate. Remember the ancient Greek passion: to see an unchanging "arche" in a world in which change is everywhere. The a-tom resolved the crisis. By rearranging a-toms, one can create all the change one wants, but the rock of our existence, the a-tom, is immutable. In chemistry, a relatively small number of atoms provide enormous choice because of the possible combinations: the carbon atom with one oxygen atom or two, hydrogen with oxygen, or chlorine or sulfur and so on. Yet the atoms of hydrogen are always hydrogen — identical one to another, immutable. But here we go, forgetting our hero Dalton.

Dalton, noting that the properties of gases can best be explained by postulating atoms, applied this idea to chemical reactions. He noticed that a chemical compound always contains the same weights of its constituent elements. For example, carbon and oxygen combine to form carbon monoxide (CO). To make CO, one always needs 12 grams of carbon and 16 grams of oxygen, or 12 pounds of carbon and 16 pounds of oxygen. Whatever units you use, the ratio is always 12 to 16. What can the explanation be? If one atom of carbon weighs 12 units and one atom of oxygen weighs 16 units, then the macroscopic weights of carbon and oxygen that disappear into CO will have this same ratio. This alone would be a weak argument for atoms. However, when you make hydrogen-oxygen compounds and hydrogen-carbon compounds, the relative weights of hydrogen, carbon, and oxygen are always 1 to 12 to 16. One begins to run out of alternative explanations. When the same logic is applied to many dozen compounds, atoms become the only sensible conclusion.

Dalton revolutionized science by declaring that the atom is the basic unit of the chemical element and that each chemical atom has its own weight. Here he is, writing in 1808:

> There are three distinctions in the kinds of bodies, or three states, which have more specially claimed the attention of philosophical chemists; namely, those which are marked by the terms elastic fluids, liquids, and solids. A very famous instance is exhibited to us in water, of a body, which, in certain circumstances, is capable of assuming all three states. In steam we recognize a perfectly elastic fluid, in water a perfect liquid, and in ice a complete solid. These observations have tacitly led to the conclusion which seems universally adopted, that all bodies of sensible magnitude, whether liquid or solid, are constituted of a vast number of extremely small particles, or atoms of matter bound together by a force of attraction, which is more or less powerful according to circumstances. . . .
>
> Chemical analysis and synthesis go no farther than to organize the separation of particles one from another and their reunion. No new creation or destruction of matter is within the reach of chemical agency. We might as well attempt to introduce a new planet into the solar system, or to annihilate one already in existence, as to create or destroy a particle of hydrogen. All the changes we can produce consist in separating particles that are in a state of cohesion or combination, and joining those that were previously at a distance.

The contrast between Lavoisier and Dalton in scientific styles is interesting. Lavoisier was a meticulous measurer. He insisted on precision, and this paid off in a dramatic restructuring of chemical methodology. Dalton had many things wrong. He used 7 instead of 8 for the relative weight of oxygen to hydrogen. He had the composition of water and ammonia wrong. Nevertheless, he made one of the profound scientific discoveries of the age: after some 2,200 years of speculation and vague hypothesis, Dalton established the reality of atoms. He presented a new view which, "if established, as I doubt not it will in time, will produce the most important changes in the system of chemistry and reduce the whole to a science of great simplicity." His apparatus was not a powerful microscope, not a particle accelerator, but some test tubes, a chemical balance, the chemical literature of his day, and creative inspiration.

What Dalton called an atom was certainly not the a-tom that Democritus envisioned. We now know that an oxygen atom, for example, is not indivisible. It has a complex substructure. But the name stuck: what we commonly call an atom today is Dalton's atom. It's a *chemical* atom, a single unit of a *chemical element,* such as hydrogen, oxygen, carbon, or uranium.

•

Headline in the *Royal Enquirer* in 1815:

> CHEMIST FINDS ULTIMATE PARTICLE,
> ABANDONS BOA CONSTRICTORS, URINE

Once in a blue moon a scientist comes along who makes an observation that is so simple and elegant that it just has to be right, an observation that appears to solve, in one swift stroke, a problem that has tormented science for thousands of years. Once in a hundred blue moons the scientist is actually right.

All you can say about William Prout is that he came very close. Prout put forward one of the great "almost correct" guesses of his century. His guess was rejected for the wrong reasons and by the fickle finger of fate. Around 1815 this English chemist thought he had found the particle from which all matter was made. It was the hydrogen atom.

To be fair, it was a profound, elegant idea, albeit "slightly" wrong. Prout was doing what a good scientist does: looking for simplicity, in the Greek tradition. He was looking for a common denominator among the twenty-five known chemical elements at the time. Frankly, Prout was a bit out of his field. To his contemporaries, his main accomplishment was writing the definitive textbook on urine. He also conducted extensive experiments with boa constrictor excrement. How this led him to atomism, I don't care to speculate.

Prout knew that hydrogen, with an atomic weight of 1, was the lightest of all the known elements. Maybe, said Prout, hydrogen is the "primary matter," and all the other elements are simply combinations of hydrogens. In the spirit of the ancients, he named this quintessence "protyle." His idea made a lot of sense, because the atomic weights of most of the elements were close to integers, multiples of the weight of hydrogen. The reason for this was that relative weights were then typically inaccurate. As the precision of atomic weights improved, the Prout hypothesis was crushed (for the wrong reason). Chlorine, for example, was found to have a relative weight of 35.5. That blew away Prout's concept because you can't have half an atom. We now know that natural chlorine is a mixture of two varieties, or isotopes. One has 35 "hydrogens" and the other has 37 "hydrogens." These "hydrogens" are really neutrons and protons, which have almost the same mass.

What Prout had really hypothesized was the existence of the nucleon (either of the particles, the proton or the neutron, that make up the nucleus) as the building block of atoms. It was a hell of a good try by Prout. The drive for a system simpler than the set of twenty-five or so elements was destined to succeed.

Not in the nineteenth century, however.

PLAYING CARDS WITH THE ELEMENTS

We end our breakneck jaunt through two hundred–plus years of chemistry with Dmitri Mendeleev (1834–1907), the Siberian-born chemist responsible for the periodic table of the elements. The table was an enormous step forward in classification and at the same time constituted progress in the search for Democritus's atom.

Even so, Mendeleev took a lot of guff in his lifetime. This odd man — he seems to have survived on a diet based on sour milk (he was testing some medical fad) — was subjected by his colleagues to considerable derision for his table. He was also a great supporter of his students at the University of St. Petersburg, and when he stood behind them during a protest late in his career, the administration booted him out.

Without students, he might never have constructed the periodic table. When first appointed to the chair of chemistry in 1867, Mendeleev couldn't find an acceptable text for his classes, so he began writing his own. Mendeleev saw chemistry as "the science of mass" — there's that concern with mass again — and in his textbook he came up with the simple idea of arranging the known elements by the order of their atomic weights.

He did so by playing cards. He wrote the symbols of the elements with their atomic weights and various properties (for example, sodium: active metal; argon: inert gas) each on a separate note card. Mendeleev enjoyed playing patience, a kind of solitaire. So he played patience with the elements, arranging the cards so that the elements were in order of increasing atomic weights. He then discovered a certain periodicity. Similar chemical properties reappeared in elements spaced eight cards apart; for example, lithium, sodium, and potassium are all chemically active metals, and their positions are 3, 11, and 19. Similarly, hydrogen (1), fluorine (9), and chlorine (17) are active gases. He rearranged the cards so that there were eight vertical columns, with the elements in each column having similar properties.

Mendeleev did something else that was unorthodox. He felt no compulsion to fill in all the slots in his grid of boxes. Just as in solitaire, he knew that some of the cards were hidden in the deck. He wanted the table to make sense not only reading across the rows but also reading down the columns. If a space called for an element with particular properties and no such element existed, he left it blank rather than trying to force an existing element into the slot. Mendeleev even named the blanks, using the prefix "eka," which is Sanskrit for "one." For example, eka-aluminum and eka-silicon were the gaps in the vertical columns beneath aluminum and silicon, respectively.

The gaps in the table were one of the reasons Mendeleev was so widely mocked. Yet five years later, in 1875, gallium was discovered and turned out to be eka-aluminum, with all the properties predicted by the periodic table. In 1886 germanium was discovered, which turned out to be eka-silicon. The game of chemical solitaire turned out to be not so nutty.

One of the factors that made Mendeleev's table possible was that chemists had become more accurate in measuring the atomic weights of the elements. Mendeleev himself had corrected the atomic weights of several elements, which did not win him many friends among those important scientists whose figures were being revised.

No one understood why the regularities appeared in the periodic table until the discovery in the following century of the nucleus and the quantum atom. In fact, the initial impact of the periodic table was to discourage scientists. There were fifty or more substances called "elements," basic ingredients of the universe that presumably could not be subdivided further — this meant more than fifty different "atoms," and the number was soon to swell to over ninety. This is a long way from an ultimate building block. Looking at the periodic table in the late 1800s should have made scientists tear their hair out. Where's the simple unity we've been seeking for over two millennia? Yet the order that Mendeleev found in this chaos pointed to a deeper simplicity. In retrospect, the organization and regularities of the periodic table cried out for an atom with some structure that repeated itself periodically. Chemists, however, were not ready to abandon the notion that their chemical atoms — hydrogen, oxygen, and so on — were indivisible. A more fruitful attack would come from a different angle.

Don't blame Mendeleev for the complexity of the periodic table, though. He was simply organizing the confusion as best he could, doing what good scientists do — looking for order in the midst of the complexity. He never was fully appreciated by his peers during his lifetime, never won the Nobel Prize, even though he was alive for several years after the founding of the Prize. At his death in 1907, however, he received the ultimate honor for a teacher. A band of students followed his funeral procession, carrying high above them the periodic table. His legacy is the famous chart of the elements that hangs in every laboratory, every high school chemistry classroom in the world.

For the final stage in the oscillating development of classical physics we swing from the study of matter and particles back to the study of a force. In this case, electricity. In the nineteenth century, electricity was considered almost a science unto itself.

It was a mysterious force. And at first appearance, it didn't seem to occur naturally, except in the frightening form of lightning. So researchers had to do an "unnatural" thing to study electricity. They had to "manufacture" this phenomenon before they could analyze it. We have come to realize that electricity is everywhere; all matter is electrical in nature. Keep this in mind when we get to the modern era, when we discuss exotic particles "manufactured" in accelerators.

Electricity was considered as exotic in the nineteenth century as quarks are today. Today electricity surrounds us, another example of how humans can alter their own environment.

There were many heroes of electricity and magnetism in this early period, many of whom left their names on various electrical units. They include Charles Augustin de Coulomb (the unit of charge), André Ampère (current), Georg Ohm (resistance), James Watt (electrical power), and James Joule (energy). Luigi Galvani gave us the galvanometer, a device for measuring currents, and Alessandro Volta gave us the volt (a unit of potential or electromotive force). Similarly C. F. Gauss, Hans Christian Oersted, and W. E. Weber all made their mark and left their names on electrical quantities calculated to generate fear and loathing in future students of electrical engineering. Only Benjamin Franklin failed to get his name on any electrical unit, despite his significant contributions. Poor Ben! Well, he has his stove and his portrait on those hundred-dollar bills. Franklin noted that there are two kinds of electricity. He could have called one Joe and the other Moe, but he chose instead plus (+) and minus (−). Franklin termed the amount of, say, negative electricity on an object "electric charge." He also introduced the concept of conservation of charge, that when electricity is transferred from one body to another, the total charge must add to zero. But the giants among all of these scientists were two Englishmen, Michael Faraday and James Clerk Maxwell.

ELECTRIC FROGS

Our story begins in the late 1700s with Galvani's invention of the battery, which was later improved by Volta, another Italian. Galvani's study of frog reflexes — he hung frog muscles on the latticework outside his window and watched them twitch during thunderstorms — demonstrated "animal electricity." This stimulated Volta's work about 1790, and a good thing too. Think of Henry Ford installing a box of frogs in each of his cars with instructions to the motorist: "Frogs must be fed every fifteen miles." What Volta found was that the frog electricity had to do with two dissimilar metals separated by some kind of frog goop, for Galvani's frogs were hung on brass hooks on an iron latticework. Volta was able to produce an electrical current sans frog by experimenting with different pairs of metals separated by pieces of leather (standing in for the frogs) soaked in brine. He soon created a "pile" of zinc and copper plates, realizing that the larger the pile, the more current he could drive through an external circuit. Crucial to this work was Volta's invention of an electrometer for measuring the current. This research yielded two important results: a laboratory tool for producing currents and a realization that electricity could be produced by chemical reactions.

Another important development was Coulomb's measurement of the strength and behavior of the electrical force between two charged balls. To make this measurement he invented the torsion balance, a device exquisitely sensitive to tiny forces. The force he was after, of course, was electricity. Using his torsion balance, Coulomb determined that the force between electrical charges varied inversely as the square of the distance between them. He also discovered that like sign charges (+ + or − −) repelled one another, whereas unlike charges (+ −) attracted. Coulomb's law, giving the *F* for electric charges, will play a crucial role in our understanding of the atom.

In a veritable frenzy of activity, there began a series of experiments on what scientists first believed to be the separate phenomena of electricity and magnetism. In a brief period of about fifty years (1820–1870) these experiments led to a grand synthesis that resulted in a unified theory that included not only electricity and magnetism but light as well.

SECRET OF THE CHEMICAL BOND: PARTICLES AGAIN

Much of our early knowledge of electricity emerged from discoveries in chemistry, specifically what is now called electrochemistry. Volta's battery taught scientists that an electrical current can flow around a circuit in a wire that reaches from one pole of the battery to the other. When the circuit is interrupted by attaching wires to pieces of metal immersed in a liquid, the current flows through the liquid. The current in the liquid, they found, creates a chemical process: decomposition. If the liquid is water, hydrogen gas appears near one piece of metal, oxygen near the other. The proportion of 2 parts hydrogen to 1 part oxygen indicates that water is being decomposed into its constituents. A solution of sodium chloride would result in a plating of sodium on one "terminal" and the appearance of the greenish gas chlorine at the other. The industry of electroplating would soon emerge.

The decomposition of chemical compounds by an electrical current indicated something profound: a connection between atomic binding and electric forces. The notion gained currency that the attractions between atoms — that is, the "affinity" one chemical has for another — were electrical in nature.

Michael Faraday began his work in electrochemistry by systematizing the nomenclature. As with Lavoisier's naming of chemicals, this helped a lot. Faraday called the metals immersed in the liquid "electrodes." The negative electrode was a "cathode," the positive an "anode." When the electricity zipped through the water, it impelled a migration of charged atoms through the liquid from cathode to anode. Normally, chemical atoms are neutral, having neither a positive nor a negative charge. But the electric current somehow charged the atoms. Faraday called these charged atoms "ions." Today we know that an ion is an atom that has become charged because it has lost or gained one or more electrons. In Faraday's time, they didn't know about electrons. They didn't know what electricity was. But did Faraday suspect the existence of electrons? In the 1830s he carried out a series of spectacular experiments that resulted in two simple summary statements known as Faraday's laws of electrolysis:

1. The mass of chemical released at an electrode is proportional to the current multiplied by the length of time it runs. That is, the released mass is proportional to the amount of electricity that passes through the liquid.
2. The mass liberated by a fixed quantity of electricity is proportional to the atomic weight of the substance multiplied by the number of atoms in the compound.

What these laws mean is that electricity is not smooth and continuous but can be divided into "chunks." Given Dalton's idea of atoms, Faraday's laws tell us that atoms in the liquid (ions) migrate to the electrode, where each ion is presented with a unit quantity of electricity that converts it to a free atom of hydrogen, oxygen, silver, or whatever. The Faraday laws thus point to an unavoidable conclusion: there are *particles of electricity.* This conclusion, however, had to wait about sixty years to be dramatically confirmed by the discovery of the electron at the end of the century.

A SHOCK IN COPENHAGEN

To continue the history of electricity — the stuff that appears in two or three slots in your electrical outlets, for a price — we have to go to Copenhagen, Denmark. In 1820 Hans Christian Oersted made a key discovery — some historians claim that it is *the* key discovery. He created an electric current in the approved manner, with wires connecting one terminal of a Voltaic contraption (battery) to the other. Electricity was still a mystery, but an electric current involved something called electric charge, moving through a wire. No surprise there, until Oersted placed a compass needle (a magnet) near the circuit. When the current flowed, the compass needle veered from pointing to the geographic North Pole (its normal job description) to taking a funny position at right angles to the wire. Oersted worried about this effect until it dawned on him that, after all, a compass is designed to detect magnetic fields. So the current in the wire must be producing a magnetic field, no? Oersted had discovered a connection between electricity and magnetism: *currents produce magnetic fields*. Magnets of course also produce magnetic fields, and their ability to attract pieces of iron (or to hold snapshots on refrigerator doors) was well studied. The news traveled across Europe and created a great stir.

Running with this information, the Parisian André Marie Ampère found a mathematical relation between current and a magnetic field. The detailed strength and direction of this field depended on the current flowing and on the shape (straight, circular, or whatever) of the wire carrying the current. By a combination of mathematical reasoning and many experiments hastily carried out, Ampère generated a one-man storm of controversy out of which emerged, in the fullness of time, a prescription for calculating the magnetic field produced by an electric current through any configuration of wire — straight, bent, formed into a circular loop, or wound densely on a cylindrical form. Since current passed through two straight wires produces two magnetic fields, these fields can push on each other; effectively, the wires exert force on each other. This discovery made possible Faraday's invention of the electric motor. The fact that a circular loop of current produces a magnetic field was also profound. Could it be that what the ancients called lodestones, natural magnets, actually are composed of atomic-scale circular currents? Another clue to the electrical nature of atoms.

Oersted, like so many other scientists, felt driven toward unification, simplification, reduction. He believed that gravity, electricity, and magnetism were all different manifestations of a single force, which is why his discovery of a direct connection between two of these forces was so exciting (shocking?). Ampère, too, looked for simplicity; he essentially tried to eliminate magnetism by considering it an aspect of electricity in motion (electrodynamics).

DÉJÀ VU ALL OVER AGAIN

Enter Michael Faraday (1791–1867). (Okay, he has already entered, but this is his formal intro. Fanfare, please.) If Faraday was not the greatest experimenter of his time, he is certainly a candidate for that title. It is said that he has generated more biographies than Newton, Einstein, or Marilyn Monroe. Why? Partly it's the Cinderella aspect of his career. Born into poverty, at times hungry (he was once given a loaf of bread as his only food for a week), Faraday was practically unschooled, with a strong religious upbringing. Apprenticed to a bookbinder at the age of fourteen, he actually managed to read some of the books he bound. He thus educated himself while developing a manual dexterity that would serve him well as an experimenter. One day a client brought in a copy of the third edition of the *Encyclopaedia Britannica* to be rebound. It had an article on electricity. Faraday read it, was hooked, and the world changed.

Think about this. Two items are received by the network news offices over the AP wires:

> FARADAY DISCOVERS ELECTRICITY,
> ROYAL SOCIETY LAUDS FEAT

and

> NAPOLEON ESCAPES FROM ST. HELENA,
> CONTINENTAL ARMIES ON THE MARCH

Which item makes the six o'clock news? Right! Napoleon. But over the next fifty years Faraday's discovery literally electrified England and set in motion as radical a change in the way people live on this planet as has ever flowed from the inventions of one human being. Now if only the gatekeepers of TV journalism had been forced to satisfy a real science requirement in college . . .

CANDLES, MOTORS, DYNAMOS

Here is what Michael Faraday accomplished. Starting his professional life as a chemist at the age of twenty-one, he discovered a number of organic compounds, including benzene. He made the transition to physics by cleaning up electrochemistry. (If those University of Utah chemists who thought they had discovered cold fusion in 1989 had understood Faraday's laws of electrolysis better, perhaps they would never have embarrassed themselves as well as the rest of us.) Faraday then went on to make major discoveries in the fields of electricity and magnetism. He:

- discovered the law (named for him) of induction, whereby a changing magnetic field creates an electric field
- was the first to produce an electric current from a magnetic field
- invented the electric motor and dynamo
- demonstrated the relation between electricity and chemical bonding
- discovered the effect of magnetism on light
- and much more!

All this without a Ph.D., M.A., B.A., or high school equivalency degree! He was also mathematically illiterate. He wrote up his discoveries not in equations but in plain descriptive language, often accompanied by pictures to explain the data.

In 1990 the University of Chicago launched a TV series called *The Christmas Lectures,* and I was honored to give the first one. I called it "The Candle and the Universe," borrowing the idea from Faraday, who started the original Christmas lectures for children in 1826. In his first talk he argued that all known scientific processes were illustrated by the burning candle. This was true in 1826, but by 1990 we had learned about a lot of processes that do *not* take place in the candle because the temperature is too low. Nevertheless, Faraday's lectures on the candle were lucid and entertaining and would make a great Christmas present for your children if some silver-voiced actor would make some CD's. So add another facet to this remarkable man — Faraday as popularizer.

We have already discussed his electrolysis research, which prepared the way for the discovery of the electrical structure of chemical atoms and, indeed, for the existence of the electron. Now I want to describe Faraday's two most remarkable contributions: electromagnetic induction and his almost mystical concept of "field."

The route to the modern understanding of electricity (more properly, electromagnetism or the electromagnetic field) is akin to the famous baseball double-play combination Tinker to Evers to Chance. In this case it's Oersted to Ampère to Faraday. Oersted and Ampère made the first steps in understanding electric currents and magnetic fields. Electric currents flowing in wires, like those in your house, make magnetic fields. Thus you can make as powerful a magnet as you want, from the tiny battery-operated magnets that drive small fans to the giant ones used in particle accelerators, by organizing currents. This understanding of electromagnets illuminates our understanding of natural magnets as containing atomic-scale current elements that cooperate to generate a magnet. Nonmagnetic materials also have these Ampèrian atomic currents, but their random orientation produces no net magnetism.

Faraday struggled for a long time to unify electricity and magnetism. If electricity can make magnetic fields, he wondered, can magnets make electricity? Why not? Nature loves symmetry. But it took him more than ten years (1820–1831) to prove it. This was probably his greatest achievement.

Faraday's experimental discovery is called electromagnetic induction, and the symmetry he sought emerged in a surprising form. The road to fame is paved with good inventions. Faraday first wondered whether a magnet could make a current-carrying wire move. Visualizing the forces, he rigged up a device that consisted of a wire connected to a battery at one end, with the other end hanging in a beaker of mercury. The electric wire hung free so that it could revolve around an iron magnet in the beaker. When the current was turned on, the wire moved in a circle around the magnet. We know this odd invention today as an electric motor. Faraday had converted electricity to motion, which could do work.

Let's jump to 1831 and another invention. Faraday wrapped a large number of turns of copper wire on one side of a soft iron doughnut, then connected the two ends of the coil to a sensitive current-measuring device called a galvanometer. He wrapped a similar length of wire on the other side of the doughnut, connecting these ends to a battery so that current could flow in the coil. This device is now called a transformer. Let's review. We have two coils wound on opposite sides of a doughnut. One, let's call it A, is connected to a battery; the other (B) is connected to a galvanometer. What happens when you turn on the juice?

The answer is important to the history of science. When the current flows in coil A, the electricity produces magnetism. Faraday reasoned that this magnetism should induce a current in coil B. But instead he got a strange effect. When he turned on the current, the needle in the galvanometer connected to coil B deflected — voilà! electricity! — but only momentarily. After the sudden jump, the needle remained pointed maddeningly to zero. When he disconnected the battery, the needle deflected briefly in the opposite direction. Increasing the sensitivity of the galvanometer had no effect. Increasing the number of turns in each coil had no effect. Using a much stronger battery had no effect. And then the Eureka moment (in England it is called the By Jove moment): Faraday figured out that current in the first coil had induced a current in the second, but only when the first current was changing. So, as the next thirty years or so of research showed, a *changing* magnetic field generates an electric field.

The technology that emerged in due course was the electric generator. By rotating a magnet mechanically, one can produce a constantly changing magnetic field, which will generate an electric field and, if connected to a circuit, an electric current. One can rotate a magnet by turning a crank, by using the force of a waterfall, or by harnessing a steam turbine. Now we had a way of generating electricity to turn night into day and to energize those electrical outlets in home and factory.

But we pure scientists . . . we are on the track of the a-tom and the God Particle; we dwell on the technology only because it would have been awfully hard to make particle accelerators without Faraday's electricity. As for Faraday, he probably wouldn't have been impressed with the electrification of the world except that now he could work at night.

Faraday built the first hand-cranked electrical generator himself; it was called a dynamo in those days. But he was too involved in the "discovery of new facts . . . being assured that the latter [practical applications] would find their full development hereafter" to figure out what to do with it. The story is often told that the British prime minister visited Faraday's laboratory in 1832 and, pointing to the funny machine, asked what use it was. "I know not, but I wager that one day your government will tax it," said Faraday. A tax on electrical generation was levied in England in 1880.

THE FIELD BE WITH YOU

Faraday's major *conceptual* contribution, crucial to our history of reductionism, was *the field.* To prepare for this, we must go back to Roger Boscovich, who published a radical hypothesis some seventy years before Faraday's time, carrying the a-tom an important step forward. How do a-toms collide? he asked. When billiard balls collide, they deform; their elastic recovery pushes the balls apart. But a-toms? Can one imagine a deformed a-tom? What would deform? What recover? Boscovich was led by such thinking to reduce a-toms to a dimensionless, structureless mathematical point. This point is the source of forces, both attractive and repulsive. He constructed a detailed geometric model that treated atomic collisions very plausibly. The point a-tom did everything that Newton's "hard, massy atom" did but offered advantages. Although it had no extension, it did have inertia (mass). Boscovich's a-tom reached out into space via forces radiating from it. This is an extremely prescient concept. Faraday also was convinced that a-toms were points, but since he could not offer proof, his support was muted. The Boscovich/Faraday view was this: matter consists of point a-toms *surrounded by forces*. Newton had said force acts on mass, so this was clearly an extension of his idea. How does this force manifest itself?

"Lets play a game," I say to the students in a large lecture hall. "When the student to your left lowers his hand, you raise and lower your hand." At the end of the row we pass the signal up one row and switch to "student on your right." We begin with the student at the extreme left of the front row, who raises her hand. Soon the "hand-up" wave travels across the room, up, back across, and so on until it peters out at the top of the hall. What we have is a disturbance propagating with some speed through a medium of students. It's the same principle as the wave, seen in football stadiums across the land. A water wave has the same properties. Although the disturbance propagates, the water particles stay put, bobbing up and down but not involved in the horizontal velocity of the disturbance. The "disturbance" is the height of the wave. The medium is water. The velocity depends on the properties of water. Sounds propagate through air in much the same way. But how does a force reach out from atom to atom through intervening empty space? Newton punted. "I frame no hypothesis," he said. Framed or not, the common hypothesis for how a force propagate was the mysterious *action-at-a-distance*, a kind of placeholder for a future understanding of how gravity is supposed to work.

Faraday introduced the concept of *field*, the ability of space to be disturbed because of a *source* somewhere. The most common example is a magnet reaching for iron nails. Faraday pictured the space around the magnet or coil as being "strained" because of the source. The field concept emerged painfully over many years in many writings, and historians enjoy differing on how, what, and when it all came out. Here is a note from Faraday in 1832: "When a magnet acts upon a distant magnet or piece of iron, the influencing cause . . . proceeds gradually from magnetic bodies and *requires time for its transmission* [emphasis mine]." Thus the concept is that a "disturbance" — for example a magnetic field strength of 0.1 tesla — can travel through space and notify a grain of iron powder not only that it is there but that it can exert a force. This is what a strong water wave does to an unwary bather. The water wave — say it's three feet high — needs water in which to propagate. We must still wrestle with what the magnetic field needs. Later.

Magnetic lines of force are revealed in the old experiment you did in school by sprinkling iron powder on a sheet of paper placed over a magnet. You gave the paper a tap to break the surface friction, and the iron powder clustered in a definite pattern of lines connecting the poles of the magnet. Faraday thought these lines were real manifestations of his field concept. The important issue is not so much Faraday's ambiguous descriptions of this alternative to action-at-a-distance but the way the concept was altered and used by our next electrician, Scotsman James Clerk (pronounced "klark") Maxwell (1831–1879).

Before we leave Faraday, we should clarify his attitude toward atoms. He left us two gemlike quotes from 1839:

> Although we know nothing of what an atom is, yet we cannot resist forming some idea of a small particle which represents it to the mind — there is an immensity of facts which justify us in believing that the atoms of matter are in some way associated with electrical powers, to which they owe their most striking qualities, and amongst them their chemical affinity [attraction of atom to atom].

and

> I must confess that I am jealous of the term atom, for although it is very easy to talk of atoms, it is very difficult to form a clear idea of their nature when compound bodies are under consideration.

Abraham Pais, citing these statements in his book *Inward Bound*, concludes: "That is the true Faraday, exquisite experimentalist, who would only accept what he was forced to believe on experimental grounds."

AT THE SPEED OF LIGHT

If the first play was Oersted to Ampère to Faraday, the next was Faraday to Maxwell to Hertz. Although Faraday the inventor changed the world, his science could not stand by itself and would have dead-ended if it were not for Maxwell's synthesis. For Maxwell, Faraday provided a semiarticulate (that is, nonmathematical) insight. Maxwell played Kepler to Faraday's Brahe. Faraday's magnetic lines of force acted as a steppingstone to the field concept, and his extraordinary comment in 1832 that electromagnetic actions are not transmitted instantaneously but require a well-defined time played a key role in Maxwell's great discovery.

Maxwell gave full credit to Faraday, even admiring his mathematical illiteracy because it forced him to express his ideas in "natural, untechnical language." Maxwell asserted that his primary motivation was to translate Faraday's view of electricity and magnetism to mathematical form. But the treatise that evolved went far beyond Faraday.

In the years 1860–1865 Maxwell's papers — models of dense, difficult, complicated mathematics (ugh!) — emerged as the crowning glory of the electrical period of science that had begun in dim history with amber and lodestones. In this final form Maxwell not only set Faraday to mathematical music (albeit atonal) but in so doing established the existence of electromagnetic waves moving through space at some finite velocity, as Faraday had predicted. This was an important point; many of Faraday and Maxwell's contemporaries thought forces were transmitted instantaneously. Maxwell specified how Faraday's field would work. Faraday had found experimentally that a changing magnetic field generates an electric field. Maxwell, struggling for symmetry and consistency in his equations, postulated that a changing electric field would generate a magnetic field. This produced, in the mathematical stuff, a surging back and forth of electric and magnetic fields, which, in Maxwell's notebooks, took off through space, speeding away from their sources at a velocity that depended on all kinds of electrical and magnetic quantities.

But there was a surprise. Not predicted by Faraday, and essentially Maxwell's major discovery, was the actual velocity of these electromagnetic waves. Maxwell pored over his equations, and after he plugged in the proper experimental numbers, out came 3×10^8 meters per second. "Gor luv a duck!" he said, or whatever Scotsmen say when they're surprised. Because 3×10^8 meters per second is the speed of light (which had been measured for the first time a few years earlier). As we learned with Newton and the mystery of the two kinds of masses, there are few real coincidences in science. Maxwell concluded that light is but one example of an electromagnetic wave. Electricity need not be confined to wires but can disseminate through space as light does. "We can scarcely avoid the inference," wrote Maxwell, "that light consists in the transverse undulations of the same medium which is the cause of electric and magnetic phenomena." Maxwell opened the possibility, which Heinrich Hertz seized, of verifying his theory by experimentally generating electromagnetic waves. It was left to others, including Guglielmo Marconi and a host of more modern inventors, to develop the second "wave" of electromagnetic technology: radio, radar, television, microwave, and laser communications.

Here is the way it works. Consider an electron at rest. Because of its electric charge, an electric field exists everywhere in space, stronger near the electron, weaker as we go farther away. The electric field "points" toward the electron. How do we know there is a field? Simple: place a positive charge anywhere, and it will feel a force pointing toward the electron. Now force the electron to accelerate up a wire. Two things happen. The electric field changes, not instantly but as soon as the information arrives at the point in space where we are measuring it. Also, a moving charge *is* a current, so a magnetic field is created.

Now apply forces on the electron (and on many of its friends) so that it surges up and down the wire at a regular cycle. The resulting *change* in electric and magnetic fields propagates away from the wire with a finite velocity — the velocity of light. This is an electromagnetic wave. We often call the wire an antenna and the force driving the electron a radio frequency signal. Thus the signal, with whatever message is contained in it, propagates away from the antenna at the speed of light. When it reaches another antenna, it will find plenty of electrons, which it will, in turn, force to jiggle up and down, creating an oscillating current that can be detected and converted to video and audio information.

Despite his monumental contribution, Maxwell was anything but an overnight sensation. Let's look at what the critics said of Maxwell's treatise:

- "A somewhat gross conception." — Sir Richard Glazebrook
- "A feeling of uneasiness, often even of mistrust is mingled with admiration . . ." — Henri Poincaré
- "Found no foothold in Germany and was scarcely even noticed." — Max Planck
- "I may say one thing about it [the electromagnetic theory of light]. I do not think it is admissible." — Lord Kelvin

With reviews like these it is hard to become a superstar. It took an experimenter to make Maxwell a legend, though not in his own time, for he died about a decade too soon.

HERTZ TO THE RESCUE

The true hero (to this highly biased student of historians) is Heinrich Hertz who, in a series of experiments spanning more than a decade (1873–1888), confirmed all the predictions of Maxwell's theory.

Waves have a wavelength, which is the distance between crests. The crests of water waves in the ocean typically may be twenty to thirty feet apart. Sound wavelengths range around inches. Electromagnetism also comes in waves. The difference between various electromagnetic waves — infrared, microwave, x-rays, radio waves — is simply a matter of their wavelengths. Visible light — blue, green, orange, red — is in the middle of the electromagnetic spectrum. Radio waves and microwaves have longer wavelengths. Ultraviolet, x-rays, and gamma rays have shorter wavelengths.

Using a high-voltage coil and a detection device, Hertz found a way to generate electromagnetic waves and measure their speed. He showed that these waves had the same reflection, refraction, and polarization properties as light waves and that they could be focused. Despite the bad reviews, Maxwell was right. Hertz, in subjecting Maxwell's theory to rigorous experiment, clarified and simplified it to a "system of four equations," which we'll get to in a moment.

After Hertz, Maxwell's ideas became generally accepted, and the old problem of action-at-a-distance was put to rest. Forces in the form of fields propagated through space with a finite velocity, the speed of light. Maxwell felt that he needed a medium to support his electric and magnetic fields, so he adapted the Faraday-Boscovich notion of an all-pervading aether in which the electric and magnetic fields vibrated. Just like Newton's discarded aether, this aether had weird properties and would soon play a crucial role in the next scientific revolution.

The Faraday-Maxwell-Hertz triumph spelled another success for reductionism. No longer did universities have to hire a professor of electricity, a professor of magnetism, and a professor of light or optics. These are all unified, and only one position is now needed (more money for the football team). A vast set of phenomena is encompassed: both things created by science and things natural — like motors and generators, transformers, and an entire electrical power industry, like sunlight and starlight, radio and radar and microwaves, and infrared and ultraviolet light and x-rays and gamma rays and lasers. The propagation of all of these is explained by Maxwell's four equations, which in their modern form, applied to electricity in free space, are written:

$$c\,\nabla \times E = -(\delta B/\delta t)$$
$$c\,\nabla \times B = (\delta E/\delta t)$$
$$\nabla \cdot B = 0$$
$$\nabla \cdot E = 0$$

In these equations, E stands for the electric field, B stands for the magnetic field, and c, the velocity of light, stands for a combination of electric and magnetic quantities that can be measured on a lab bench. Note here the symmetry of E and B. Never mind the incomprehensible doodles; for our purposes it's not important to explain the workings of these equations. The point is, this is the scientific summons: "Let there be light!"

Physics and engineering students the world over wear T-shirts sporting these four crisp equations. Maxwell's original equations, however, looked nothing like the above. These simple versions are the work of Hertz, a rare example of someone who was more than the usual experimenter with only a working grasp of theory. He was exceptional in both areas. Like Faraday, he was aware of, but uninterested in, the immense practical importance of his work. He left that to lesser scientific minds, such as Marconi and Larry King. Hertz's theoretical work consisted largely of cleaning up Maxwell, reducing and popularizing his theory. Without Hertz's efforts, physics students would have to lift weights so they could wear triple-extra-large T-shirts in order to accommodate Maxwell's clumsy mathematics.

True to our tradition and our promise to Democritus, who recently faxed us a reminder, we have to interview Maxwell (or his estate) on atoms. Of course he believed. He was also the author of a very successful theory that treated gases as an assembly of atoms. He believed, correctly, that chemical atoms were not just tiny rigid bodies, but had some complex structure. This belief came out of his knowledge of optical spectra, which became important, as we shall see, in the development of quantum theory. Maxwell believed, incorrectly, that his complex atoms were uncuttable. He said it so beautifully in 1875: "Though in the course of ages catastrophies have occurred and may yet occur in the heavens, though ancient systems may be dissolved and new systems evolved out of their ruins, the [atoms] out of which these systems [earth, solar system, and so on] are built — the foundation stones of the material universe — remain unbroken and unworn." If only he had used the terms "quarks and leptons" instead of "atoms."

The ultimate judgment on Maxwell comes again from Einstein, who stated that Maxwell made the single most important contribution of the nineteenth century.

THE MAGNET AND THE BALL

We have glossed over some important details in our story. How do we know that fields propagate at a fixed speed? How did physicists in the nineteenth century even know what the speed of light was? And what is the difference between instantaneous action-at-a-distance and time-delayed response?

Consider a very powerful electromagnet at one end of a football field and, at the other end, an iron ball suspended by a thin wire from a very high support. The ball tilts ever so slightly toward the faraway magnet. Now suppose we *very rapidly* turn the current off in the electromagnet. Precise observations of the ball and its wire would record a response as the ball relaxes back to its equilibrium position. But is the response instantaneous? Yes, say the action-at-a-distance folk. The connection between magnet and iron ball is tight and, when the magnet disappears, the ball *instantaneously* begins to move back to zero tilt. "No!" say the finite-velocity people. The information "magnet is turned off, you can relax now" travels across the gridiron with a definite velocity, so the ball's response is delayed.

Today we know the answer. The ball has to wait, not very long because the information travels at the speed of light, but there is a measurable delay. But in Maxwell's time this problem was at the heart of a raging debate. At stake was the validity of the field concept. Why didn't scientists just do an experiment and settle the issue? Because light is so fast that it takes only one millionth of a second to cross the football field. In the 1800s that was a difficult delay to measure. Today it is trivial to measure time intervals a thousand times shorter, so the finite propagation of electromagnetic happenings is easily gauged. For example, we bounce laser signals off a new reflector on the moon to measure the distance between earth and moon. The round trip takes about 1.0 second.

An example on a larger scale: On February 23, 1987, at exactly 7:36 UT Greenwich mean time, a star was observed to explode in the southern sky. This supernova event took place in the Large Magellanic Cloud, a cluster of stars and dust located 160,000 light-years away. In other words, it took 160,000 years for the electromagnetic information from the supernova to arrive at planet earth. And Supernova 87A was a relatively near neighbor. The most distant object observed is about 8 billion light-years old. Its light set out for our telescope rather close to the Beginning.

The velocity of light was first measured in an earthbound laboratory by Armand-Hippolyte-Louis Fizeau, in 1849. Lacking oscilloscopes and crystal-controlled clocks, he used an ingenious arrangement of mirrors (to extend the length of the light path) and a rapidly rotating toothed wheel. If we know how fast the wheel is turning, and we know the radius of the wheel, we can calculate the time it takes for a gap to be replaced by a tooth. We can adjust the rotation speed so that this time is precisely the time a light beam takes to proceed from gap to distant mirror and back to gap, and then through gap to the eyeball of M. Fizeau. *Mon dieu! I see it!* Now gradually speed up the wheel (shorten the time) until the light is blocked. There. Now we know the distance the beam traveled — from light source through gap to mirror and back to wheel tooth — and we know the time it took. Fiddling with this arrangement gave Fizeau the famous number 300 million (3×10^8) meters per second or 186,000 miles per second.

I am continually surprised at the philosophical depth of all these guys during this electromagnetic renaissance. Oersted believed (contrary to Newton) that all forces of nature (at the time: gravity, electricity, and magnetism) were different manifestations of one primordial force. This is s-o-o-o modern! Faraday's efforts to establish the symmetry of electricity and magnetism invokes the Greek heritage of simplicity and unification, 2 of the 137 goals at Fermilab in the 1990s.

TIME TO GO HOME?

In these past two chapters we've covered more than three hundred years of classical physics, from Galileo to Hertz. I've left out some good people. The Dutchman Christiaan Huygens, for example, told us a lot about light and waves. The Frenchman René Descartes, the founder of analytical geometry, was a leading advocate of atomism, and his sweeping theories of matter and cosmology were imaginative but unsuccessful.

We've looked at classical physics from an unorthodox point of view, that of searching for Democritus's a-tom. Usually the classical era is viewed as an examination of forces: gravity and electromagnetism. As we've seen, gravitation derives from the attraction between masses. In electricity Faraday recognized a different phenomenon; matter is irrelevant here, he said. Let's look at force fields. Of course, once you have a force you must still invoke Newton's second law ($F = ma$) to find the resultant motion, and here inertial matter really matters. Faraday's matter-doesn't-matter approach was derived from the intuition of Boscovich, a pioneer in atomism. And, of course, Faraday provided the first hints about "atoms of electricity." Perhaps one isn't supposed to look at science history this way, as a search for a concept, the ultimate particle. Yet, it's there beneath the surface in the intellectual lives of many of the heroes of physics.

By the late 1890s, physicists thought they had it all together. All of electricity, all of magnetism, all of light, all of mechanics, all moving things, as well as cosmology and gravity — all were understood by a few simple equations. As for atoms, most chemists felt that the subject was pretty much closed. There was the periodic table of the elements. Hydrogen, helium, carbon, oxygen et al. were indivisible elements, each with its own invisible, indivisible atom.

There *were* some mysterious cracks in the picture. For example, the sun was puzzling. Using then-current beliefs in chemistry and atomic theory, the British scientist Lord Rayleigh calculated that the sun should have burned up all its fuel in 30,000 years. Scientists knew that the sun was a lot older than that. This aether business was also troubling. Its mechanical properties would have to be bizarre indeed. It had to be totally transparent, capable of slipping between atoms of matter without disturbing them, yet it had to be as rigid as steel to support the huge velocity of light. Still, it was hoped that these and other mysteries would be solved in due time. Had I been teaching back in 1890, I might have been tempted to send my physics students home, advising them to find a more interesting major. All the big questions had been answered. Those issues that were not understood — the sun's energy, radioactivity, and a number of other puzzles — well, everybody believed that sooner or later they would yield to the power of the Newton-Maxwell theoretical juggernaut. Physics had been neatly wrapped up in a box and tied with a bow.

Then suddenly, at the end of the century, the whole package began to unravel. The culprit, as usual, was experimental science.

THE FIRST TRUE PARTICLE

During the nineteenth century, physicists fell in love with the electrical discharges produced in gas-filled glass tubes when the pressure was lowered. A glass blower would fashion an exquisite three-foot-long glass tube. Metal electrodes were sealed into the tube. The experimenter would pump out the air as best he could, then bleed in a desired gas (hydrogen, air, carbon dioxide) at low pressure. Wires from each electrode were attached to an external battery. Large electrical voltages were applied. Then, in a darkened room, experimenters were awed as splendid glows appeared, changing shape and size as the pressure decreased. Anyone who has seen a neon sign is familiar with this kind of glow. At low enough pressure, the glow turned into a ray, which traveled from the cathode, the negative terminal, toward the anode. Logically, it was dubbed a cathode ray. These phenomena, which we now know to be rather complex, fascinated a generation of physicists and interested laypersons all over Europe.

Scientists knew a few controversial, even contradictory details about these cathode rays. They carried a negative charge. They traveled in straight lines. They could spin a fine paddle wheel sealed into the glass. Electric fields *didn't* deflect them. Electric fields *did* deflect them. A magnetic field would cause a narrow beam of cathode rays to bend into a circular arc. The rays were stopped by thick metal but could penetrate thin metal foils.

Interesting facts, but the critical mystery remained: what were these rays? In the late nineteenth century, there were two guesses. Some researchers thought cathode rays were massless electromagnetic vibrations in the aether. Not a bad guess. After all, they glowed like a light beam, another kind of electromagnetic vibration. And obviously, electricity, which is a form of electromagnetism, had something to do with the ray.

Another camp thought the rays were a form of matter. A good guess was that they were composed of gas molecules in the tubes that had picked up a charge from the electricity. Another guess was that cathode rays were composed of a new form of matter, small particles never before isolated. For a variety of reasons, the idea that there is a basic carrier of electric charge was "in the air." We'll let the cat out of the bag right now. Cathode rays weren't electromagnetic vibrations and they weren't gas molecules.

If Faraday had been alive in the late 1800s, what would he have said? Faraday's laws strongly implied that there were "atoms of electricity." As you'll recall, he did some similar experiments, except that he passed electricity through liquids rather than gases, ending up with ions, charged atoms. As early as 1874 George Johnstone Stoney, an Irish physicist, had coined the term "electron" for the unit of electricity that is lost when an atom becomes an ion. Had Faraday witnessed a cathode ray, perhaps he would have known in his heart that he was watching electrons at work.

Some scientists in this period may have strongly suspected that cathode rays were particles; maybe some thought they had finally found the electron. How do you find out? How do you prove it? In the intense period before 1895, many prominent researchers in England, Scotland, Germany, and the United States were studying gaseous discharges. The one who struck pay dirt was an Englishman named J. J. Thomson. There were others who came close. We'll take a look at two of them and what they did, just to show how heartbreaking scientific life is.

The guy who came nearest to beating Thomson was Emil Weichert, a Prussian physicist, who demonstrated his technique to a lecture audience in January 1887. His glass tube was about fifteen inches long and three inches in diameter. The illuminated cathode rays were easily visible in a partially darkened room.

If you're trying to corral a particle, you must describe its charge (e) and mass (m). At the time, the particle in question was too small to weigh. To get around this problem many researchers independently seized upon a clever technique: subject the ray to known electric and magnetic forces and measure its response. Remember $F = ma$. If indeed the rays were composed of electrically charged particles, the force experienced by the particles would vary with the quantity of charge (*e*) they carried. The response would be muted by their inertial mass (*m*). Unfortunately, therefore, the effect that could be measured was the quotient of these two quantities, the ratio *e/m*. In other words, researchers couldn't find individual values for *e* or *m*, just a number equal to one value divided by the other. Let's look at a simple example. You are given the number 21 and told that it is the quotient of two numbers. The 21 is a clue only. The two numbers you're looking for might be 21 and 1, 63 and 3, 7 and 1/3, 210 and 10, ad infinitum. But if you have an inkling of what one number is, you can deduce the second.

To go after *e/m*, Weichert put his tube into the gap of a magnet, which bent the beam into an arc. The magnet pushes on the electric charge of the particles; the slower the particles, the easier it is for the magnet to bend them into a circular arc. Once he figured out the speed, the deflection of particles by the magnet gave him a fair value for *e/m*.

Weichert understood that if he made an informed guess as to the value of the electric charge, he could deduce the approximate mass of the particle. He concluded: "We are not dealing with atoms known from chemistry because the mass of these moving [cathode ray] particles turns out to be 2,000 to 4,000 times smaller than the lightest known chemical atom, the hydrogen atom." Weichert almost hit the bull's-eye. He knew he was looking at some new kind of particle. He was damn close to the mass. (The electron's mass turned out to be 1,837 times smaller than that of hydrogen.) So why is Thomson famous and Weichert a footnote? Because he simply assumed (guessed) the value of the electric charge; he had no evidence for it. Weichert was also distracted by a job change and a competing interest in geophysics. He was a scientist who reached the right conclusion but didn't have all of the data. No cigar, Emil!

The second runner-up was Walter Kaufmann, in Berlin. He got to the finish line in April 1897, and his shortcoming was the opposite of Weichert's. The book on him was good data, bad thinking. He also derived *e/m* using magnetic and electric fields, but he took the experiment a significant step further. He was especially interested in how the value of *e/m* might change with changes in pressure and with the gas used in the tube — air, hydrogen, carbon dioxide. Unlike Weichert, Kaufmann thought that cathode ray particles were simply charged atoms of the gas in the tube, so they should have a different mass for each gas used. Surprise — he discovered that *e/m* does *not* change. He always got the same number no matter what gas, what pressure. Kaufmann was stumped and missed the boat. Too bad, because his experiments were quite elegant. He got a better value for *e/m* than the champ, J. J. It's one of the cruel ironies of science that he missed what his data were screaming at him: your particles are a new form of matter, dummkopf! And they are the universal constituents of all atoms; that's why *e/m* doesn't change.

Joseph John Thomson (1856–1940) started out in mathematical physics and was surprised when, in 1884, he was appointed professor of experimental physics at the famous Cavendish Laboratory at Cambridge University. It would be nice to know whether he really wanted to be an experimentalist. He was famous for his clumsiness with experimental apparatus but was fortunate in having excellent assistants who could do his bidding and keep him away from all that breakable glass.

In 1896 Thomson sets out to understand the nature of the cathode ray. At one end of his fifteen-inch glass tube the cathode emits its mysterious rays. These head for an anode with a hole that permits some of the rays (read electrons) to pass through. The narrow beam thus formed goes on to the end of the tube, where it strikes a fluorescent screen, producing a small green spot. Thomson's next surprise is to insert into the glass tube a pair of plates about six inches long. The cathode beam passes through the gap between these plates, which Thomson connects to a battery, creating an electric field perpendicular to the cathode ray. This is the deflection region.

If the beam moves in response to the field, that means it is carrying an electric charge. If, on the other hand, the cathode rays are photons — light particles — they will ignore the deflection plates and continue on their way in a straight line. Thomson, using a powerful battery, sees the spot on the fluorescent screen move down when the top plate is negative, up when the top plate is positive. He thus proves that the rays are charged. Incidentally, if the deflection plates carry an alternating voltage (varying rapidly plus-minus-plus-minus), the green spot will move up and down rapidly, creating a green line. This is the first step in making a TV tube and seeing Dan Rather on the CBS nightly news.

But it is 1896, and Thomson has other things on his mind. Because the force (the strength of the electric field) is known, it is easy, using simple Newtonian mechanics, to calculate how far the spot should move *if* one can figure out the velocity of the cathode rays. Now Thomson uses a trick. He places a magnetic field around the tube in such a direction that the magnetic deflection exactly cancels the electric deflection. Since this magnetic force depends on the unknown velocity, he has merely to read the strength of the electric field and the magnetic field in order to derive a value for the velocity. With the velocity determined, he can now go back to testing the deflection of the ray in electric fields. What emerges is a precise value for *e/m*, the ratio of the electric charge on a cathode ray particle divided by its mass.

Fastidiously, Thomson applies fields, measures deflections, cancels deflections, measures fields, and gets numbers for *e/m*. Like Kaufmann, he double-checks by changing the cathode material — aluminum, platinum, copper, tin — and repeating the experiment. All give the same number. He changes the gas in the tube: air, hydrogen, carbon dioxide. Same result. Thomson does not repeat Kaufmann's mistake. He concludes that the cathode rays are not charged gas molecules but fundamental particles that must be part of all matter.

Not satisfied that he has enough proof, he hits on using the idea of conservation of energy. He captures the cathode rays in a metal block. Their energy is known; it is simply the electrical energy given to the particles by the voltage from the battery. He measures the heat generated by the cathode rays, and notes that in relating the energy acquired by the hypothetical electrons to the heat generated in the metal block, the ratio *e/m* appears. In a long series of experiments, Thomson gets a value for *e/m* (2.0×10^{11} coulombs per kilogram), that is not very different from his first result. In 1897 he announces the result: "We have in the cathode rays matter in a new state, a state in which the subdivision of matter is carried very much further than in the ordinary gaseous state." This "subdivision of matter" is an ingredient in all matter and is part of the "substance from which the chemical elements are built up."

What to call this new particle? Stoney's term "electron" was handy, so electron it became. Thomson lectured and wrote about the corpuscular properties of cathode rays from April to August 1897. This is known as marketing your results.

There was still one puzzle to be solved: the separate values of *e* and *m*. Thomson was in the same fix as Weichert a few years earlier. So he did something clever. Noting that the *e/m* of this new particle was about a thousand times bigger than that of hydrogen, the lightest of all the chemical atoms, he realized that either the electron's *e* was much bigger or its *m* was much smaller. What's it to be: big *e* or little *m*? Intuitively, he went with little *m* — a brave choice, for he was guessing that this new particle had a tiny mass, far smaller than that of hydrogen. Remember, most physicists and chemists still thought that the chemical atom was the indivisible a-tom. Thomson now said that the glow in his tube was evidence of a universal ingredient, a smaller constituent of all chemical atoms.

In 1898 Thomson went on to measure the electric charge of his cathode rays, thus indirectly measuring the mass as well. He used a new technique, the cloud chamber, invented by his Scottish student C. T. R. Wilson in order to study the properties of rain, not a rare commodity in Scotland. Rain happens when water vapor condenses on dust to form drops. When the air is clear, electrically charged ions can stand in for dust, and that's what happens in a cloud chamber. Thomson measured the total charge in the chamber using an electrometric technique and determined the individual charge on each droplet by counting them and dividing the total.

I had to build a Wilson cloud chamber as part of my Ph.D. thesis, and I've hated the technique ever since, hated Wilson, hated anyone who had anything to do with this contrary and mulelike device. That Thomson got the correct value of *e* and hence a measurement of the mass of the electron is miraculous. And that's not all. During the whole process of pinning down this particle, his dedication had to be unwavering. How does he know the electric field? Does he read the label on the battery? No labels. How does he know the precise value of his magnetic field in order to measure velocity? How does he measure the current? Reading a pointer on a dial has its problems. The pointer is a bit thick. It may shiver and shake. How is the scale calibrated? Is it meaningful? In 1897 absolute standards were not catalogue items. Measuring voltages, currents, temperatures, pressures, distances, time intervals were all formidable problems. Each required a detailed knowledge of the workings of the battery, the magnets, the meters.

Then there was the political problem: how to convince the powers that be to give you the resources to do the experiment in the first place. Being the boss, as Thomson was, really helped. And I left out the most crucial problem of all: how to decide which experiment to do. Thomson had the talent, the political know-how, the stamina, to carry through where others had failed. In 1898 he announced that electrons are components of the atom and that cathode rays are electrons that have been separated from the atom. Scientists thought the chemical atom was structureless, uncuttable. Thomson had torn it apart.

The atom was split, and we had found our first elementary particle, our first a-tom. Do you hear that giggle?

· 5 ·

THE NAKED ATOM

There's something happening here.
What it is ain't exactly clear.

— Buffalo Springfield

ON NEW YEAR'S EVE 1999, while most of the world prepares for the last blowout of the century, physicists from Palo Alto to Novosibirsk, from Cape Town to Reykjavik, will be resting, having exhausted themselves almost two years earlier celebrating the one hundredth anniversary (in 1998) of the discovery of the electron — the first truly elementary particle. Physicists love to celebrate. They'll celebrate any particle's birthday, no matter how obscure. But the electron, wow! They'll be dancing in the streets.

After its discovery, the electron was frequently toasted in its birthplace, the Cavendish Laboratory at Cambridge University, with: "To the electron, may it remain forever useless!" Fat chance. Today, less than a century later, our entire technological superstructure is based upon that little fellow.

Almost as soon as the electron was born, it began causing problems. It still perplexes us today. The electron is "pictured" as a sphere of electric charge that spins rapidly around an axis, creating a magnetic field. J. J. Thomson struggled mightily to measure the electron's charge and mass, but now both quantities are known to a high degree of precision.

Now for the spooky features. In the curious world of the atom, the radius of the electron is generally taken to be zero. This gives rise to some obvious problems:

- If the radius is zero, what spins?
- How can it have mass?
- Where is the charge?

- How do we know the radius is zero in the first place?
- Can I get my money back?

Here we meet the Boscovich problem face to face. Boscovich solved the problem of "atoms" colliding by making them into points, things with no dimension. His points were literal mathematicians' points, but he allowed these point particles to have conventional properties: mass and something we call charge, the source of a field of force. Boscovich's points were theoretical, speculative. But the electron is real. Probably a point particle, but with all other properties intact. Mass, yes. Charge, yes. Spin, yes. Radius, no.

Think of Lewis Carroll's Cheshire Cat. Slowly the Cheshire Cat disappears until all that's left is its smile. No cat, just smile. Imagine the radius of a spinning glob of charge slowly shrinking until it disappears, leaving intact its spin, charge, mass, and smile.

This chapter is about the birth and development of the quantum theory. It is the story of what happens inside the atom. I begin with the electron, because a particle with spin and mass but no dimension is counterintuitive to us humans. Thinking about such stuff is a kind of mental pushup. It might hurt the brain a bit because you'll have to use certain obscure cerebral muscles that may not have had much use.

Still, the idea of the electron as point mass, point charge, and point spin does raise conceptual problems. The God Particle is intimately tied to this structural difficulty. A deep understanding of mass still escapes us, and the electron in the 1930s and '40s was the harbinger of these difficulties. Measuring the size of the electron became a cottage industry, generating Ph.D.'s galore, from New Jersey to Lahore. Through the years, increasingly sensitive experiments gave smaller and smaller numbers, all consistent with zero radius. It's as if God took the electron in Her hand and squeezed it as small as She could. With the large accelerators built in the 1970s and '80s, the measurements became ever more precise. In 1990 the radius was measured at *less than* .000000000000000001 inches or, scientifically, 10^{-18} centimeters. This is the best "zero" physics can supply . . . so far. If I had a good experimental idea as to how to add a zero I'd drop everything to try to get it approved.

Another interesting property of the electron is its magnetism, which is described by a number called the g-factor. Using quantum theory, the electron's g-factor is calculated to be:

$$2 \times (1.001159652190)$$

And what calculations! It took skilled theorists years and impressive amounts of supercomputer time to come up with this number. But this was theory. For verification, experimenters devised ingenious methods for measuring the g-factor with equivalent precision. The result, obtained by Hans Dehmelt of the University of Washington:

$$2 \times (1.001159652193)$$

As you can see, we have verification to almost twelve places. This is a spectacular agreement of theory and experiment. The point here is that the calculation of the g-factor is an outgrowth of quantum theory, and at the heart of quantum theory lies what are known as the Heisenberg uncertainty principles. In 1927 a German physicist proposed a startling idea: that it is impossible to measure *both* the speed and the position of a particle to arbitrary precision. This impossibility is independent of the brilliance and the budget of the experimenter. It is a fundamental law of nature.

And yet, despite the fact that uncertainty is woven into the very fabric of quantum theory, it churns out predictions, such as the g-factor above, that are accurate to eleven decimal places. Quantum theory is a prima facie scientific revolution that forms the base rock on which twentieth-century science flourishes . . . and it starts with a confession of uncertainty.

How did the theory come about? It's a good detective story, and as in any mystery, there are clues — some valid, some false. There are butlers all over the place to confuse the detectives. The city cops, the state police, the FBI collide, argue, cooperate, fall apart. There are many heroes. There are coups and countercoups. I'll give a very partial view, hoping to convey a sense of the evolution of ideas from 1900 until the 1930s, when the very mature revolutionaries put the "finishing" touches on the theory. But be forewarned! The microworld is counterintuitive: point masses, point charges, and point spins are experimentally consistent properties of particles in the atomic world, but they are not quantities we can see around us in the normal macroscopic world. If we are to survive together as friends through this chapter, we have to learn to recognize hangups derived from our narrow experience as macro-creatures. So forget about normal; expect shock, disbelief. Niels Bohr, one of the founders, said that anyone who isn't shocked by quantum theory doesn't understand it.

Richard Feynman asserted that no one understands quantum theory. ("So, what do you want from us?" say my students.) Einstein, Schrödinger, and other good scientists never accepted the implications of the theory, yet in the 1990s, elements of quantum spookiness are considered crucial to an understanding of the origin of the universe.

The armory of intellectual weapons that the explorers carried with them into the new world of the atom included Newtonian mechanics and Maxwell's equations. All macroscopic phenomena seemed to be subject to these powerful syntheses. But the experiments of the 1890s began to trouble the theorists. We've already discussed cathode rays, which led to the discovery of the electron. In 1895 Wilhelm Roentgen discovered x-rays. In 1896 Antoine Becquerel accidentally discovered radioactivity, when he stored photographic plates near some uranium in a desk drawer. Radioactivity soon led to a concept of *lifetimes.* Radioactive stuff decayed over characteristic times whose average could be measured, but the decay of a particular atom was unpredictable. What did this mean? No one knew. Indeed, all of these phenomena defied explanation by classical means.

WHEN THE RAINBOW ISN'T ENOUGH

Physicists were also beginning to look closely at light and its properties. Newton, using a glass prism, had shown that he could replicate the rainbow by spreading white light out into its spectral composition, the colors going from red at one end of the spectrum to violet at the other, one color graduating smoothly into another. In 1815 Joseph von Fraunhofer, a skilled craftsman, greatly refined the optical system used to observe the colors emanating from the prism. Now when one squinted through a small telescope, the spread-out colors appeared in exquisite focus. With this instrument — bingo! — Fraunhofer made a discovery. The splendid colors of the sun's spectrum were overlaid by a series of fine dark lines, seemingly irregularly spaced. Fraunhofer eventually recorded some 576 of these lines. What did they mean? In Fraunhofer's time light was known to be a wave phenomenon. Later James Clerk Maxwell would show that light waves are electric and magnetic fields and that a crucial parameter is the distance between wave crests, the wavelength, which determines color.

Knowing wavelengths, we can assign a numerical scale to the band of colors. Visible light ranges from deep red, at 8,000 angstrom units

(.00008 cm), to deep violet, at about 4,000 angstrom units. Using such a scale, Fraunhofer could locate precisely each of the fine dark lines. For example, one famous line known as H_α, or "aitch-sub-alpha" (if you don't like aitch-sub-alpha, call it Irving), has a wavelength of 6,562.8 angstrom units, in the green, close to the middle of the spectrum.

Why do we care about these lines? Because by 1859 the German physicist Gustav Robert Kirchhoff had found a deep connection between these lines and the chemical elements. This fellow heated up various elements — copper, carbon, sodium, and so on — by putting them in a hot flame until they glowed. He energized various gases in tubes and used even more improved viewing apparatus to examine the spectra of light emitted by these glowing gases. He discovered that each element emitted a characteristic series of very sharp, bright-colored lines superimposed on a darker glow of continuous colors. Inside the telescope was an engraved scale, calibrated in wavelengths, so that the location of each bright line could be pinpointed. Because the line spacings were different for each element, Kirchhoff and his accomplice, Robert Bunsen, were able to fingerprint elements by their *spectral lines*. (Kirchhoff needed someone to help him heat up the elements; who better than the man who invented the Bunsen burner?) With some skill, researchers could identify small impurities of one chemical element embedded in another. Science now had a tool to examine the composition of anything that gives off light — for example, the sun, and indeed, in time, the distant stars. By finding spectral lines not previously recorded, scientists discovered a lode of new elements. In the sun a new element called helium was identified in 1878. It wasn't until seventeen years later that this star-born element was discovered on earth.

Think of the thrill of discovery when the light from the first bright star was analyzed . . . and was found to be made of the same stuff we have here on earth! Since starlight is very faint, it took great telescopic and spectroscopic skill to study its patterns of colors and lines but the conclusion was unavoidable: the sun and stars are made of the same stuff as the earth. In fact, we've yet to find an element in space that we don't have here on earth. We are all star material. For any overarching concept about the world in which we live, this discovery is clearly of incredible significance. It reinforces Copernicus: we are not special.

Ah, but why was Fraunhofer, the guy who started all this, finding those dark lines in the sun's spectrum? The explanation was soon forthcoming. The hot core of the sun (white, white hot) emitted light of all wavelengths. But as this light filtered through the relatively cool gases at the sun's surface, those gases absorbed the light of just those wavelengths that they like to emit. So Fraunhofer's dark lines represented *absorption*. Kirchhoff's bright lines were light *emissions*.

Here we are in the late 1800s, and what do we make of all this? The chemical atoms are supposed to be hard, massy, structureless, uncuttable a-toms. But each one seems to be capable of emitting or absorbing its own characteristic series of sharp lines of electromagnetic energy. To some scientists, this screamed one word, "structure!" It was well known that mechanical objects have structures that resonate to regular impulses. Piano or violin strings vibrate to make musical notes in their crafted instruments, and wineglasses shatter when a large tenor sings the perfect note. Bridges could be set into violent motion by the unfortunate beat of marching soldiers. Light waves are just that, impulses with a "beat" equal to the velocity divided by the wavelength. These mechanical examples raised the question: if atoms had no internal structure, how could they display resonant properties such as spectral lines?

And if atoms had a structure, what would Newton's and Maxwell's theories say about it? X-rays, radioactivity, the electron, and spectral lines had one thing in common. They couldn't be explained by classical theory (although many scientists tried). On the other hand, none of these phenomena flatly contradicted classical Newton/Maxwell theory either. They just couldn't be explained. But as long as there was no smoking gun, there was hope that some smartass kid eventually could find a way to save classical physics. That never happened. Instead, the smoking gun finally materialized. In fact, there were at least three smoking guns.

SMOKING GUN NO. 1: THE ULTRAVIOLET CATASTROPHE

The first observational evidence that flatly contradicted classical theory was "black body radiation." All objects radiate energy. The hotter they are, the more energy they radiate. A living, breathing human emits about 200 watts of radiation in the invisible infrared region of the spectrum. (Theorists emit 210 watts and politicians go to 250.)

All objects also absorb energy from their surroundings. If their temperature is higher than the surroundings, they cool because they radiate more energy than they absorb. "Black body" is the technical term for an ideal absorber, one that absorbs 100 percent of the radiation hitting it. Such an object, when cold, appears black because it reflects no light. Experimenters like to use a black body as a standard for measuring emitted radiation. What is interesting about the radiation from such an object — a piece of coal, an iron horseshoe, a toaster wire — is the color spectrum of the light: how much light it gives off at the various wavelengths. As we heat these objects, our eyes perceive a dull red glow at first, then, as the objects get hotter, bright red, then yellow, then blue-white, then (lots of heat!) bright white. Why do we end up with white?

The shift of the color spectrum means that the peak intensity of the light is moving, as the temperature is raised, from infrared to red to yellow to blue. As the peak moves, the distribution of light among the wavelengths broadens. By the time the peak is at the blue, so many of the other colors are being radiated that the hot body appears to our eyes as white. White hot, we say. Today astrophysicists are studying the black body radiation left over from the most incandescent radiation in the history of the universe — the Big Bang.

But I digress. In the late 1890s, the data on black body radiation were getting better and better. What did Maxwell's theory say about these data? Catastrophe! It was just wrong. Classical theory predicted the wrong shape for the curve of distribution of light intensity among the various colors, the various wavelengths. In particular, it predicted that the peak quantity of light would always be emitted at the shortest wavelengths, toward the violet end of the spectrum and even into the invisible ultraviolet. This is not what happens. Hence "the ultraviolet catastrophe," and hence the smoking gun.

Initially, it was believed that this failure of the application of Maxwell's equations would be solved by a better understanding of how electromagnetic energy was generated by the radiating matter. The first physicist to appreciate the significance of this failure was Albert Einstein in 1905, but the stage was set for the master by another theorist.

Enter Max Planck, a Berlin theorist in his forties, who had had a long career in physics and was an expert on the theory of heat. He was smart, and he was professorial. Once, when he forgot which room he was supposed to lecture in, he stopped by the department office and asked, "Please tell me in which room does Professor Planck lecture today?" He was told sternly, "Don't go there, young fellow.

You are much too young to understand the lectures of our learned Professor Planck."

In any case, Planck was close to the experimental data, much of which had been acquired by colleagues in his Berlin laboratory, and he was determined to understand them. He made an inspired guess at a mathematical expression that would fit the data. Not only did it fit the distribution of light intensity at any given temperature, but it agreed with the way the curve (the distribution of wavelengths) changed as the temperature changed. For future events it is important to emphasize that a given curve allows one to calculate the temperature of the body emitting the radiation. Planck had reason to be proud of himself. "Today I made a discovery as important as that of Newton," he boasted to his son.

Planck's next problem was to tie his lucky educated guess to some law of nature. Black bodies, so the data insisted, emitted very little radiation at short wavelengths. What "law of nature" could result in a suppression of the short wavelengths so beloved by classical Maxwell theory? A few months after publishing his successful equation, Planck hit on a possibility. Heat is a form of energy, and thus the energy content of the radiating body is limited by its temperature. The hotter the object, the more energy available. In classical theory this energy is distributed equally among the different wavelengths. BUT (get goose pimples, damn it, we are about to discover quantum theory) suppose the amount of energy depends on the wavelength. Suppose short wavelengths "cost" more energy. Then, as we try to radiate shorter wavelengths, we run out of available energy.

Planck found that to justify his formula (now called the Planck law of radiation) he had to make two explicit assumptions. He said, first, that the energy radiated is related to the wavelength of the light, and second, that discreteness is inextricably linked to this phenomenon. Planck could justify his formula and keep peace with the laws of heat by assuming that the radiation was emitted in discrete bundles or "packets" of energy or (here it comes) "quanta." Each bundle's energy is related to the frequency via a simple connection: $E = hf$. A quantum of energy E is equal to the frequency, f, of the light times a constant, h. Since frequency is inversely related to wavelength, the short wavelengths (or high frequencies) cost more energy. At any given temperature, only so much energy is available, so high frequencies are suppressed. This discreteness was essential to get the right answer. Frequency is the speed of light divided by the wavelength.

The constant that Planck introduced, h, was determined by the data. But what is h? Planck called it the "quantum of action," but history calls it Planck's constant, and it will forevermore symbolize the revolutionary new physics. Planck's constant has a value, 4.11×10^{-15} eV-second, for what it's worth. Don't memorize. Just note that it's a very small number, thanks to the 10^{-15} (15 places past the decimal point).

This — the introduction of the notion of a quantum or bundle of light energy — is the turning point, although neither Planck nor his colleagues understood the depth of this discovery. The exception was Einstein, who did recognize the true significance of Planck's quanta, but for the rest of the scientific community it took twenty-five years to sink in. Planck's theory disturbed him; he didn't want to see classical physics destroyed. "We have to live with quantum theory," he finally conceded. "And believe me, it will expand. It will not be only in optics. It will go in all fields." How right he was!

As a final comment, in 1990 the Cosmic Background Explorer (COBE) satellite transmitted back to its delighted astrophysicist masters data on the spectral distribution of the cosmic background radiation that pervades all of space. The data, of unprecedented precision, fit the Planck formula for black body radiation exactly. Remember, the curve of distribution of light intensity allows one to define the temperature of the body emitting the radiation. Using the data from the COBE satellite and Planck's equation, the researchers were able to calculate the average temperature of the universe. It's cold here: 2.73 degrees above absolute zero.

SMOKING GUN NO. 2: THE PHOTOELECTRIC EFFECT

Now we zip over to Albert Einstein, working as a clerk in the Swiss Patent Office in Bern. The year is 1905. Einstein obtained his Ph.D. in 1903 and spent the next year brooding about the system and the meaning of life. But 1905 was a good year for him. He managed to solve three of the outstanding problems of physics that year: the photoelectric effect (our topic), the theory of Brownian motion (look it up!), and, oh yes, the special theory of relativity. Einstein understood that Planck's guess meant that light, electromagnetic energy, was being emitted in discrete globs of energy, hf, rather than in the classical idyll of emission, one wavelength continuously and smoothly changing to another.

This perception must have given Einstein the idea of explaining an experimental observation of Heinrich Hertz, who was generating radio waves to verify Maxwell's theory. Hertz did this by striking sparks between two metal balls. In the course of this work he noticed that sparks would jump across the gap more readily if the balls were freshly polished. He suspected that the polishing enabled the electric charge to leave the surface. Being curious, he spent some time studying the effect of light on metal surfaces. He noticed that the blue-violet light of the spark was essential in drawing charges out of the metal surface. These charges fueled the cycle by aiding the formation of sparks. Hertz reasoned that polishing removes oxides, which interfere with the interaction of light with a metal surface.

The blue-violet light was stimulating electrons to pour out of the metal, which at the time seemed a bizarre effect. Experimenters systematically studied the phenomenon and came up with these curious facts:

1. Red light is incapable of releasing electrons, even if the light is extraordinarily intense.
2. Violet light, even if relatively faint, releases electrons easily.
3. The shorter the wavelength (the more violet the light), the higher the energy of the released electrons.

Einstein realized that Planck's idea that light appears in bundles could be the key to understanding the photoelectric mystery. Imagine an electron, minding its own business in the metal of one of Hertz's highly polished balls. What kind of light can give that electron enough energy to jump out of the surface? Einstein, using Planck's equation, noted that if the wavelength of light is short enough, the electron receives enough energy to breach the surface of the metal and escape. Either the electron swallows the entire bundle of energy or it doesn't, reasoned Einstein. Now, if the wavelength of the bundle swallowed is too long (not energetic enough), the electron cannot escape; it doesn't have enough energy. Drenching the metal with impotent (long-wavelength) bundles of light energy doesn't do any good. Einstein said that what's important is the energy of the bundle, not how many bundles you have.

Einstein's idea works perfectly. In the photoelectric effect the light quanta, or *photons,* are absorbed rather than, as with the Planck theory, emitted. Both processes seem to demand quanta with energy $E = hf$. The quantum concept was gaining. The photon idea wasn't convincingly proven until 1923, when American physicist Arthur Compton succeeded in demonstrating that a photon could collide with an electron much as two billiard balls collide, changing direction, energy, and momentum and acting in every way like a particle — but a very special particle somehow connected with a vibration frequency or wavelength.

Here was a ghost arisen. The nature of light was an old battleground. Recall that Newton and Galileo held that light consisted of "corpuscles." The Dutch astronomer Christiaan Huygens argued for a wave theory. This historic battle of Newton's corpuscles and Huygen's waves had been settled in favor of waves by Thomas Young's double-slit experiment (which we will review soon) early in the nineteenth century. In quantum theory, the corpuscle was resurrected, in the form of the photon, and the wave-corpuscle dilemma was revived with a surprise ending.

But there was even more trouble ahead for classical physics, thanks to Ernest Rutherford and his discovery of the nucleus.

SMOKING GUN NO. 3: WHO LIKES PLUM PUDDING?

Ernest Rutherford is one of those characters almost too good to be true, as if he were delivered to the scientific community by Central Casting. A big, gruff New Zealander with a walrus moustache, Rutherford was the first foreign research student admitted to the famed Cavendish Laboratory, run at the time by J. J. Thomson. Rutherford arrived just in time to witness the discovery of the electron. Good with his hands (unlike his boss, J.J.), he was an experimenter's experimenter, a worthy rival to Faraday as the best ever. He was known for his profound belief that swearing at an experiment made it work better, a notion backed up by experimental results, if not theory. In evaluating Rutherford one must especially add in his students and postdocs, who, under his baleful eye, carried out great experiments. There were many: Charles D. Ellis (discoverer of beta decay), James Chadwick (discoverer of the neutron), and Hans Geiger (of counter fame), among others. Don't think it's easy to supervise some fifty graduate students. For one thing, one must read their papers. Listen to one of my best students begin his thesis: "This field of physics is so virginal that no human eyeball has ever set foot in it." But back to Ernest.

Rutherford had ill-concealed contempt for theorists, though, as you'll see, he wasn't such a bad one himself. And it's a good thing there wasn't the press coverage of science at the turn of the century that there is today. Rutherford was so quotable he'd have skewered himself out of a truckful of grants. Here are a few Rutherfordisms that have leaked down to us over the decades.

- "Don't let me catch anyone talking about the universe in my department."
- "Oh that stuff [relativity]. We never bother with that in our work."
- "All science is either physics or stamp collecting."
- "I've just been reading some of my early papers and, you know, when I finished, I said to myself, 'Rutherford, my boy, you used to be a damned clever fellow.' "

This damned clever fellow put in his time with Thomson, then crossed the Atlantic to work at McGill University in Montreal, then trekked back to England for a post at Manchester University. By 1908 he had won a Nobel Prize for his work with radioactivity. That would seem a fitting climax to a career for most men, but not for Rutherford. Now his work began in Ernest.

One cannot talk about Rutherford without talking about the Cavendish Lab, created in 1874 as the research laboratory of Cambridge University. The first director was Maxwell (a theorist as lab director?). The second was Lord Rayleigh, followed by Thomson in 1884. Rutherford arrived from the boonies of New Zealand as a special research student in 1895 at a fantastic time for rapid developments. One of the major ingredients for professional success in science is luck. Without this, forget it. Rutherford had it. His work on the newly discovered radioactivity — Becquerel rays they were called — honed him for his most important discovery, the atomic nucleus, in 1911. He made that discovery at the University of Manchester, then returned in triumph to the Cavendish, where he succeeded Thomson as director.

You'll recall that Thomson had seriously complicated the issue of matter by discovering the electron. The chemical atom, thought to be the indivisible particle put forth by Democritus, now had little guys running around inside. These electrons had a negative charge, which presented a problem. Matter is neutral, neither positive nor negative. So what offsets the electrons?

The dramatic story begins quite prosaically. The boss comes into the lab. There sit a postdoc, Hans Geiger, and an undergraduate hanger-on, Ernest Marsden. They are engaged in alpha-particle scattering experiments. A radioactive source — for example, radon 222 — naturally and spontaneously emits alpha particles. The alpha particles, it turns out, are nothing but helium atoms without their electrons — that is, helium nuclei, as Rutherford discovered in 1908. The radon source is placed in a lead case with a narrow hole that aims the alpha particles at a piece of extremely thin gold foil. As the alphas pass through the foil, their paths are deflected by the gold atoms. The angles of these deflections are the subject of the study. Rutherford had set up what became the historical prototype of a scattering experiment. You shoot particles at a target and see where they bounce. In this case the alpha particles were little probes whose purpose was to find out how atoms are structured. The gold-foil target is surrounded on all sides — 360 degrees — by zinc sulfide screens. When a zinc sulfide molecule is struck by an alpha particle, it emits a flash of light, which allows the researchers to measure the angle of deflection. The alpha particle zips into the gold foil, hits an atom, and is deflected into one of the zinc sulfide screens. Flash! Most of the alpha particles are deflected only slightly and strike the zinc sulfide screen directly behind the gold foil. It was a tough experiment to do. They had no particle counter — Geiger hadn't invented it yet — so Geiger and Marsden were forced to sit in a dark room for several hours to adapt their eyesight to see the flashes. Then they had to spot and catalogue the number and positions of the little sparks.

Rutherford — who didn't have to sit in dark rooms because he was the boss — said: "See if any of the alpha particles are *reflected* from the foil." In other words, see if any of the alphas hit the gold foil and bounce back toward the source. Marsden recalled, "To my surprise I was able to observe the effect. . . . I told Rutherford when I met him later, on the steps leading to his room."

The data, later published by Geiger and Marsden, recorded that one in 8,000 alpha particles was reflected from the metal foil. Rutherford's now-famous reaction to this news: "It was quite the most incredible event that ever happened to me in my life. It was as if you fired a fifteen-inch artillery shell at a piece of tissue paper and it came back and hit you."

This was May 1909. Early in 1911 Rutherford, acting now as a theoretical physicist, cracked the problem. He greeted his students with a broad smile. "I know what the atom looks like and I understand the strong backward scattering." In May of that year, his article declaring the existence of the nuclear atom was published. This was the end of an era. The chemical atom was now seen, correctly, as complex, not simple, and as cuttable, not at all a-tomlike. It was the beginning of a new era, the era of nuclear physics, and it marked the demise of classical physics, at least inside the atom.

Rutherford took at least eighteen months to think through a problem that is now solved by physics majors in their junior year. Why was he so puzzled by the ricocheting alpha particles? It had to do with how scientists at the time viewed the atom. Here is the massive, positively charged alpha particle charging into a gold atom and bouncing backward. The 1909 consensus was that the alpha should have blasted right through, like an artillery shell through tissue paper, to use Rutherford's metaphor.

The tissuelike model of the atom went back to Newton, who said forces have to cancel out if one is to have mechanical stability. Thus the electrical forces of attraction and repulsion had to be balanced in a stable atom that you could trust. The theorists of that turn-of-the-century epoch went in for a frenzy of model making, trying to arrange the electrons to make a stable atom. Atoms were known to have lots of negatively charged electrons. Therefore they had to have an equal amount of positive charge distributed in some unknown way. Since the electrons are very light and the atom is heavy, either an atom must have thousands of electrons (to make the weight) or the weight must reside in the positive charge. Out of the many models proposed, by 1905 the leading model of the day had been postulated by none other than J. J. Thomson, Mr. Electron. It was called the plum-pudding model because it had the positive charge spread out in a sphere covering the entire atom, with the electrons embedded throughout like plums in a pudding. Such an arrangement appeared to be mechanically stable and even allowed the electrons to vibrate around equilibrium locations. But the nature of the positive charge was a complete mystery.

Rutherford, on the other hand, calculated that the only configuration capable of knocking an alpha particle backward was one in which the entire mass and positive charge were concentrated in a very small volume in the center of a relatively huge (atom-size) sphere. The nucleus! The electrons would be spaced throughout the sphere. In time and with better data, Rutherford's theory was refined. The central positive charge (nucleus) occupies a volume no more than one trillionth of the volume of the atom. According to the Rutherford model, matter is predominantly empty space. When we pound on a table, it feels solid, but it is the interplay of electrical forces (and quantum rules) among atoms and molecules that creates the illusion of solidity. The atom is mostly void. Aristotle would be appalled.

Rutherford's surprise at the alpha particles bouncing backward may be appreciated if we abandon his artillery shell and think instead of a bowling ball thundering down the alley toward an array of tenpins. Picture the bowler's shock if the ball were stopped by the pins and then rebounded, careening back to the bowler, who would then run for her life. Could this happen? Well, suppose somewhere in the middle of the triangular array of pins there was a special "fat pin" made out of solid iridium, the densest metal known. This pin is heavy! It weighs fifty times more than the ball. A sequence of time-lapse photos would show the ball impinging on the fat pin, deforming it but coming to rest. Then, as the pin re-forms to its original shape and, indeed, recoils just a little bit, it imparts a resounding force to the ball, which reverses its velocity. This is what happens in any elastic collision, say of a billiard ball and cushion. Rutherford's more picturesque military shell metaphor was derived from his preconception, and that of most physicists of his day, that the atom was a sphere of pudding tenuously spread over a large volume. For a gold atom, this was an "enormous" sphere of radius 10^{-9} meters.

To get a sense of the Rutherford atom, if we picture the nucleus as the size of a green pea (about a quarter inch in diameter) the atom is a sphere of radius 300 feet, something that can surround six football fields, packed into a rough square. Rutherford's luck held here too. His radioactive source just happened to produce alphas with an energy of about 5 million electron volts (we write it 5 MeV), which was ideal for discovering the nucleus. The energy was low enough that the alpha particle never got too close to the nucleus but was turned back by its strong positive electric charge. The surrounding cloud of electrons had much too little mass to have any appreciable effect on the alpha. If the alpha had had much more energy, it would have penetrated the nucleus, sampling the strong nuclear force (we'll learn about this later) and greatly complicating the pattern of scattered alpha particles. (The vast majority of alphas pass through the atom so far from the nucleus that their deflections are small.) As it was, the pattern of scattered alpha particles, as subsequently measured by Geiger and Marsden and eventually by a host of continental competitors, was mathematically equivalent to what would be expected if the nucleus were a point. We know now that nuclei are not points, but if the alpha particles don't get too close, the arithmetic is the same.

Boscovich would have been pleased; the Manchester experiments backed up his vision. The outcome of a collision is determined by the force fields around the "point" things. Rutherford's experiment had implications beyond the discovery of the nucleus. It established that very large deflections imply small "pointlike" concentrations, a crucial idea experimenters eventually employed when going after quarks, the real points. In the slowly emerging view of the structure of the atom, Rutherford's model was a clear milestone. It was very much a miniature solar system: a dense, positively charged central nucleus with a number of electrons in various orbits such that the total negative charge exactly canceled the positive nuclear charge. Maxwell and Newton were duly invoked. The orbiting electron, like the planets, obeyed Newton's commandment, $F = ma$. F was now the electrical force (Coulomb's law) between charged particles. Since this is also an inverse-square force like gravity, one would assume at first glance that stable, planetary orbits would follow. And there you have it, the nice neat solar system model of the chemical atom. Everything was fine.

Well, it was fine until the arrival in Manchester of a young Danish physicist of theoretical persuasion. "Name's Bohr, Niels Henrik David Bohr, Professor Rutherford. I'm a young theoretical physicist and I'm here to help you." We can only imagine the reaction of the gruff, earthy New Zealander.

THE STRUGGLE

The evolving revolution known as quantum theory didn't spring fully grown from the foreheads of theorists. It was slowly induced from data that emerged from the chemical atom. One can look at the struggle to understand this atom as practice for the real contest, understanding the sub-atom, the subnuclear jungle.

This slow unfolding of the real world is probably a blessing. What would Galileo or even Newton have done if the full data emerging from Fermilab had somehow been revealed to them? A colleague of mine at Columbia, a very young, very bright, articulate, enthusiastic professor, was given a unique teaching assignment. Take the forty or so freshmen who had declared physics as their major and give them two years of intensive instruction: one professor, forty aspiring physicists, two years. The experiment turned out to be a disaster. Most of the students switched to other fields. The reason came later from a graduating mathematics major: "Mel was terrific, best teacher I ever had. In those two years not only did we get through the usual — Newtonian mechanics, optics, electricity and so on — but he opened a window on the world of modern physics and gave us a glimpse of the problems he was facing in his own research. I felt there was no way I could ever handle such a difficult set of problems, so I switched to mathematics."

This raises a deeper question, whether the human brain will ever be prepared for the mysteries of quantum physics, which in the 1990s continues to disturb some of the very best physicists. Theoretician Heinz Pagels (who died tragically a few years ago in a mountain-climbing accident) suggested, in his fine book *The Cosmic Code,* that the human brain may not have evolved enough to understand quantum reality. Perhaps he's right, although a few of his colleagues seem convinced that they have evolved much more than the rest of us.

The overriding point is that quantum theory, as a highly refined and dominant theory of the 1990s, works. It works in atoms. It works in molecules. It works in complex solids, metals, insulators, semiconductors, superconductors, and anywhere it has been applied. The success of the quantum theory accounts for a significant fraction of the industrial world's total gross national product (GNP). But more important for us, it is the only tool we have as we proceed down into the nucleus, into its constituents and down, down into the vast minuteness of primordial matter — where we will confront the a-tom and the God Particle. And it is there that quantum theory's conceptual difficulties, dismissed by most working physicists as mere "philosophy," may play a significant role.

BOHR: ON THE WINGS OF A BUTTERFLY

Rutherford's discovery, coming after several experimental results that contradicted classical physics, was the last nail in the coffin. In the ongoing contest between experiment and theory, this would have been a good time to rub it in: "How clear do we experimenters have to make it before you theorists are convinced you need a new thing?" It appears that Rutherford didn't realize how much havoc his new atom was going to wreak on classical physics.

And then along came Niels Bohr, who would play Maxwell to Rutherford's Faraday, Kepler to his Brahe. Bohr's first position in England was at Cambridge, where he went to work for the great J. J., but the twenty-five-year-old kept irritating the master by finding mistakes in his book. While studying at the Cavendish Lab, on a Carlsberg Beer fellowship no less, Bohr heard Rutherford lecture about his new model of the atom in the fall of 1911. Bohr's thesis had been a study of "free" electrons in metals, and he was aware that all was not well with classical physics. He knew of course about Planck and about Einstein's more dramatic deviation from classical orthodoxy. And the spectral lines emitted by certain elements when they were heated was another clue to the quantum nature of the atom. Bohr was so impressed by Rutherford's lecture, and his atom, that he arranged to go to Manchester for a four-month visit in 1912.

Bohr saw the real significance in the new model. He realized that to satisfy Maxwell's equations, the electrons in circular orbits around a central nucleus would have to radiate energy, just like an electron accelerating up and down an antenna. To satisfy the laws of energy conservation, the orbits would shrink, and in no time flat the electron would spiral into the nucleus. If all these conditions were met, matter would be unstable. The model was a classical disaster! Yet there really was no alternative.

Bohr had no choice but to try something very new. The simplest atom of all is hydrogen. So Bohr studied the available data, such as how alpha particles slow down in hydrogen gas, and concluded that hydrogen has one electron in a Rutherford orbit around a positively charged nucleus. In facing up to a break with classical theory, Bohr was encouraged by other curiosities. He noted that nothing in classical physics determines the radius of the electron's orbit in the hydrogen atom. In fact, the solar system is a good example of a variety of planetary orbits. According to Newton's laws, any planetary orbit can be imagined; all it needs is to be started off properly. Once a radius is fixed, the planet's speed in orbit and its period (the year) are determined. But all hydrogen atoms, it would seem, are exactly alike. The atom shows none of the variety exhibited by the solar system. Bohr made the sensible but absolutely anticlassical assertion that only certain orbits are allowed in atoms.

Bohr also proposed that in these very special orbits the electron *doesn't* radiate. This, in historical context, was an incredibly audacious hypothesis. Maxwell rotated in his grave, but Bohr was simply trying to make sense of the facts. One important fact concerned the spectral lines that Kirchhoff had found shining out of atoms decades earlier. Glowing hydrogen, like other elements, emits a distinctive series of spectral lines. To get spectral lines, Bohr realized he must allow the electron to have the option of a number of different orbits corresponding to different energy contents. So he gave hydrogen's single electron a set of allowed radii representing a set of states of higher and higher energy. To explain spectral lines, he postulated (out of the blue) that radiation occurs when an electron "jumps" from one energy level to a lower one; the energy of the radiating photon is the difference of the two energy levels. He then proposed an absolutely outrageous rule for these special radii that determine the energy levels. Orbits are allowed, he said, in which the angular moment, a well-known quantity that measures the rotational momentum of the electron, takes on only integer values when measured in a new quantum unit. Bohr's quantum unit was nothing but Planck's constant, h. Bohr later said that "it was in the air to try to use the preexisting quantum ideas."

Now what is Bohr doing in his attic room late at night in Manchester with a sheaf of blank paper, a pencil, a sharp knife, slide rule, and some reference books? He is searching for nature's rules, rules that will correspond to the facts listed in his reference books. What right does he have to make up rules for the behavior of invisible electrons orbiting the nucleus (also invisible) of the hydrogen atom? His legitimacy is ultimately established by his success in explaining the data. He starts with the simplest atom, hydrogen. He understands that his rules ultimately have to emerge from some deep principle, but first the rules. This is how theorists work. Bohr in Manchester was, in the words of Einstein, trying to know the mind of God.

Bohr soon returned to Copenhagen to allow his seminal idea to germinate. Finally, in three papers published in April, June, and August of 1913 (the great trilogy), he presented his quantum theory of the hydrogen atom — a mixture of classical laws and totally arbitrary assertions (hypotheses) clearly designed *to get the right answer.* He manipulated his model of the atom so that it would explain the known spectral lines. Tables of these spectral lines, a series of numbers, had been painstakingly compiled by the followers of Kirchhoff and Bunsen, checked in Strasbourg and Göttingen, in London and Milan. What kind of numbers? Here are a few from hydrogen: λ_1 = 4,100.4, λ_2 = 4,339.0, λ_3 = 4,858.5, λ_4 = 6,560.6. (Sorry you asked?

Don't worry. No need to memorize them.) How do these spectral vibrations come about? And why only these, no matter how the hydrogen is energized? Oddly, Bohr later minimized the importance of spectral lines: "One thought that spectra are marvelous. But it is not possible to make progress there. Just as if you had the wing of a butterfly, then certainly it is very regular with its colors and so on. But nobody thought that you could get the basis of biology from the coloring of the wing of a butterfly." And yet it turned out that the spectral lines of hydrogen, the wing of the butterfly, provided a crucial clue.

Bohr's theory was crafted to give the numbers for hydrogen that were on the books. Crucial to his analyses was the overriding concept of *energy,* a term that was precisely defined in Newton's time, then evolved and enlarged. An understanding of it is necessary for the educated person. So let's take two minutes for energy.

TWO MINUTES FOR ENERGY

In high school physics, an object with a certain mass and a certain velocity is said to have kinetic energy (energy by virtue of motion). Objects have energy also by virtue of where they are. A steel ball on top of the Sears Tower has potential energy because someone worked hard to get it up there. If you drop it off the tower, it will, in falling, trade in its potential energy for kinetic energy.

The only thing that makes energy interesting is that it is conserved. Picture a complex system of billions of atoms in a gas, all in rapid motion, colliding with the walls of the vessel and with one another. Some atoms may gain energy; others lose it. But the total energy never changes. It wasn't until the eighteenth century that scientists discovered that heat is a form of energy. Chemicals can release energy via reactions such as the burning of coal. Energy can and does continually change from one form to another. Today we recognize mechanical, thermal, chemical, electrical, and nuclear energy. We know that mass can be converted to energy via $E = mc^2$. In spite of these complexities, we are still a hundred percent convinced that in complex reactions the total energy (including mass) always remains constant. Example: slide a block along a smooth plane. It stops. Its kinetic energy was changed into heat in the ever so slightly warmer plane. Example: you fill your car with gasoline, knowing that you have bought 12 gallons of chemical energy (measured in joules), which you can use to give your Toyota a certain kinetic energy. The gasoline goes away, but its energy can be accounted for — 320 miles, from Newark to North Hero. The energy is conserved. Example: a waterfall crashes onto the rotor of an electric generator, converting natural potential energy to electrical energy to warm and illuminate a distant town. In nature's ledger it all has to add up. You end up with what you brought.

SO?

Okay, how does this relate to the atom? In Bohr's picture, the electron must confine itself to specific orbits, with each orbit defined by its radius. Each of the allowed radii corresponds to a well-defined energy state (or energy level) of the atom. The smallest radius corresponds to the lowest energy, which is called the ground state. If we pour energy into a sample of hydrogen gas, some of it is used in shaking up the atoms so that they move faster. Some of the energy, however, is absorbed by the electron in a very specific bundle (remember the photoelectric effect), which allows the electron to reach another of its energy levels, or radii. The levels are numbered 1, 2, 3, 4, . . . , and each one has its energy, E_1, E_2, E_3, E_4, and so on. Bohr constructed his theory to include the Einstein idea that the energy of a photon determines its wavelength.

If photons of all wavelengths rain down on a hydrogen atom, the electron will eventually swallow the appropriate photon (light bundle of some particular energy) and jump up from E_1 to E_2 or E_3, say. In this way electrons populate the higher energy levels of the atom. This is what happens, for example, in a discharge tube. When electrical energy goes in, the tube glows with the characteristic colors of hydrogen. The energy induces some electrons in the trillions of atoms to jump to higher energy states. If the input electrical energy is large enough, many of the atoms will have electrons occupying essentially all possible higher energy states.

In Bohr's picture the electrons in higher energy states spontaneously jump down to lower levels. Now remember our little lecture on the conservation of energy. If electrons jump down, they lose energy, and that lost energy has to be accounted for. Bohr said, "No problem." A dropping electron emits a photon of energy equal to the difference in energy of the orbits. If the electron jumps from level 4 to level 2, for example, the photon's energy is equal to E_4 minus E_2. There are lots of jump possibilities, such as $E_2 \rightarrow E_1$, $E_3 \rightarrow E_1$, or $E_4 \rightarrow E_1$. Multilevel jumps are also allowed, such as $E_4 \rightarrow E_2$, then $E_2 \rightarrow E_1$. Every change of energy results in the emission of a corresponding wavelength, and a series of spectral lines is observed.

Bohr's ad hoc, quasi-classical explanation of the atom was a virtuoso, if unorthodox, performance. He used Newton and Maxwell when they were convenient. He discounted them when they weren't. He used Planck and Einstein when they worked. It was outrageous. But Bohr was smart, and he got the right answer.

Let's review. Thanks to the work of Fraunhofer and Kirchhoff back in the nineteenth century, we knew about spectral lines. We knew that atoms (and molecules) emit and absorb radiation at specific wavelengths and that each atom has its own characteristic pattern of wavelengths. Thanks to Planck, we knew that light is emitted in quanta. Thanks to Hertz and Einstein, we knew that it is also absorbed in quanta. Thanks to Thomson, we knew there are electrons. Thanks to Rutherford, we knew that the atom has a dense little nucleus, lots of void, and electrons scattered throughout. Thanks to my mother and father, I got to learn this stuff. Bohr put this data — and much more — together. The electrons are allowed only certain orbits, said Bohr. They absorb energy in quanta, which forces them to jump to higher orbits. When they drop back down to lower orbits, they emit photons, quanta of light. Scientists observe these quanta as specific wavelengths — the spectral lines peculiar to each element.

Bohr's theory, developed between 1913 and 1925, is now referred to as the "old quantum theory." Planck, Einstein, and Bohr had each taken a step to flout classical physics. All had firm experimental data that told them they were right. Planck's theory beautifully agreed with the black body spectrum, Einstein's with detailed measurements of photoelectrons. In Bohr's mathematical formula one finds such quantities as the electron's charge and mass, Planck's constant, some π's, numbers like 3, and an important integer (the quantum number) that enumerated the energy states. All of these, when factored together, provided a formula from which all the spectral lines of hydrogen could be calculated. It was in remarkable agreement with the data.

Rutherford loved Bohr's theory but raised the question of when and how the electron decides to jump from one state to another — something Bohr didn't discuss. Rutherford remembered a previous puzzle: when does a radioactive atom decide to decay? In classical physics, every action has a cause. In the atomic domain that kind of causality doesn't seem to appear. Bohr recognized the crisis (which wasn't really solved until Einstein's 1916 work on "spontaneous transitions") and pointed out a direction. But the experimenters, still exploring the phenomena of the atomic world, found a number of things that Bohr hadn't counted on.

When the American physicist Albert Michelson, a precision fanatic, examined the spectral lines more closely he noticed that each of the hydrogen lines was actually two narrowly spaced lines — two wavelengths that were very close together. This doubling of lines means that when an electron is ready to jump down, it has a choice of two lower energy states. Bohr's model didn't predict the doubling, which was called "fine structure." Arnold Sommerfeld, a contemporary and associate of Bohr, noticed that the velocity of electrons in the hydrogen atom is a significant fraction of the velocity of light and should be treated in accordance with Einstein's 1905 theory of relativity. He made the first step toward joining the two revolutions, quantum theory and relativity. When he included the effects of relativity, he noted that where the Bohr theory predicted one orbit, the new theory predicted two closely spaced orbits. This explained the doubling of the lines. In carrying out this calculation, Sommerfeld introduced a "new abbreviation" of some constants that frequently appeared in his equations. It was $2\pi e^2/hc$, which he abbreviated with the Greek letter alpha (α). Don't worry about the equation. The interesting thing is this: when one plugs in the known numbers for the electron's charge, e, Planck's constant, h, and the velocity of light, c, out pops $\alpha = 1/137$. There's that 137 again, a pure number.

Experimenters continued to add pieces to the Bohr model of the atom. In 1896, before the discovery of the electron, a Dutchman, Pieter Zeeman, put a Bunsen burner between the poles of a strong magnet and placed a lump of table salt in the burner. He examined the yellow light from sodium with a very precise spectrometer he had constructed. Sure enough, in the magnetic field the yellow spectral lines became broader, meaning that the magnetic field actually splits the lines. This effect was confirmed by more precise measurements up through 1925, when two Dutch physicists, Samuel Goudsmit and George Uhlenbeck, came up with a bizarre suggestion that the effect could be explained only by giving the electrons the property of "spin." In a classical object, say a top, spin is the rotation of the top around its axis of symmetry. Electron spin is the quantum analogue.

All of these new ideas, though valid by themselves, were rather ungracefully tacked on to the 1913 Bohr atomic model, like products picked up at a custom-car shop. With these accoutrements, the now greatly aggrandized Bohr theory, like an old Ford retrofitted with air conditioning, spinner hubcaps, and fake tailfins, could account for a very impressive amount of precise and brilliantly achieved experimental data.

There was only one problem with the model. It was wrong.

A PEEK UNDER THE VEIL

The crazy-quilt theory initiated by Niels Bohr in 1912 was running into increasing difficulties when a French graduate student in 1924 uncovered a crucial clue. This clue, revealed in an unlikely source, the turgid prose of a doctoral dissertation, would, in three dramatic years, yield a totally new conception of reality in the microworld. The author was a young aristocrat, Prince Louis-Victor de Broglie, sweating out his Ph.D. in Paris. De Broglie was inspired by a paper by Einstein, who in 1909 was mulling over the significance of his light quanta. How could light behave like a swarm of energy bundles — that is, like particles — and at the same time exhibit all the behaviors of waves, such as interference, diffraction, and other properties that require a wavelength?

De Broglie thought that this curious dual character of light might be a fundamental property of nature that could be applied to material objects such as electrons as well. In his photoelectric theory, following Planck, Einstein had assigned a certain energy to a quantum of light, related to its wavelength or frequency. De Broglie then invoked a new symmetry: if waves can be particles, then particles (electrons) can be waves. He devised a way to assign electrons a wavelength related to their energy. His idea immediately hit pay dirt when he applied it to electrons in the hydrogen atom. An assigned wavelength gave an explanation for Bohr's mysterious ad hoc rule that only certain radii are allowed to the electron. It's totally obvious! It is? Sure. If in a Bohr orbit the electron has a wavelength of some teensy fraction of a centimeter, then only those orbits are allowed in which an integral (whole) number of wavelengths can fit around the circumference. Try this crude visualization. Go get a nickel and a handful of pennies. Place the nickel (the nucleus) on a table and arrange a number of pennies in a circle (the electron orbit) around the nickel. You'll find you need seven pennies to make the smallest orbit. This defines a radius. If you want to use eight pennies you are forced to make a bigger circle, but not *any* bigger circle; only one radius will do it. Larger radii will permit nine, ten, eleven, or more pennies. You can see from this dumb example that if you restrict yourself to whole pennies — or whole wavelengths — only certain radii are allowed. To get circles in between requires overlapping the pennies, and if they represent wavelengths, the waves wouldn't connect up smoothly around the orbit. De Broglie's idea was that the wavelength of the electron (the diameter of the penny) determines the allowed radii. Key to his concept was the idea of assigning a wavelength to the electron.

De Broglie, in his dissertation, speculated as to whether electrons would demonstrate other wavelike effects such as interference and diffraction. His faculty advisers at the University of Paris, though impressed by the young prince's virtuosity, were nonplused by the notion of particle waves. One of his examiners, wanting an outside opinion, sent a copy to Einstein, who wrote back this compliment about de Broglie: "He has lifted a corner of the great veil." His Ph.D. thesis was accepted in 1924 and eventually earned him a Nobel Prize, making de Broglie the only physicist up to that time to win the Prize on the basis of a dissertation. The biggest winner, though, was Erwin Schrödinger, who saw the real potential in de Broglie's work.

Now comes an interesting pas de deux of theory-experiment. De Broglie's idea had no experimental support. An electron wave? What does it mean? The necessary support appeared in 1927, in, of all places, New Jersey — not a Channel island but an American state near Newark. Bell Telephone Laboratories, the famous industrial research institution, was engaged in a study of vacuum tubes, an ancient electronic device used before the dawn of civilization and the invention of transistors. Two scientists, Clinton Davisson and Lester Germer, were bombarding various oxide-coated metal surfaces with streams of electrons. Germer, working under Davisson's direction, noticed that a curious pattern of electrons was reflected from certain metal surfaces that had no oxide coating.

In 1926 Davisson traveled to a meeting in England and learned about de Broglie's idea. He rushed back to Bell Labs and began to analyze his data from the point of view of wave behavior. The patterns he observed fit precisely with the theory of electrons behaving as waves whose wavelength was related to the energy of the bombarding particles. He and Germer rushed to publish. They were none too soon. In the Cavendish Laboratory, George P. Thomson, son of the famous J. J., was carrying out similar research. Davisson and Thomson shared the 1938 Nobel Prize for first observing electron waves.

The filial affection of J. J. and G. P. is, incidentally, amply documented in their warm correspondence. In one of his more emotional letters, G. P. gushed:

> Dear Father,
> Given a spherical triangle with sides ABC . . .
> [And, after three densely written pages of the same]
> Your son, George

So now a wave is associated with an electron whether it is imprisoned in an atom or traveling in a vacuum tube. But what is there about this electron that waves?

THE MAN WHO DIDN'T KNOW BATTERIES

If Rutherford was the prototypical experimenter, Werner Heisenberg (1901–1976) qualified as his theoretical counterpart. He would have fit I. I. Rabi's definition of a theorist as one who "couldn't tie his own shoelaces." One of the most brilliant students in Europe, Heisenberg almost failed his Ph.D. orals at the University of Munich when one of his examiners, Wilhelm Wien, a pioneer in the study of black body radiation, took a dislike to him. Wien started asking practical questions, like how does a battery work? Heisenberg had no idea. Wien, after grilling him with more questions about experimentation, wanted to flunk him. Cooler heads prevailed, and Heisenberg got off with the equivalent of a gentleman's C.

His father was a professor of Greek at Munich, and as a teenager Heisenberg had read the *Timaeus,* which includes all of Plato's atomic theory. Heisenberg thought Plato was nuts — his "atoms" were little cubes and pyramids — but he was fascinated with Plato's basic tenet that one can never understand the universe until the smallest components of matter are known. Young Heisenberg decided to devote his life to studying the smallest particles of matter.

Heisenberg tried hard to picture the Rutherford-Bohr atom in his mind and kept coming up empty. Bohr's electron orbits were like none he could imagine. The cute little atom that would become the Atomic Energy Commission's logo for so many years — a nucleus with electrons circling in "magic" radii without radiating — just didn't make any sense. Heisenberg realized that Bohr's orbits were merely constructs that made the numbers come out right and got rid of or (better) finessed classical objections to the Rutherford model of the atom. But real orbits? No. Bohr's quantum theory didn't go far enough in discarding the baggage of classical physics. The unique way in which space in the atom permitted only certain orbits required a more radical proposition. Heisenberg came to realize that this new atom was fundamentally not visualizable. He developed a firm guide: do not deal with anything that can't be measured. Orbits can't be measured. Spectral lines, however, *can* be. Heisenberg wrote a theory called "matrix mechanics," based on mathematical forms called matrices. His methods were difficult mathematically, and even more difficult to visualize, but it was clear that he had made a major improvement in Bohr's old theory. In time, matrix mechanics repeated all the successes of the Bohr theory without the arbitrary magic radii. Heisenberg's matrices went on to new successes where the old theory failed. But physicists found the matrices hard to use.

And then came the most famous vacation in the history of physics.

MATTER WAVES AND THE LADY IN THE VILLA

A few months after Heisenberg completed his matrix formulation, Erwin Schrödinger decided he needed a holiday. It was about ten days before Christmas in the winter of 1925. Schrödinger was a competent but undistinguished professor of physics at the University of Zurich, and all college teachers deserve a Christmas holiday. But this was no ordinary vacation. Leaving his wife at home, Schrödinger booked a villa in the Swiss Alps for two and a half weeks, taking with him his notebooks, two pearls, and an old Viennese girlfriend. Schrödinger's self-appointed mission was to save the patched-up, creaky quantum theory of the time. The Viennese-born physicist placed a pearl in each ear to screen out any distracting noises. Then he placed the girlfriend in bed for inspiration. Schrödinger had his work cut out for him. He had to create a new theory *and* keep the lady happy. Fortunately he was up to the task. (Don't become a physicist unless you are prepared for such demands.)

Schrödinger had begun his career as an experimenter but had switched to theory rather early on. He was old for a theorist, thirty-eight that Christmas. Obviously, there are lots of middle-aged, even elderly, theorists around. But they usually do their best work in their twenties, then retire, intellectually speaking, in their thirties to become "elder statesmen" of physics. This shooting-star phenomenon was especially true during the heyday of quantum theory, which saw Paul Dirac, Werner Heisenberg, Wolfgang Pauli, and Niels Bohr all crafting their finest theories as very young men. When Dirac and Heisenberg went to Stockholm to accept their Nobel Prizes, they were, in fact, accompanied by their mothers. Dirac once wrote:

> Age is of course a fever chill
> That every physicist must fear.
> He's better dead than living still
> When once he's past his thirtieth year.

(He won his Nobel for physics, not for literature.) Fortunately for science, Dirac didn't take his own verse to heart, living well into his eighties.

One of the items Schrödinger took with him on vacation was de Broglie's paper on particles and waves. Working feverishly, he extended the quantum concept even further. He didn't just treat electrons as particles with wave characteristics. He came up with an equation in which electrons *are* waves, matter waves. A key actor in Schrödinger's famous equation is the Greek symbol psi, or ψ. Physicists are fond of saying that the equation thus reduces everything to psi's (sighs). ψ is known as the *wave function,* and it contains all we know or can know about the electron. When Schrödinger's equation is solved, it gives ψ as it varies in space and changes with time. Later the equation was applied to systems of many electrons and eventually to any system requiring a quantum treatment. In other words, the Schrödinger equation, or "wave mechanics," applies to atoms, molecules, protons, neutrons, and, especially important to us today, clusters of quarks, among other particles.

Schrödinger was out to rescue classical physics. He insisted that electrons were truly classical waves, like sound waves, water waves, or Maxwell's electromagnetic light and radio waves, and that their particle aspect was illusory. They were *matter waves*. Waves were well understood, simple to visualize, unlike the electrons in the Bohr atom, jumping willy-nilly from orbit to orbit. In Schrödinger's interpretation, ψ (really the square of ψ, or ψ^2) described the density distribution of this matter wave. His equation described these waves under the influence of the electrical forces in the atom. For example, in the hydrogen atom, Schrödinger's waves clump in places where the old

Bohr quantum theory talked orbits. The equation gave the Bohr radii automatically, with no adjustments, and provided the spectral lines, not only for hydrogen but for the other elements as well.

Schrödinger published his wave equation within weeks after he left the villa. It was an immediate sensation, one of the most powerful mathematical tools ever devised to deal with the structure of matter. (By 1960, more than 100,000 scientific papers had been published based on the application of Schrödinger's equation.) He wrote five more papers in quick succession; all six papers were published in a six-month period that was among the greatest bursts of creativity in scientific history. J. Robert Oppenheimer called the theory of wave mechanics "perhaps one of the most perfect, most accurate, and most lovely man has discovered." Arthur Sommerfeld, the great physicist and mathematician, said Schrödinger's theory "was the most astonishing among all the astonishing discoveries of the twentieth century."

For all of this, I personally forgive Schrödinger for his romantic dalliances which, after all, are of concern only to biographers, sociological historians, and envious colleagues.

A WAVE OF PROBABILITY

Physicists loved Schrödinger's equation because they could solve it and it worked. Although Heisenberg's matrix mechanics and Schrödinger's equation both seemed to give the correct answers, most physicists seized on the Schrödinger method since this was a good old differential equation, a warm and familiar form of mathematics. A few years later it was shown that the physical ideas and numerical consequences of Heisenberg's and Schrödinger's theories were identical. They were just written in different mathematical languages. Today a mixture of the most convenient aspects of both theories is used.

The only problem with Schrödinger's equation was that his interpretation of the "wave" was wrong. It turned out that the ψ thing could not represent matter waves. There was no doubt it represented some sort of wave, but the question was, what's waving?

The answer was provided by the German physicist Max Born, still in that eventful year 1926. Born insisted that the only consistent interpretation of Schrödinger's wave function is that ψ^2 represents the *probability* of finding a particle, the electron, at various locations. ψ varies in space and time. Where ψ^2 is large, the electron has a large probability of being found. Where $\psi = 0$, the electron is never found. The wave function is a wave of probability.

Born was influenced by experiments in which a stream of electrons is directed toward some sort of energy barrier. This could be, for example, a wire screen connected to the negative terminal of a battery, say at −10 volts. If the electrons have an energy of only 5 volts, they should be effectively repelled by the "10-volt barrier" in the classical view. If an electron's energy is larger than that of the barrier, it will penetrate the barrier like a ball thrown over a wall. If its energy is less than that of the barrier, the electron is reflected, like a ball thrown against the wall. However, Schrödinger's quantum equation indicates that some of the ψ-wave penetrates and some of the wave is reflected. This is typical light behavior. Pass a store window and you see the goodies displayed, but you also see a dim image of yourself. Light waves are both transmitted through and reflected by the glass. Schrödinger's equation predicts similar results. But we never see a fraction of an electron!

The experiment goes as follows: we send 1,000 electrons toward the barrier. Geiger counters find that 550 penetrate the barrier and 450 are reflected, but in every case, it is an entire electron that is detected. The Schrödinger waves, when properly squared, give 550 and 450 as a statistical prediction. If we accept the Born interpretation, a single electron has a 55 percent probability of penetrating and a 45 percent chance of being reflected. Since a single electron never divides, Schrödinger's wave cannot be the electron. It can be only a probability.

Born, along with Heisenberg, was part of the Göttingen school, a group of some of the brightest physicists of the age whose professional and intellectual lives revolved around the University of Göttingen in Germany. Born's statistical interpretation of Schrödinger's psi came from the Göttingen school's conviction that electrons are particles. They make Geiger counters click. They leave sharp tracks in Wilson cloud chambers. They collide with other particles and bounce off. So here is Schrödinger's equation, which gives correct answers but describes electrons as waves. How can it be converted to a particle equation?

Irony is a constant companion of history, and the idea that changed everything was given (again!) by Einstein in a speculative paper of 1911 on the relationship of photons to Maxwell's classical field equations. Einstein had suggested that the field quantities guided the photons to places of high probability. Born's resolution of the particle-wave conflict is simply this: the electron (and its friends) act like particles at least when they are being detected, but their distribution in space between measurements follows the wavelike probability patterns that emerge from the Schrödinger equation. In other words, the Schrödinger psi quantity describes the probable location of the electrons. And this probability can behave like a wave. Schrödinger did the hard part, crafting the equation that lies at the heart of the theory. But it was Born, inspired by Einstein's paper, who figured out what the equation was actually predicting. The irony is that it was Born's probability interpretation of the wave function that Einstein never accepted.

WHAT THIS MEANS, OR THE PHYSICS OF CLOTH CUTTING

The Born interpretation of the Schrödinger equation is the single most dramatic and major change in our world view since Newton. It is not surprising that Schrödinger found the idea totally unacceptable and regretted inventing an equation that would involve such foolishness. However, Bohr, Heisenberg, Sommerfeld, and others accepted it with little fuss because "probability was in the air." Born's paper made the eloquent assertion that the equation can only predict probability but that the mathematical form of probability is developed along perfectly predictable paths.

In this new interpretation, the equation deals with probability waves, ψ, which predict what the electron is doing, what its energy is, where it will be, and so on. However, these predictions are in the form of probabilities. What "waves" about the electron is just these probability predictions. These wavelike solutions to the equations can pile up in one place to add up to high probability and cancel in other places to yield low probability. When one puts these predictions to the test, one in effect does the experiment a huge number of times. Indeed, in most of the trials, the electron ends up where the equation says the probability is high; only very rarely does it end up where the probability is low. There is quantitative agreement. What is shocking is that for two apparently identical experiments one can get two quite different results.

The Schrödinger equation with Born's probability interpretation of the wave function has been enormously successful. It is the key to understanding hydrogen and helium and, given a big enough computer, uranium. It was used to understand how two elements combine to make a molecule, putting chemistry on a far more scientific footing. It allows one to design electron microscopes and even proton microscopes. In the period 1930–1950 it was carried into the nucleus and was found to be as productive there as in the atom.

The Schrödinger equation predicts with a high degree of accuracy, but, again, what it predicts is probability. What does that mean? Probability in physics is similar to probability in life. It's a billion-dollar business, as executives from insurance companies, clothing manufacturers, and a good fraction of the Fortune 500 industries will assure you. Actuaries can tell us that the average white American nonsmoking male born in, say, 1941, will live to be 76.4 years old. But they can't tell you diddly about your brother Sal, who was born that same year. For all they know, he could be run over by a truck tomorrow or die of an infected toenail in two years.

In one of my classes at the University of Chicago, I play garment-center mogul for my students. Being a success in the rag trade is similar to making a career in particle physics. In either case, you need a strong grasp of probability and a working knowledge of tweed jackets. I ask the students to sing out their heights while I plot each student's height on a graph. I have two students at 4 foot 8 inches, one at 4 foot 10, four at 5 foot 2, and so on. One guy is 6 foot 6, way outside the others. (If Chicago only had a basketball team!) The average is 5 foot 7. After polling 166 students I have a nice, bell-shaped set of steps going up to 5 foot 7 and then stepping down toward the 6-foot-6 anomaly. Now I have a "distribution curve" of college freshman heights, and if I'm reasonably sure that choosing physics to fulfill the science requirement does not distort the curve, I have a representative sample of student heights at the University of Chicago. I can read percentages using the vertical scale; for example, I can figure out what percentage of students is between 5 foot 2 and 5 foot 4. With my graph I can also read that there is a 26 percent probability that the next student who shows up will be between 5 foot 4 and 5 foot 6, if this is something I want to know.

Now I'm ready to make suits. If these students are my market (an unlikely prospect if I'm in the suit business), I can estimate what percentage of my suits should be size 36, 38, and so on. If I don't have a graph of heights, I have to guess, and if wrong, at the end of the season I have 137 size-46 suits left unsold (which I have to blame on my partner Jake, the schlemiel!).

The Schrödinger equation, when solved for any situation involving atomic processes, generates a curve analogous to the distribution-of-student-heights curve. However, the shape may be quite different. If we want to know where the electron hangs out in the hydrogen atom — how far it is from the nucleus — we'll find some distribution that drops off sharply at about 10^{-8} centimeters, with about an 80 percent probability of finding the electron within the sphere of radius 10^{-8} centimeters. This is the ground state. If we excite the electron to the next energy level, we'll get a bell curve with a mean radius that's about four times as big. We can compute probability curves of other processes as well. Here we must clearly differentiate *probability* predictions from *possibilities.* The possible energy levels are very precisely known, but if we ask which energy state the electron will be found in, we calculate only a probability, which depends on the history of the system. If the electron has more than one choice as to which lower energy state to jump to, we can again predict probabilities; for example, an 82 percent probability of jumping to E_1, 9 percent into E_2, and so on. Democritus said it best when he proclaimed, "Everything existing in the universe is the fruit of chance and necessity." The various energy states are the necessities, the only conditions that are possible. But we can only predict the probabilities of the electron being in any of these possible states. That's a matter of chance.

Concepts of probability are well known to actuarial experts today. But they were upsetting to physicists trained in classical physics in the early part of the century (and remain upsetting to many people today). Newton described a deterministic world. If you threw a rock, launched a rocket, or introduced a new planet to a solar system, you could predict where it would go with total certainty, at least in principle, as long as you knew the forces and the initial conditions. Quantum theory said no: initial conditions are inherently uncertain. You get only probabilities for predictions of whatever you want to measure: a particle's location, its energy, velocity, or whatever. The Born interpretation of Schrödinger was unsettling to physicists, who in the three centuries since Galileo and Newton had come to accept determinism as a way of life. Quantum theory threatened to transform them into high-level actuaries.

A SURPRISE ON A MOUNTAINTOP

In 1927 the English physicist Paul Dirac was trying to extend quantum theory, which at the time appeared to be at odds with Einstein's special theory of relativity. The two theories had already been introduced to each other by Sommerfeld. Dirac, intent on making the two theories happily compatible, supervised the marriage and its consummation. In doing so, he found an elegant new equation for the electron (curiously, we call it the Dirac equation). Out of this powerful equation comes the postdictum that electrons must have spin and must produce magnetism. Recall the g-factor from the beginning of the chapter. Dirac's calculations showed that the strength of the electron's magnetism as measured by g was 2.0. (It was much later that refinements led to the precise value given earlier.) More! Dirac (age twenty-four or so) found that in obtaining the electron-wave solution to his equation, there was another solution with bizarre implications. There had to be another particle with properties identical to those of the electron but with opposite electric charge. Mathematically, this is a simple concept. As every little kid knows, the square root of four is plus two, but it is also minus two because minus two times minus two is also four: $2 \times 2 = 4$, and $-2 \times -2 = 4$. So there are two solutions. The square root of four is plus *or* minus two.

The problem was that the symmetry implied by Dirac's equation meant that for every particle there must exist another particle with the same mass but opposite charge. So Dirac, a conservative gentleman who was so uncharismatic as to have generated legends, struggled with his negative solution and eventually predicted that nature must contain positive electrons as well as negative electrons. Someone coined the word *antimatter.* This antimatter should be all over the place, yet no one had ever spotted any.

In 1932, a young Cal Tech physicist named Carl Anderson built a cloud chamber designed to register and photograph subatomic particles. A powerful magnet surrounded his apparatus to bend the path of the particles, giving a measure of their energy. Anderson bagged a bizarre new particle — or, rather, the track of one — in the cloud chamber. He called this strange new object a *positron,* because it was identical to an electron except that it had a positive charge instead of a negative charge. Anderson's publication made no reference to Dirac's theory, but the connection was soon made. He had found a new form of matter, the antiparticle that had popped out of the Dirac equation a few years earlier. The tracks were made by cosmic rays, radiation from particles that strike our atmosphere from the far reaches of our galaxy. Anderson, to get even better data, transported his apparatus from Pasadena to the top of a mountain in Colorado, where the air is thin and the cosmic rays are more intense.

A front-page photograph of Anderson in the *New York Times,* announcing the discovery, was an inspiration to the young Lederman, his first exposure to the romantic adventure of schlepping equipment to the top of a high mountain to make important scientific measurements. Antimatter turned out to be a very big deal, inextricably involved in the lives of particle physicists, and I promise to say more about it in later chapters. Another quantum-theory success.

UNCERTAINTY AND ALL THAT

In 1927 Heisenberg invented his uncertainty relations, which put the cap on the great scientific revolution we call quantum theory. In truth, quantum theory wasn't wrapped up until the 1940s. Indeed, in its quantum field theory version, its evolution continues today, and the theory will not be complete until it is fully combined with gravitation. But for our purposes the uncertainty principle is a good place to end. Heisenberg's uncertainty relations are a mathematical consequence of the Schrödinger equation. They could also have been the logical postulates, or assumptions, of the new quantum mechanics. Since Heisenberg's ideas are crucial to understanding just how new the quantum world is, we need to dwell a bit here.

Quantum designers insist that only measurements, dear to the hearts of experimenters, count. All we can ask of a theory is to predict the results of events that can be measured. This sounds like an obvious point, but forgetting it leads to the so-called paradoxes that popular writers without culture are fond of exploiting. And, I should add, it is in the theory of measurement that the quantum theory meets its past, present, and no doubt future critics.

Heisenberg announced that our *simultaneous* knowledge of a particle's location and its motion is limited and that the combined uncertainty of these two properties must exceed . . . nothing other than Planck's constant, h, which we first met in the formula $E = hf$. Our measurements of the particle's location and its motion (actually, its momentum) are reciprocally related to each other. The more we know about one, the less we know about the other. The Schrödinger equation gives us probabilities for these factors. If we devise an experiment that pinpoints the location of the electron — say it's at some coordinate with an extremely small uncertainty of position — the spread in the possible values of the momentum is correspondingly large according to Heisenberg's relation. The product of the two uncertainties (we can assign them numbers) is always greater than Planck's ubiquitous *h*. Heisenberg's relations dispose, once and for all, of the classical picture of orbits. The very concept of location or place is now less definite. Let's go back to Newton and to something we can visualize.

Suppose we have a straight road on which a Hyundai is tooling along at some respectable speed. We decide that we are going to measure its location at some instant of time as it whizzes past us. We also want to know how fast it is going. In Newtonian physics, pinpointing the position and velocity of an object at a specific time allows one to predict precisely where it will be at any future time. However, when we assemble our rulers and clocks, our flashbulbs and cameras, we find that the more carefully we measure the position, the poorer our ability to measure the speed and vice versa. (Recall that the speed is the change of position divided by the time.) However, in classical physics we can continually improve on our accuracy in both quantities to arbitrary precision. We simply ask some government agency for more funds to build better equipment.

In the atomic domain, by contrast, Heisenberg proposed a basic unknowability that cannot be reduced by any amount of equipment, ingenuity, or federal funding. He proposed that it is a fundamental property of nature that the product of the two uncertainties always exceeds Planck's constant. Strange as this may sound, there is a firm physical basis for this uncertainty in measurability of the microworld. For example, let's try to nail down the position of an electron. To do so, you must "see" it. That is, you have to bounce light, a beam of photons, off the electron. Okay, there! Now you see the electron. You know its location at a moment in time. But a photon glancing off the electron changes the electron's state of motion. One measurement undermines the other. In quantum mechanics, measurement inevitably produces change because you are dealing with atomic systems, and your measuring tools cannot be any smaller, gentler, or kinder. Atoms are one ten-billionth of a centimeter in radius and weigh a millionth of a billion-billionth of a gram, so it doesn't take much to influence them profoundly. By contrast, in a classical system, one can make sure that the act of measuring barely influences the system being measured. Suppose we want to measure water temperature. We don't change the temperature of a lake, say, by dipping a small thermometer into it. But dipping a fat thermometer into a thimble of water would be stupid since the thermometer would change the temperature of the water. In atomic systems, quantum theory says, we must include the measurement as part of the system.

THE AGONY OF THE DOUBLE SLIT

The most famous and most instructive example of the counterintuitive nature of quantum theory is the double-slit experiment. This experiment was first carried out by Thomas Young, a physician, in 1804 and was heralded as experimental proof of the wave nature of light. The experimenter aimed a beam of, say, yellow light at a wall in which he had cut two very fine parallel slits a very short distance apart. A distant screen caught the light that squirted through the slits. When Young covered one of the slits, a simple, bright, slightly broadened image of the other slit was projected on the screen. But when both slits were uncovered, the result was surprising. A careful examination of the light area on the screen revealed a series of equally spaced bright and dark fringes. Dark fringes are places where no light arrives.

The fringes are proof, said Young, that light is a wave. Why? They are part of an interference pattern, which occurs when waves of any kind bump into each other. When two water waves, for example, collide crest to crest, they reinforce each other, creating a bigger wave. When they collide trough to crest, they cancel each other out. The wave flattens.

Young's interpretation of the double-slit experiment was that at certain locations the wavelike disturbances from the two slits arrive on the screen in just the right phases to cancel each other out: a peak of the light wave from slit one arrives exactly at a trough of light from slit two. A dark fringe results. Such cancellations are quintessential indicators of wave interference. When two peaks or two troughs coincide at the screen, we get a bright fringe. The fringe pattern was accepted as proof that light was a wave phenomenon.

Now in principle the same experiment can be carried out with electrons. In a way this is what Davisson did at Bell Labs. Using electrons, the experiment also results in an interference pattern. The screen is covered with tiny Geiger counters, which click when an electron hits. The Geiger counter detects *particles*. To check that the counters are working, we put a thick piece of lead over slit two: no electrons can penetrate. Now all Geiger counters click if we wait long enough for some thousands of electrons to pass through the remaining open slit. But when two slits are open, some columns of Geiger counters never click!

Wait a minute. Hold it. When one slit is closed, the electrons, squirting through the other slit, spread out, some going to the left, some straight, some to the right, causing a roughly uniform pattern of clicks across the screen, just as Young's yellow light resulted in a broad bright line in his one-slit experiment. In other words, the electrons behave, logically enough, like particles. But if we remove the lead and let some of the electrons go through slit two, the pattern changes and no electrons reach those columns of Geiger counters corresponding to the dark fringe locations. Now the electrons are acting like waves. Yet we know they are particles because the counters are clicking.

Maybe, you might argue, two or more electrons are passing simultaneously through the slits and simulating a wave interference pattern. To verify that no two electrons are passing simultaneously through the slits, we reduce the rate of electrons to one per minute. Same patterns. Conclusion: electrons going through slit one "know" that slit two is open or closed because they change their patterns in each case.

How do we come up with this idea of "smart" electrons? Put yourself in the place of the experimenter. You have an electron gun, so you know you're shooting particles at the slits. You also know that you end up with particles at the destination, the screen, because the Geiger counters click. A click means particle. So, whether we have one slit or two slits open, we begin and end with particles. However, where the particles land depends on whether one or two slits are open. So a particle going through slit one seems to know whether slit two is open or closed, because it appears to change its path depending on that information. If slit two is closed, it says to itself, "Okay, I can land anywhere on the screen." If slit two is open, it says, "Uh-oh, I have to avoid certain bands on the screen in order to create a fringe pattern." Since particles can't "know," our wave-particle ambiguity has created a logical crisis.

Quantum mechanics says we can predict the probability of the electrons' passage through slits and subsequent arrival at the screen. The probability is a wave, and waves exhibit two-slit interference patterns. When both slits are open, the ψ probability waves can interfere to result in zero probability ($\psi = 0$) at certain places on the screen. The anthropomorphic complaint of the previous paragraph is a classical hangover; in the quantum world, "How does the electron know which slit to go through?" is not a question that can be answered by measurement. The detailed point-by-point trajectory of the electron is not being observed, and therefore the question "Which slit did the electron go through?" is not an operational question. Heisenberg's uncertainty relations also solve our hangup by pointing out that if you try to measure the electron's trajectory between the electron gun and the wall, you totally change the motion of the electron and destroy the experiment. We can know the initial conditions (electron fired from gun); we can know the results (electron hits some position on screen); we *cannot* know the path from A to B unless we are prepared to screw up the experiment. This is the spooky nature of the new world in the atom.

The quantum mechanics solution, that is, Don't worry! We can't measure it, is logical enough, but not satisfying to most human minds, which strive to understand the details of the world around us. For some tortured souls, this quantum unknowability is still too high a price to pay. Our defense: this is the only theory we know now that works.

NEWTON VS. SCHRÖDINGER

A new intuition must be cultivated. We spend years teaching physics students classical physics, then turn around and teach them quantum theory. It takes graduate students two or more years to develop quantum intuition. (You, lucky reader, are expected to perform this pirouette in the space of just one chapter.)

The obvious question is, which is correct? Newton's theory or Schrödinger's? The envelope, please. And the winner is . . . Schrödinger! Newton's physics was developed for big things; it doesn't work inside the atom. Schrödinger's theory was designed for microphenomena. Yet when the Schrödinger equation is applied to macroscopic situations it gives results identical to Newton's.

Let's look at a classic example. The earth orbits the sun. An electron orbits — to use the old Bohr language — a nucleus. The electron, however, is constrained to specific orbits. Are there only certain allowable quantum orbits for the planet earth around the sun? Newton would say no, the planet can orbit wherever it wants. But the correct answer is yes. We can apply the Schrödinger equation to the earth-sun system. Schrödinger's equation would give the usual discrete set of orbits, but there would be a huge number of them. In using the equation, you'd plug the mass of the earth (instead of the mass of the electron) into the denominator, so the orbital spacings out where the earth is, say, 93 million miles from the sun, would end up so small — say, one every billionth of a billionth of an inch — as to be in effect continuous. For all practical purposes, you end up with the Newtonian result that *all* orbits are allowed. When you take the Schrödinger equation and apply it to macro objects, it changes in front of your very eyes to . . . $F = ma$! Or thereabouts. It was Roger Boscovich, by the way, in the eighteenth century who surmised that Newton's formulas were simply approximations that were good over large distances but wouldn't survive in the microworld. So our graduate students do not have to discard their mechanics books. They may get a job with NASA or the Chicago Cubs, plotting rocket reentry trajectories or pop-ups with good old Newtonian equations.

In quantum theory, the concept of orbits, or of what the electron is doing in the atom or in a beam, is not useful. What matters is the result of a measurement, and here quantum methods can only predict the probability of any possible result. If you measure where the electron is, say in the hydrogen atom, your result could be a number, the distance of the electron from the nucleus. You do this, not by measuring a single electron but by repeating the measurement many times. You get a different result each time, and finally you draw a curve graphing all the results. It is this graph that can be compared to the theory. The theory cannot predict the result of any given measurement. It is a statistical thing. Going back to my cloth-cutter analogy, if we know that the average height of freshmen at the University of Chicago is 5 foot 7, the next new freshman might still be 5 foot 3 or 6 foot 1. We cannot predict the height of the next freshman; we can only draw a kind of actuarial curve.

Where it gets spooky is in predictions of a particle's passage through a barrier or the decay time of a radioactive atom. We prepare an *identical* setup many times. We shoot a 5.00 MeV electron at a 5.50 MeV potential barrier. We predict that 45 times out of 100 it will penetrate. But we can't ever be sure what a given electron will do. One gets through; the next one, identical in every way, does not. Identical experiments have different results. That's the quantum world. In classical science we stress the importance of replicating experiments. In the quantum world, we can replicate everything except the result.

In the same way, take the neutron, which has a "half-life" of 10.3 minutes, meaning that if you start with 1,000 neutrons, half have disintegrated in 10.3 minutes. But a given neutron? It can decay in 3 seconds or 29 minutes. Its exact time of decay is unpredictable. Einstein hated this idea. "God does not play dice with the universe," he said. Other critics said, suppose there is, in each neutron or each electron, some mechanism, some spring, some "hidden variable" that makes each neutron different, like human beings, who also have an average lifetime. In the case of humans, there are plenty of not-so-hidden things — genes, clogged arteries, and so on — which in principle can be used to predict an individual's day of demise, barring falling elevators, disastrous love affairs, or an out-of-control Mercedes.

The hidden-variable hypothesis has been essentially disproven for two reasons: no such variables have shown up in all the billions of experiments done on electrons and new, improved theories related to quantum-mechanics experiments have ruled them out.

THREE THINGS TO REMEMBER ABOUT QUANTUM MECHANICS

Quantum mechanics can be said to have three remarkable qualities: (1) it is counterintuitive; (2) it works; and (3) it has aspects that made it unacceptable to the likes of Einstein and Schrödinger and that have made it a source of continuing study in the 1990s. Let's touch on each of these.

1. It is counterintuitive. Quantum mechanics replaces continuity with discreteness. Metaphorically, instead of a liquid being poured into the glass, it is very fine sand. The smooth hum you hear is the beating of huge numbers of atoms on your eardrums. Then there is the spookiness of the double-slit experiment, already discussed.

Another counterintuitive phenomena is "tunneling." We talked about sending electrons toward an energy barrier. The classical analogue is rolling a ball up a hill. If you give the ball enough initial push (energy), it will go over the top. If the initial energy is too low, the ball will come back down. Or picture a roller coaster with the car stuck in a trough between two terrifying rises. Suppose the car rolls halfway up one rise and loses power. It will slide back down, then almost halfway up the other side, then oscillate back and forth, trapped in the trough. If we could remove friction, the car would oscillate forever, imprisoned between the two insurmountable rises.

In quantum atomic theory such a system is known as a bound state. However, our description of what happens to electrons aimed at an energy barrier or an electron trapped between two barriers must include probabilistic waves. It turns out that some of the wave can "leak" through the barrier (in atomic or nuclear systems the barrier is either an electrical or a strong force), and therefore there is a finite probability that the trapped particle will appear outside the trap. This was not only counterintuitive, it was considered a major paradox, since the electron on its way through the barrier would have negative kinetic energy — a classical absurdity. But with evolving quantum intuition one responds that the condition of the electron "in the tunnel" is not observable and therefore not a question for physics. What one does observe is that it does get out. This phenomenon, called tunneling, was used to explain alpha-radioactivity. It is the basis of an important solid state electronic device known as a tunnel diode. Spooky as it is, this tunnel effect is essential to modern computers and other electronic devices.

Point particles, tunneling, radioactivity, double slit anguish — all of these contributed to the new intuitions that quantum physicists needed as they fanned out in the late 1920s and '30s with their new intellectual armaments to seek unexplained phenomena.

2. It works. As a result of the events of 1923–1927, the atom was understood. Even so, in those pre-computer days, only simple atoms — hydrogen, helium, lithium, and atoms in which some electrons are removed (ionized) — could be properly analyzed. A breakthrough was made by Wolfgang Pauli, one of the *wunderkinder,* who understood the theory of relativity at the age of nineteen and became the "enfant terrible" of physics as an elder statesman.

A digression on Pauli is unavoidable at this point. Noted for his high standards and irascibility, Pauli was the conscience of physics in his time. Or was he just candid? Abraham Pais reports that Pauli once complained to him that he had trouble finding a challenging problem to work on: "Perhaps it's because I know too much." Not a brag, just a statement of fact. You can imagine that he was tough on assistants. When one new young assistant, Victor Weisskopf, a future leading theorist, reported to him at Zurich, Pauli looked Weisskopf over, shook his head, and muttered, "Ach, so young and already you are unknown." After some months, Weisskopf presented Pauli with a theoretical effort. Pauli took one glance and said, "Ach, that isn't even wrong!" To one postdoc he said, "I don't mind your thinking slowly. I mind your publishing faster than you think." No one was safe from Pauli. In recommending a fellow to be assistant to Einstein, who was, in his later years, deep into the mathematical exotica of his fruitless quest for a unified field theory, Pauli wrote: "Dear Einstein, This student is good, but he does not clearly grasp the difference between mathematics and physics. On the other hand, you, dear Master, have long lost this distinction." That's our boy Wolfgang.

In 1924 Pauli proposed a fundamental principle that explained the Mendeleev periodic table of the elements. The problem: we build up the atoms of the heavier chemical elements by adding positive charge to the nucleus and electrons to the various allowed energy states of the atom (orbits, in the old quantum theory). Where do the electrons go? Pauli announced what has become known as the Pauli exclusion principle: no two electrons can occupy the same quantum state. Originally an inspired guess, the principle turned out to be a consequence of a deep and lovely symmetry.

Let's see how Santa, in his workshop, makes the chemical elements. He has to do this right because he works for Her, and She is tough. Hydrogen is easy. He takes one proton — the nucleus. He adds an electron, which occupies the lowest possible energy state — in the old Bohr theory (which is still useful pictorially) the orbit with the smallest allowed radius. Santa doesn't have to be careful; he just drops the electron anywhere near the proton and it "jumps" eventually to this lowest "ground" state, emitting photons on the way. Now helium. He assembles the helium nucleus, which has two plus charges. So he needs to drop in two electrons. And with lithium it takes three electrons to form the electrically neutral atom. The issue is, where do these electrons go? In the quantum world, only certain states are allowed. Do they all crowd into the ground state, three, four, five . . . electrons? This is where the Pauli principle comes in. No, says Pauli, no two electrons can be in the same quantum state. In helium, the second electron *is* allowed to join the first electron in the lowest energy state *only* if it spins in the opposite sense to its partner. When we add the third electron, for the lithium atom, it is excluded from the lowest energy level and must go into the next lowest level. This turns out to have a much larger radius (again à la Bohr theory), thus accounting for lithium's chemical activity — namely, the ease with which it can use this lone electron to combine with other atoms. After lithium we have the four-electron atom, beryllium, in which the fourth electron joins the third in its "shell," as the energy levels are called.

As we proceed merrily along — beryllium, boron, carbon, nitrogen, oxygen, neon — we add electrons until each shell is filled. No more in that shell, says Pauli. Start a new one. Briefly, the regularity of chemical properties and behaviors all comes out of this quantum buildup via the Pauli principle. Decades earlier, scientists had derided Mendeleev's insistence on lining the elements up in rows and columns according to their characteristics. Pauli showed that this periodicity was precisely tied to the various shells and quantum states of electrons: two can be accommodated in the first shell, eight in the second, eight in the third, and so on. The periodic table did indeed contain a deeper meaning.

Let's summarize this important idea. Pauli invented a rule for how the chemical elements change their electronic structure. This rule accounts for the chemical properties (inert gas, active metal, and so on), tying them to the numbers and states of the electrons, especially those in the outermost shells, where they are most readily in contact with other atoms. The dramatic implication of the Pauli principle is that if a shell is filled, it is impossible to add an additional electron to that shell. The resistive force is huge. This is the real reason for the impenetrability of matter. Although atoms are way more than 99.99 percent empty space, I have a real problem in walking through a wall. Probably you share this frustration. Why? In solids, where atoms are locked together via complicated electrical attractions, the imposition of your body's electrons on the system of "wall" atoms meets Pauli's prohibition on having electrons too close together. A bullet is able to penetrate a wall because it ruptures the atom-atom bonds and, like a football blocker, makes room for its own electrons. Pauli's principle also plays a crucial role in such bizarre and romantic systems as neutron stars and black holes. But I digress.

Once we understand atoms, we solve the problem of how they combine to make molecules, for example, H_2O or NaCl. Molecules are formed via the complex of forces among electrons and nuclei in the combining atoms. The arrangement of the electrons in their shells provides the key to creating a stable molecule. Quantum theory gave chemistry a firm scientific base. Quantum chemistry today is a thriving field, out of which has come new disciplines like molecular biology, genetic engineering, and molecular medicine. In materials science, quantum theory helps us explain and control the properties of metals, insulators, superconductors, and semiconductors. Semiconductors led to the discovery of the transistor, whose inventors fully credit the quantum theory of metals as their inspiration. And out of that discovery came computers and microelectronics and the revolution in communications and information. And then there are masers and lasers, which are complete quantum systems.

When our measurements reached into the atomic nucleus — a scale 100,000 times smaller than the atom — the quantum theory was an essential tool in that new regime. In astrophysics, stellar processes produce such exotic objects as suns, red giants, white dwarfs, neutron stars, and black holes. The life story of these objects is based on quantum theory. From the point of view of social utility, as we have estimated, quantum theory accounts for over 25 percent of the GNP of all the industrial powers. Just think, here are these European physicists obsessed with how the atom works, and out of their efforts come trillions of dollars of economic activity. If only wise and prescient governments had thought to put a 0.1 percent tax on quantum technological products, set aside for research and education . . . Anyway, it does indeed work.

3. It has problems. This issue has to do with the wave function (psi, or ψ) and what it means. In spite of the great practical and intellectual success of quantum theory, we cannot be sure we know what the theory means. Our uneasiness may be intrinsic to the mind of man, or it may be that some genius will eventually come up with a conceptual scheme that makes everyone happy. If it makes you queasy, don't worry. You're in good company. Quantum theory has made many physicists unhappy, including Planck, Einstein, de Broglie, and Schrödinger.

There is a rich literature on the objections to the probabilistic nature of quantum theory. Einstein led the battle, and in a long series of efforts (not easy to follow) to undermine the uncertainty relations, he was continually thwarted by Bohr, who had established what is now called the "Copenhagen interpretation" of the wave function. Bohr and Einstein really went at it. Einstein would invent a thought experiment that was an arrow to the heart of the new quantum theory, and Bohr, usually after a long weekend of hard work, would find the flaw in Einstein's logic. Einstein was the bad boy, the needler in these debates. Like a troublemaking kid in catechism class ("If God is all-powerful, can She build a rock so heavy that not even She can lift it?"), Einstein kept coming up with paradoxes in the quantum theory. Bohr was the priest who kept countering Einstein's objections.

The story is told that many of their discussions took place during walks in the forest. I can see what happened when they encountered a huge bear. Bohr immediately drew a pair of $300 Reebok Pump running shoes out of his backpack and began lacing them up. "What are you doing, Niels? You know you can't outrun a bear," Einstein logically pointed out. "Ah, I don't have to outrun the bear, dear Albert," responded Bohr. "I only have to outrun you."

By 1936 Einstein had reluctantly agreed that quantum theory correctly describes all possible experiments, at least those that can be imagined. He then switched gears and decided that quantum mechanics cannot be a complete description of the world, even though it does correctly give the probability for various measurement outcomes. Bohr's defense was that the incompleteness that worried Einstein was not a fault of the theory but a quality of the world in which we live. These two debated quantum mechanics into the grave, and I'm quite sure they are still at it unless the "Old One," as Einstein called God, out of misplaced concern settled the problem for them.

Einstein and Bohr's debate requires books to tell, but I will try to illustrate the problem with one example. A reminder about Heisenberg's fundamental tenet: no attempt to make a simultaneous measurement of where a particle is and where it is going can ever be entirely successful. Design a measurement to locate the atom, and there it is, as precise as you like. Design a measurement to see how fast it is going — presto, we get its speed. But we can't have both. The reality that these measurements reveal depends on the strategy that the experimenter adopts. This subjectivity challenges our cherished beliefs in cause and effect. If an electron starts at point A and is seen to arrive at point B, it seems "natural" to assume it took a particular path from A to B. Quantum theory denies this, saying the path is unknowable. All paths are possible, and each has its probability.

To expose the incompleteness of this ghostly-trajectory idea, Einstein proposed a crucial experiment. I cannot do justice to his concept, but I'll try to get across the basic idea. It's called the EPR thought experiment, for Einstein, Podolsky, and Rosen, the three inventors. They proposed a two-particle experiment, in which one particle's fate is tied to the other's. There are ways of creating a pair of particles flying apart from each other so that if one spins up the other must spin down, or if one spins right the other must go left. We send one particle speeding off to Bangkok, the other to Chicago. Einstein said, okay, let's accept the idea that we can't know anything about a particle until we measure it. So we measure particle A, in Chicago, and discover that it spins right. Ergo, we now know about particle B, in Bangkok, whose spin is about to be measured. Before the Chicago measurement, the probability of spin left versus spin right was 50 percent. Now, after Chicago, we know that particle B spins left. But how does particle B know the result of the Chicago experiment? Even if it carries a little radio, radio waves travel at the speed of light, and it would take some time for the message to arrive. What is this communicating mechanism, that doesn't even have the courtesy to travel at the velocity of light? Einstein called this "spooky action at a distance." The EPR conclusion is that the only way to understand the connection of A happenings (the decision to measure at A) with the outcome of B is to provide more details, which quantum theory cannot do. Aha! cried Albert, quantum mechanics isn't complete.

When Einstein hit him with EPR, even traffic in Copenhagen stopped while Bohr pondered this problem. Einstein was trying to finesse the Heisenberg uncertainty relation by measuring an accomplice particle. Bohr's eventual rejoinder was that one cannot separate the A and B events, that the system must include A, B, and the observer who decides when to make the measurements. This holistic response was thought to have some ingredients of Eastern religious mysticism, and (too many) books have been written about these connections. The issue is whether the A particle and the A observer, or detector, have a real Einsteinian existence or are irrelevant intermediate ghosts before measurement. This particular issue was resolved by a theoretical breakthrough and (aha!) a brilliant experiment.

Thanks to a theorem developed in 1964 by a particle theorist named John Bell, it became clear that a modified form of the EPR thought experiment could actually be done in the lab. Bell devised an experiment that would predict different amounts of long-distance correlation between A and B particles depending on whether Einstein's or Bohr's point of view was right. Bell's theorem has almost a cult following today, partly because it fits on a T-shirt. For example, there's at least one women's club, probably in Springfield, that meets every Thursday afternoon to discuss Bell's theorem. Much to Bell's chagrin, his theorem was heralded by some as "proof" of paranormal and psychic phenomena.

Bell's idea resulted in a series of experiments, the most successful of which was carried out by Alan Aspect and colleagues in 1982 in Paris. The experiment in effect measured the number of times detector A results correlated with detector B results, that is, left spin and left spin or right spin and right spin. Bell's analysis enabled one to predict this correlation using the Bohr interpretation of a "complete-as-can-be" quantum theory as opposed to the Einstein notion that there must be hidden variables that determine the correlation. The experiment clearly showed that Bohr's analysis was correct and Einstein's wrong. Apparently these long-distance correlations between particles are the way nature works.

Did this end the debate? No way. It rages today. One of the more intriguing places where quantum spookiness has arisen is in the very creation of the universe. In the earliest phase of creation, the universe was of subatomic dimensions, and quantum physics applied to the entire universe. I may be speaking for the masses of physicists in saying that I'll stick to my accelerator research, but I'm mighty glad someone is still worrying about the conceptual foundations of quantum theory.

For the rest of us, we are heavily armed with Schrödinger, Dirac, and the newer quantum field theory equations. The road to the God Particle — or at least its beginning — is now very clear.

· Interlude B ·

THE DANCING MOO-SHU MASTERS

DURING THE ENDLESS PROCESS of raising, and reraising, enthusiasm for the construction of the SSC (the Superconducting Super Collider), I was visiting the Washington office of Senator Bennett Johnston, a Louisiana Democrat whose support was important to the fate of the Super Collider, which is expected to cost $8 billion. Johnston is a curious kind of guy for a U.S. senator. He likes to talk about black holes, time warps, and other phenomena. As I entered his office, he stood up behind his desk and shook a book in my face. "Lederman," he pleaded, "I have a lot of questions for you about this." The book was *The Dancing Wu Li Masters* by Gary Zukav. During our talk, he kept extending my "fifteen minutes" until we had spent an hour talking physics. I kept looking for an opening, a pause or a phrase I could use as a segue into my pitch for the Super Collider. ("Speaking of protons, I have this machine . . .") But Johnston was relentless. He talked physics nonstop. When his appointments secretary had interrupted for the fourth time, he smiled and said, "Look, I know why you came. Had you given me your pitch I would have promised to 'do what I can.' But this was much more fun! And I'll do what I can." Actually, he did a great deal.

To me it was a little disturbing that this U.S. senator, hungry for knowledge, had satisfied his curiosity with Zukav's book. There has been a spate of books over the past several years — *The Tao of Physics* is another example — that attempt to explain modern physics in terms of Eastern religion and mysticism. The authors are apt to conclude rapturously that we are all part of the cosmos and the cosmos is part of us. We are all one! (Though, inexplicably, American Express bills us separately.) My concern was that a senator might get some anxious ideas from such books just before an important vote for an $8 billion–plus machine to be run by physicists. Of course, Johnston is science-literate and knows a lot of scientists.

The inspiration for such books is usually quantum theory and its inherent spookiness. One book, which shall go nameless, presents sober explanations of the Heisenberg uncertainty relations, the Einstein-Podolsky-Rosen thought experiment, and Bell's theorem, then launches into a rapturous discussion of LSD trips, poltergeists, and a long-dead entity named Seth who communicated his ideas by taking over the voice and writing hand of an Elmira, New York, housewife. Evidently one premise of this book, and of many like it, is that quantum theory is spooky, so why not accept other strange stuff as scientific fact also?

Normally, one wouldn't care about such books if they were found in the religion, paranormal, or poltergeist sections of bookstores. Unfortunately, they are often placed in the science category, probably because words like "quantum" and "physics" are used in their titles. Too much of what the reading public knows about physics, it knows from reading these books. We'll pick on just two here, the most prominent of the lot: *The Tao of Physics* and *The Dancing Wu Li Masters,* both published in the 1970s. To be fair, *Tao*, by Fritjof Capra, who holds a Ph.D. from the University of Vienna, and *Wu Li,* by Gary Zukav, a writer, have introduced many people to physics, which is good. And there's certainly nothing wrong with finding parallels between the new quantum physics and Hinduism, Buddhism, Taoism, Zen, or Hunan cuisine, for that matter. Capra and Zukav have also gotten a lot of things right. There is some good physics writing in both of these books, which gives them a feeling of credibility. Unfortunately, the authors jump from solid, proven concepts in science to concepts that are outside of physics and to which the logical bridge is extremely shaky or nonexistent.

In *Wu Li,* for example, Zukav does a nice job of explaining Thomas Young's famous double-slit experiment. But his analysis of the results is rather bizarre. As already discussed, because one gets different patterns of photons (or electrons) depending on whether one slit or two slits are open, an experimenter might ask herself, "How does the particle 'know' how many slits are open?" This, of course, is a whimsical phrasing of a question on mechanisms. The Heisenberg uncertainty principle, a concept which is the basis of quantum theory, says that one cannot determine which slit the particle slithers through without destroying the experiment. By the curious but effective rigor of quantum theory, such questions are not relevant.

But Zukav gets a different message from the double-slit experiment: the particle *does* know whether one or two slits are open. Photons are smart! Wait, it gets better. "We have little choice but to acknowledge," Zukav writes, "that photons, which are energy, do appear to process information and to act accordingly, and that therefore, strange as it may sound, they seem to be organic." This is fun, perhaps even philosophical, but we have departed from science.

Paradoxically, while Zukav is ready to ascribe consciousness to photons, he refuses to accept the existence of atoms. He writes, "Atoms were never 'real' things anyway. Atoms are hypothetical entities constructed to make experimental observations intelligible. No one, not one person, has ever seen an atom." There's our lady in the audience again, challenging us with the question "Have you ever seen an atom?" To give the lady credit, she was willing to listen to the answer. Zukav has already answered the question, in the negative. Even on a literal level, he is now way off the mark. Since his book was published, many people have seen atoms, thanks to the scanning tunneling microscope, which takes beautiful pictures of the little fellows.

As for Capra, he's much cleverer, hedging his bets and his language, but essentially he's another nonbeliever. He insists that the "simple mechanistic picture of building blocks" should be abandoned. Starting with reasonable descriptions of quantum physics, he constructs elaborate extensions, totally bereft of the understanding of how carefully experiment and theory are woven together and how much blood, sweat, and tears go into each painful advance.

If the casual disregard of such writers turns me off, the true charlatans positively disconnect me. In fact, *Tao* and *Wu Li* constitute a relatively respectable middle ground between good science books and a lunatic fringe of fakes, charlatans, and crazies. These folks guarantee eternal life if you restrict your diet to sumac roots. They give firsthand evidence of the visit of extraterrestrials. They expose the fallacy of relativity in favor of a Sumerian version of the *Farmer's*

Almanac. They write for the "*New York Inquirer*" and contribute to the crackpot mail of all prominent scientists. Most of these people are harmless, like the seventy-year-old woman who reported to me, in eight closely spaced handwritten pages, her conversation with small green space visitors. Not all are harmless, however. A secretary of the *Physical Review*, a journal, was shot to death by a man whose incoherent article was refused publication.

The important point, I believe, is this: all disciplines, all fields of endeavor, have an "establishment," be it the collection of aging physics professors in the prestige universities, the tycoons of the fast-food business, the senior officials of the American Bar Association, or the elder statesmen of the Fraternal Order of Postal Workers. In science the road to advancement is most rapid when giants are overturned. (I knew I'd get a good mixed metaphor out of this.) Thus, iconoclasts, rebels with (intellectual) bombs, are sought after zealously — even by the science establishment itself. Of course, no theorist enjoys having his theory trashed, and some may even react — momentarily and instinctively — like the political establishment in the face of a rebellion. But the tradition of overthrow is too strongly ingrained. The nurturing and rewarding of the young and creative is a sacred obligation of the science establishment. (The saddest report one can get about so-and-so is that it is not enough to be young.) This ethic — that we should remain open to the young, the unorthodox, and the rebellious — creates an opening for the charlatans and the misguided, who can prey upon scientifically illiterate and careless journalists, editors, and other gatekeepers of the media. Some fakes have had remarkable success, such as the Israeli magician Uri Geller or the writer Immanuel Velikovsky or even some Ph.D.'s in science (a Ph.D. is even less a guarantee of truth than a Nobel Prize) who push totally off-the-wall things like "seeing hands," "psychokinesis," "creation science," "polywater," "cold fusion," and so many other fraudulent ideas. Usually the claim is that the revealed truth is being suppressed by the ensconced establishment, intent on preserving the status quo with all the rights and privileges.

Sure, that can happen. But in our discipline, even members of the establishment rail against the establishment. Our patron saint, Richard Feynman, in the essay "What Is Science?" admonished the student: "Learn from science that you must doubt the experts. . . . Science is the belief in the ignorance of experts." And later: "Each generation that discovers something from its experience must pass that on, but it must pass that on with a delicate balance of respect and disrespect, so that the race . . . does not inflict its errors too rigidly on its youth, but it does pass on the accumulated wisdom plus the wisdom that it may not be wisdom."

This eloquent passage expresses the deep training in all of us who have labored in the vineyard of science. Of course, not all scientists can summon the critical juices, the mixture of passion and perception that Feynman could bring to an issue. That's what differentiates scientists, and it is also true that many great scientists take themselves too seriously. They are then handicapped in applying their critical powers to their own work or, worse still, to the work of the kids who are challenging them. No discipline is perfect. But what is rarely understood by the lay public is how ready, how eager, how desperately the collective science community in a given discipline welcomes the intellectual iconoclast — if he or she has the goods.

The tragedy in all this is not the sloppy pseudoscience writers, not the Wichita insurance salesman who knows exactly where Einstein went wrong and publishes his own book on it, not the faker who will say anything to make a buck — not the Gellers or Velikovskys. It is the damage done to the gullible and science-illiterate general public, which can so easily be duped. This public will buy pyramids, pay a fortune for monkey gland injections, chew apricot pits, go anywhere and do anything to follow the huckster who, having progressed from the back of the wagon to the prime-time TV channel, sells ever more flagrant palliatives in the name of "science."

Why are we, meaning we the public, so vulnerable? One possible answer is that the lay public is uncomfortable with science, unfamiliar with the way it evolves and progresses. The public sees science as some monolithic edifice of unbending rules and beliefs, and — thanks to the media's portrayal of scientists as uptight nerds in white coats — sees scientists as stodgy old artery-hardened defenders of the status quo. In truth, science is a much more flexible thing. Science is not about status quo. It's about revolution.

THE RUMBLES OF REVOLUTION

Quantum theory becomes a ready target for writers who declare it akin to some sort of religion or mysticism. Classical Newtonian physics is often portrayed as safe, logical, intuitive. Quantum theory, counterintuitive and spooky, comes along and "replaces" it. It's hard to understand. It's threatening. One solution — the solution in some of the books discussed above — is to think of quantum physics as a religion. Why not consider it a form of Hinduism (or Buddhism, etc.)? That way we can simply abandon logic altogether.

Another way is to think of quantum theory as, well, science. And don't be taken in by this idea of its "replacing" what went before. Science doesn't toss out centuries-old ideas willy-nilly — especially if those ideas have worked. It is worth a short digression here to explore how revolutions in physics come about.

New physics doesn't necessarily vanquish old physics. Revolutions in science tend to be executed conservatively and cost-effectively. They may have staggering philosophical consequences, and they may seem to abandon the conventional wisdom about how the world works. But what really happens is that the established dogma is extended to a new domain.

Take that old Greek Archimedes. In 100 B.C. he summarized the principles of statics and hydrostatics. Statics is the study of the stability of structures like ladders, bridges, and arches — usually things that man has devised to make himself more comfortable. Archimedes' work on hydrostatics had to do with liquids and what floats and what sinks, with what floats upright and what rolls over, principles of buoyancy, why you scream "Eureka!" in a bathtub, and so on. These issues and Archimedes' treatment of them are as valid today as they were two thousand years ago.

In 1600 Galileo examined the laws of statics and hydrostatics, but extended his measurements to moving objects, objects rolling down inclined planes, balls tossed from towers, weighted lute strings swinging back and forth in his father's workshop. Galileo's work included Archimedes' work but explained much more. Indeed, his work extended to the features of the lunar surface and the moons of Jupiter. Galileo did not vanquish Archimedes. He engulfed him. If we were to represent their work pictorially, it would look like this:

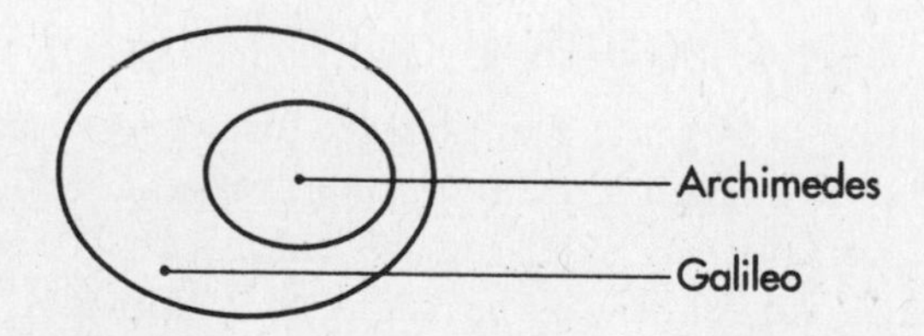

Newton reached far beyond Galileo. By adding causation he was able to examine the solar system and diurnal tides. Newton's synthesis included new measurements of the motion of planets and their moons. Nothing in the Newtonian revolution threw any doubt on the contributions of Galileo or Archimedes, but Newton's revolution extended the regions of the universe that are subject to this grand synthesis.

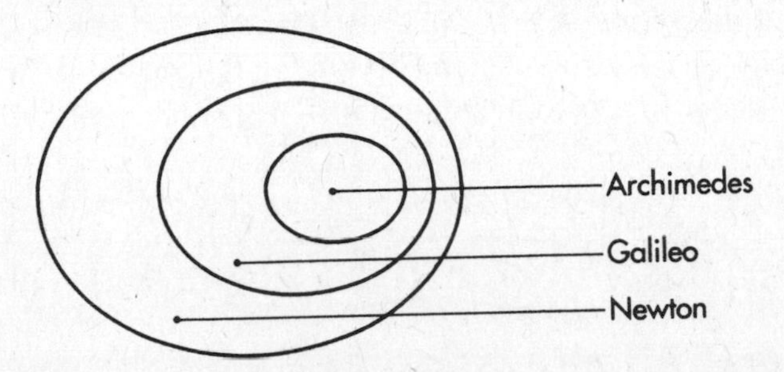

In the eighteenth and nineteenth centuries, scientists began to study a phenomenon that was outside normal human experience. It was called electricity. Except for the frightening occurrence of lightning flashes, electrical phenomena had to be contrived to be studied (just as some particles must be "manufactured" in our accelerators). Electricity was then as exotic as quarks are today. Slowly, currents and voltages, electrical and magnetic fields, were understood and even controlled. The laws of electricity and magnetism were extended and codified by James Maxwell. As Maxwell and then Heinrich Hertz and then Guglielmo Marconi and then Charles Steinmetz and many others put these ideas to use, the human environment changed. Electricity surrounds us, communications crackle in the air we breathe. But Maxwell's respect for all who went before was flawless.

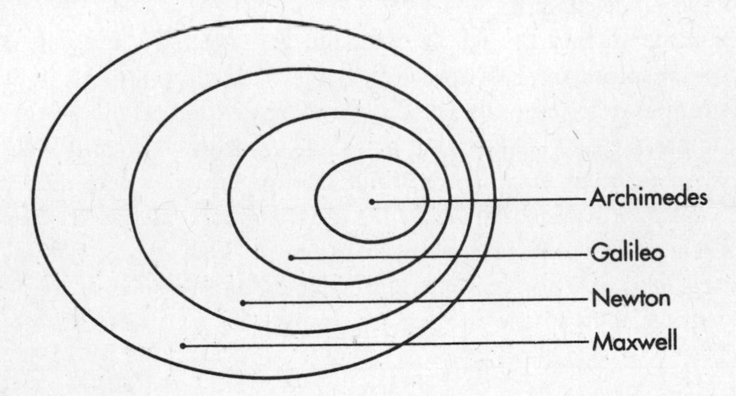

There wasn't much out beyond Maxwell and Newton — or was there? Einstein focused his attention at the rim of the Newtonian universe. His conceptual ideas went very deep; aspects of Galilean and Newtonian assumptions troubled him and eventually drove him to bold new premises. However, the domain of his observations now included things that moved with extraordinary speeds. Such phenomena were irrelevant to observers of the pre-1900 era, but as humans examined atoms, devised nuclear instruments, and began to look at happenings in the earliest moments of the universe's existence, Einstein's observations became important.

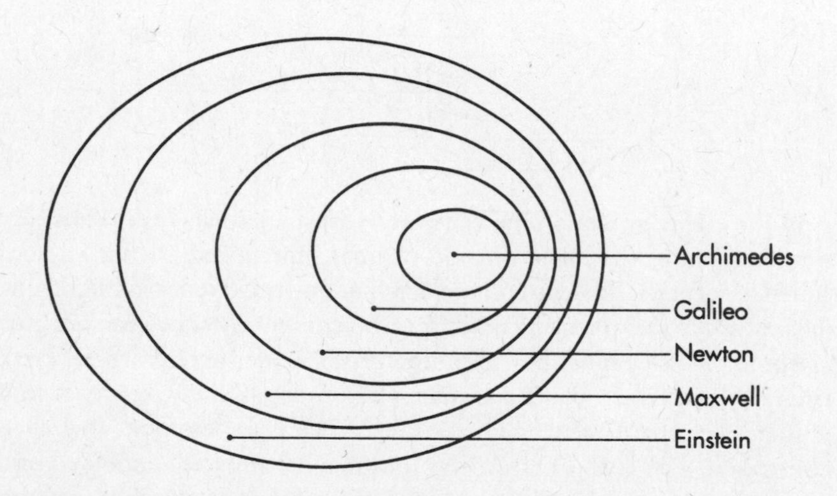

Einstein's theory of gravity also went beyond Newton's to include the dynamics of the universe (Newton believed in a static universe) and its expansion from an initial cataclysmic happening. Yet when Einstein's equations are aimed at the Newtonian world, they give Newtonian results.

So now we had the whole schmeer, no? No! We had yet to look inside the atom, and when we did, we needed concepts far beyond Newton (and unacceptable to Einstein) that extended the world down to the atom, the nucleus, and, as far as we know, beyond. (Within?) We needed quantum physics. Still, nothing in the quantum revolution cashed in Archimedes, sold out Galileo, skewered Newton, or defiled Einstein's relativity. Rather, a new domain had been sighted, new phenomena encountered. Newton's science was found inadequate, and in the fullness of time a new synthesis was discovered.

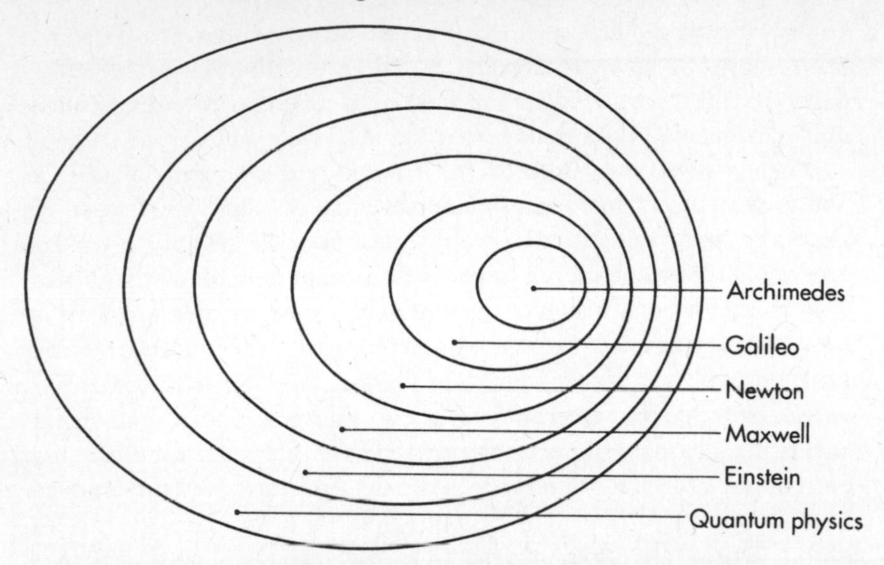

Remember we said in Chapter 5 that the Schrödinger equation was created to deal with electrons and other particles, but when applied to baseballs and other large objects, it transforms itself in front of our eyes to Newton's $F = ma$, or thereabouts. Dirac's equation, the one that predicted antimatter, was a "refinement" of the Schrödinger equation, designed to deal with "fast" electrons, which move at a significant fraction of the speed of light. Yet when the Dirac equation is applied to slow-moving electrons, out pops . . . the Schrödinger equation, but magically revised to include the spin of the electron. But discard Newton? No way.

If this march of progress sounds wonderfully efficient, it's worth pointing out that it generates a good deal of waste as well. As we open new areas to observation with our inventions and our unquenchable curiosity (and plenty of federal grants), the data usually stimulate a cornucopia of ideas, theories, and suggestions, most of which are wrong. In the contest for control of the frontier there is, in terms of concepts, only one winner. The losers vanish into the debris of history's footnotes.

How does a revolution happen? During any period of intellectual tranquility, such as occurred in the late nineteenth century, there is always a set of phenomena that are "not yet explained." The experimental scientists hope their observations will kill the reigning theory. Then a better theory will take its place and reputations will have been made. More often, either the measurements are wrong or a clever application of the current theory turns out to explain the data. But not always. Since there are always three possibilities — (1) wrong data, (2) old theory resilient, and (3) need new theory — experiment makes science a lively métier.

When a revolution does occur, it extends the domain of science, and it may also have a profound influence on our world view. An example: Newton created not only the universal law of gravitation but also a deterministic philosophy that caused theologians to place God in a new role. Newtonian rules established mathematical equations that determined the future of any system if the initial conditions were known. In contrast, quantum physics, applicable to the atomic world, softens the deterministic view, allowing individual atomic events the pleasures of uncertainty. In fact, developments indicate that even outside the subatomic world, the deterministic Newtonian order is really too idealized. The complexities that compose the macroscopic world are so prevalent that for many systems, the most insignificant change in the initial conditions produces huge changes in the outcome. Systems as simple as the flow of water down a hill or a pair of dangling pendulums will exhibit "chaotic" behavior. The science of nonlinear dynamics, or "chaos," tells us that the real world is not nearly as deterministic as was once thought.

Which doesn't mean that science and the Eastern religions have suddenly discovered a lot in common. Still, if the religious metaphors offered up by the authors of texts comparing the new physics to Eastern mysticism help you in some way to appreciate the modern revolutions in physics, then by all means use them. But metaphors are only metaphors. They are crude maps. And to borrow an old expression: we must never mistake the map for the territory. Physics is not religion. If it were, we'd have a much easier time raising money.

· 6 ·

ACCELERATORS: THEY SMASH ATOMS, DON'T THEY?

SENATOR JOHN PASTORE: Is there anything connected with the hopes of this accelerator that in any way involves the security of this country?

ROBERT R. WILSON: No sir. I don't believe so.

PASTORE: Nothing at all?

WILSON: Nothing at all.

PASTORE: It has no value in that respect?

WILSON: It has only to do with the respect with which we regard one another, the dignity of men, our love of culture. It has to do with, are we good painters, good sculptors, great poets? I mean all the things we really venerate and honor in our country and are patriotic about. It has nothing to do directly with defending our country except to make it worth defending.

We have a tradition at Fermilab. Every June 1, rain or shine, at 7 A.M., the staff is invited to jog the four miles around the main ring of the accelerator on the surface road, which doubles as a jogging path. We always run in the direction that the antiprotons accelerate. My last unofficial time around the ring was 38 minutes. The current director of Fermilab, my successor John Peoples, put up a sign his first summer in the job, inviting the staff to run on June 1 with "a younger, faster director." Swifter he was, but neither of us was fast enough to beat the antiprotons. They complete the circuit in about 22 millionths of a second, which means that each antiproton laps me about 100 million times.

The Fermilab staff continues to be humiliated by the antiprotons. We get even, though, because we get to design the experiments. We steer the antiprotons head-on into protons racing just as fast in the opposite direction. The process of getting particles to collide is the essence of this chapter.

Our discussion of accelerators will be a bit of a departure. We've been racing through century after century of scientific progress like a runaway truck. Let's slow down the pace. Here we'll talk not so much about discoveries or even physicists but about machines. Instruments have been inextricably tied to scientific progress, from Galileo's inclined plane to Rutherford's scintillation chamber. Now an instrument takes center stage. One cannot understand the physics of the past several decades without understanding the nature of the accelerator and its accompanying array of particle detectors, the dominant tools in the field for the past forty years. By understanding the accelerator, one also learns much of the physics, for this machine embodies many principles that physicists have labored centuries to perfect.

I sometimes think about the tower at Pisa as the first particle accelerator, a (nearly) vertical linear accelerator that Galileo used in his studies. However, the real story starts much later. The development of the accelerator stems from our desire to go down into the atom. Galileo aside, the history begins with Ernest Rutherford and his students, who became masters of the art of exploiting the alpha particle to explore the atom.

The alpha particle is a gift. When some naturally radioactive materials spontaneously disintegrate, they shoot out these heavy, energetic particles. An alpha particle typically has an energy of 5 million electron volts. An electron volt (eV) is the amount of energy a single electron would receive if it crossed from the case (negative) of a 1-volt flashlight battery to the battery's terminal (positive). By the time you finish the next couple of chapters, the electron volt will be as familiar as the inch, the calorie, or the megabyte. Here are four abbreviations you should know before we go on:

KeV: thousand electron volts (K for kilo)
MeV: million electron volts (M for mega)
GeV: billion electron volts (G for giga)
TeV: trillion electron volts (T for tera)

Beyond TeV we resort to powers-of-ten notation, 10^{12} eV being one TeV. Beyond 10^{14}, our foreseen technology runs out, and we are in the domain of cosmic ray particles, which bombard the earth from outer space. The numbers of cosmic ray particles are small, but their energies go all the way up to 10^{21} eV.

In particle physics terms, 5 MeV isn't very much. Rutherford's alphas barely managed to break up the nucleus of a nitrogen atom in perhaps the first on-purpose nuclear collision. And only tantalizing hints of what was to be learned emerged from those collisions. Quantum theory tells us that the smaller the object being studied, the more energy you need — equivalent to sharpening the Democritan knife. To cut the nucleus effectively we need energies of many tens or even hundreds of MeV. The more the better.

IS GOD MAKING THIS UP AS SHE GOES ALONG?

A philosophical digression. As I will describe, particle scientists went along cheerfully building ever more powerful accelerators for all the reasons any of us *sapiens* does anything — curiosity, ego, power, greed, ambition . . . Every so often, a group of us in quiet contemplation over a beer would speculate about whether God Herself knows what our next machine — for example, the 30 GeV "monster" that was nearing completion at Brookhaven in 1959 — will produce. Are we just inventing puzzles for ourselves by achieving these new, unheard-of energies? Does God, in Her insecurity, look over the shoulder of Gell-Mann or Feynman or one of Her other favorite theorists to find out what to do at those huge energies? Does She call together a committee of resident angels — Reb Newton, Einstein, Maxwell — to suggest what 30 GeV should do? This point of view is occasionally encouraged by the jumpy nature of theoretical history — as if She is making it up as we go along. However, progress in astrophysics and cosmic ray research quickly assures us that this is just Friday-evening-before-the-sabbath nonsense. Our colleagues who look up tell us with assurance that the universe is very much concerned with 30 GeV, with 300 GeV, indeed with 3 billion GeV. Space is awash with particles of astronomical (ouch!) energies, and what is today a rare and exotic happening at an infinitesimal collision point on Long Island or Batavia or Tsukuba was, just after the birth of the universe, an ordinary, everyday, garden-variety happening.

And now back to the machines.

WHY SO MUCH ENERGY?

The most powerful accelerator today, Fermilab's Tevatron, produces collisions at about 2 TeV, or 400,000 times the energy created by Rutherford's alpha particle collisions. The yet-to-be-built Superconducting Super Collider is designed to operate at about 40 TeV.

Now 40 TeV sounds like a great deal of energy, and it is indeed when invested in a single collision of two particles. But we should put this into perspective. When we strike a match, we involve about 10^{21} atoms in the reaction, and each process releases about 10 eV, so the total energy is roughly 10^{22} eV, or about 10 billion TeV. At the Super Collider there will be 100 million collisions per second, each one releasing 40 TeV, for a total of 4 billion TeV — not too different from the energy released in lighting a match! But the key is that the energy is concentrated in a few particles rather than in the billions and billions and billions of particles contained in any speck of visible matter.

We can look at the entire accelerator complex — from the oil-fired power station through the electrical power lines to the lab where transformers ship the electrical energy to magnets and radio-frequency cavities — as a giant device for concentrating, with extremely low efficiency, the chemical energy of oil into a measly billion or so protons per second. If the macroscopic quantity of oil was heated so that each of the constituent atoms had 40 TeV, the temperature would be 4×10^{17} degrees, four hundred thousand trillion degrees on the Kelvin scale. The atoms would melt into their constituent quarks. Such was the state of the entire universe less than a millionth of a billionth of a second after creation.

So what do we do with all this energy? Quantum theory demands more and more powerful accelerators to study smaller and smaller things. Here's a table showing the approximate energy one needs to crack open various interesting structures:

ENERGY (approximate)	SIZE OF STRUCTURE
0.1 eV	Molecule, large atom, 10^{-8} meters
1.0 eV	Atom, 10^{-9} m
1,000 eV	Atomic core, 10^{-11} m
1 MeV	Fat nucleus, 10^{-14} m
100 MeV	Nuclear core, 10^{-15} m
1 GeV	Neutron or proton, 10^{-16} m
10 GeV	Quark effects, 10^{-17} m
100 GeV	Quark effects, 10^{-18} m (more detail)
10 TeV	God Particle? 10^{-20} m

Note how predictably the required energy goes up as the size goes down. Note also that you need only 1 eV to study atoms, but 10 billion eV to begin to study quarks.

Accelerators are like the microscopes used by biologists to study ever smaller things. Ordinary microscopes use light to illuminate the structure of, say, red corpuscles in blood. Electron microscopes, beloved by the microbe hunters, are more powerful precisely because the energy of the electrons is higher than that of the light in an optical microscope. The electrons' shorter wavelengths allow biologists to "see" the molecules from which a corpuscle is constructed. It is the wavelength of the bombarding object that determines the size of what you can "see" and study. In quantum theory we know that as the wavelength gets shorter the energy increases; our chart simply demonstrates the connection.

In 1927, Rutherford, in an address to Britain's Royal Society, expressed the hope that one day scientists would find a way to accelerate charged particles to energies higher than that provided by radioactive decay. He foresaw the invention of machines capable of generating many millions of volts. There was a motivation for such machines beyond pure power. Physicists needed to be able to hurl more projectiles at a given target. Alpha sources provided by nature were less than bountiful: fewer than a million particles could be directed toward a 1-square-centimeter target per second. A million sounds like a lot, but nuclei occupy only one hundredth of a millionth of the target area. You need at least a thousand times more accelerated particles (a billion) and, as mentioned, a lot more energy — many millions of volts (physicists weren't sure how many millions) — to probe the nucleus. This seemed like a daunting task in the late 1920s, yet physicists in many laboratories began to work on the problem. What ensued was a race to create machines that would accelerate the requisite huge number of particles to at least one million volts. Before we discuss the advances in the technology of accelerators, we should talk about some basics.

THE GAP

The physics of particle acceleration is simple to explain (watch out!). Connect the terminals of a DieHard battery to two metal plates (also called terminals), positioned, say, a foot apart. This arrangement is called the Gap. Seal the two terminals into a can from which the air is removed. Organize the equipment so that an electrically charged particle — electrons and protons are the prime projectiles — can move freely across the gap. A negatively charged electron will gladly rush toward the positive terminal, gaining an energy of (look at the label on the battery) 12 eV. Thus the Gap produces an acceleration. If the positive metallic terminal is made of wire screen instead of a solid plate, most of the electrons will pass through it, creating a directed beam of 12 eV electrons. Now an electron volt is an extremely small amount of energy. What we need is a billion-volt battery, but Sears doesn't handle such an item. To achieve high voltages requires moving beyond chemical devices. But no matter how big the accelerator, whether we're talking about a 1920s Cockcroft-Walton device or the fifty-four-mile-around Super Collider, the basic mechanism remains the same — the Gap, across which particles gain energy.

The accelerator takes normal, law-abiding particles and gives them extra energy. Where do we get the particles? Electrons are easy. We heat a wire to incandescence, and electrons pour out. Protons are easy, too. The proton is the nucleus of the hydrogen atom (hydrogen nuclei have no neutrons), so all we need is commercially available hydrogen gas. Other particles can be accelerated, but they must be stable — that is, have long lifetimes — because the acceleration process is time consuming. And they must be electrically charged, since the Gap obviously wouldn't work on a neutral particle. The leading candidates for acceleration are protons, antiprotons, electrons, and positrons (anti-electrons). Heavier nuclei such as deuterons and alpha particles can also be accelerated, and they have their special uses. An unusual machine under construction on Long Island in New York will accelerate uranium nuclei to billions of electron volts.

THE PONDERATOR

What does the acceleration process do? The easy but incomplete answer is that it speeds up the lucky particles. In the early days of accelerators, this explanation worked fine. A better description is that it raises the *energy* of the particles. As accelerators got more powerful, they soon were able to achieve speeds close to the ultimate: the velocity of light. Einstein's 1905 special theory of relativity asserts that nothing can travel faster than light. Because of relativity, "velocity" is not a very useful concept. For example, one machine may accelerate protons, say, to 99 percent of the velocity of light, while a much more expensive one, built ten years later, can achieve 99.9 percent of the velocity of light. Big deal. Go explain this to the congressman who voted all that dough just to achieve another 0.9 percent!

It's not speed that sharpens the Democritan knife and yields new domains of observation. It's energy. A 99-percent-of-the-velocity-of-light proton has an energy of about 7 GeV (the Berkeley Bevatron, 1955), whereas a 99.95 percent proton has 30 GeV (Brookhaven AGS, 1960), and a 99.999 percent proton has 200 GeV (Fermilab, 1972). So Einstein's relativity, which rules the changes in velocity and energy, makes it silly to talk about speed. What is important is energy. A related attribute is momentum, which, for a high-energy particle, can be considered directed energy. Incidentally, the particle being accelerated also gets heavier because of $E = mc^2$. In relativity a particle at rest still has the energy given by $E = m_0c^2$, where m_0 is defined as the "rest mass" of the particle. When the particle is accelerated its energy, E, and hence its mass increase. The closer to the velocity of light we get, the heavier the object becomes, and consequently the more difficult it is to increase its speed. But the energy keeps going up. Conveniently, a proton's rest mass is about 1 GeV. So the mass of a 200 GeV proton is more than two hundred times that of the proton resting comfortably in the hydrogen gas bottle. Our accelerator is actually a "ponderator."

MONET'S CATHEDRAL, OR THIRTEEN WAYS OF LOOKING AT A PROTON

Now, how do we use these particles? Simply said, we cause them to make collisions. Since this is the core process by which we learn about matter and energy, we must go into detail. It's okay to forget the various particulars about the machinery and how the particles are accelerated, interesting as these may be. But remember this part. The whole point of the accelerator is the collision.

Our technique of observing and eventually comprehending the abstract world of the subnuclear domain is similar to how we comprehend anything — a tree, for example. What is the process? First, we need light. Let's use sunlight. The photons from the sun stream toward the tree and reflect off leaves and bark, twigs and branches, and some fraction of these photons is collected by our eyeball. The photons, we can say, are scattered by the object toward the detector. The lens of the eye focuses the light on the retina at the back of the eye. The retina detects the photons and sorts out the various qualities: color, shade, intensity. This information is organized and sent to the on-line processor, the occipital lobe of the brain, which specializes in visual data. Eventually, the off-line processor comes to a conclusion: "By Jove, a tree! How lovely."

The information coming to the eye may be filtered through spectacles or sunglasses, adding to the distortion that the eye has already introduced. It's up to the brain to correct these distortions. Let's replace the eye with a camera, and now, a week later, with a greater degree of abstraction, the tree is seen projected in a family slide show. Or a video recorder can convert the data provided by scattered photons into digital electronic information: zeroes and ones. To enjoy this, one plays it through the TV, which converts the digital information back to analog — a tree shows up on the screen. If one wanted to send "tree" to our scientist colleagues on the planet Ugiza, the digital information might not be converted to analog, but it could convey, with maximum precision, the configuration that earthlings call a tree.

Of course, it's not so simple in an accelerator. Different kinds of particles are used in different ways. Still, we can push the metaphor for nuclear collisions and scattering another step. The tree looks different in the morning, at noon, in the setting sun. Anyone who has seen Monet's numerous paintings of the entrance to the cathedral at Rouen at different times of the day knows what a difference the quality of light makes. What is the truth? To the artist the cathedral has many truths. Each shimmers in its own reality — the hazy morning light, the stark contrasts of the noontime sunshine, or the rich glow of the late afternoon. In each of these lights a different aspect of truth is exhibited. Physicists work with the same bias. We need all the information we can get. The artist employs the sun's changing light. We employ different particles: a stream of electrons, a stream of muons or neutrinos — at ever-changing energies.

Here's how it works.

What is *known* about a collision is what goes in and what comes out — and how it comes out. What happens in that tiny volume of the collision? The maddening truth is that we can't see. It's as if a black box covers the collision region. The inner mechanistic details of the collision are not observable — are hardly even capable of being imagined — in the spooky, shimmering quantum world. What we do have is a model for the forces at play and, where relevant, for the structure of the colliding objects. We see what goes in and what comes out, and we ask if the patterns are predictable by our model of what is in the box.

In a Fermilab education program for ten-year-olds, we confront them with this problem. We give them an empty square box to look at, shake, weigh. Then we put something in the box, such as a wooden block or three steel balls. Then we ask the students again to weigh, shake, tilt, listen, and to tell us everything they can about the objects: size, shape, weight . . . It's an instructive metaphor for our scattering experiments. You'd be surprised how often the kids get it right.

Let's switch to grownups and particles. Let's say we want to find out the size of protons. So we take a tip from Monet. We look at them in different forms of "light." Could protons be points? To find out, physicists hit protons with other protons at very low energy to explore the electromagnetic force between the two charged objects. Coulomb's law says that this force reaches out to infinity, decreasing in strength as the square of the distance. The target proton and the accelerated proton are, of course, both positively charged, and since like charges repel, the slow proton is readily repelled by the target proton. It never gets very close. In this kind of "light," the proton does in fact look like a point, a point of electric charge. So we increase the energy of the accelerated protons. Now the deviations in the patterns of scattered protons indicate that the penetrations are getting deep enough to touch what's called the strong force, the force that we now know holds the proton's constituents together. The strong force is a hundred times stronger than the Coulomb electrical force, but unlike the electrical force, its range is anything but infinite. The strong force reaches out only to a distance of about 10^{-13} centimeters, then fades quickly to zero.

By increasing the energy of the collision, we unearth more and more details about the strong force. As the energy increases, the wavelength of the protons (remember de Broglie and Schrödinger) shrinks. And, as we have seen, the smaller the wavelength, the more detail that can be discerned in the particle being studied.

Some of the best "pictures" of the proton were taken in the 1950s by Robert Hofstadter of Stanford University. There the "light" used was a beam of *electrons* rather than protons. Hofstadter's team aimed a well-organized beam of, say, 800 MeV electrons at a small vat of liquid hydrogen. The electrons bombarded the protons in the hydrogen, resulting in a scattering pattern, the electrons emerging in a variety of directions relative to their original motion. Not too different from what Rutherford did. Unlike the proton, the electron does not respond to the strong nuclear force. It responds only to the electric charge in the proton, so the Stanford scientists were able to explore the shape of the charge distribution in the proton. In effect, this revealed the proton's size. It was clearly not a point. The radius was measured to be 2.8×10^{-13} centimeters, with the charge piling up at the center and fading out at the edges of what we call a proton. Similar results were obtained when the experiments were repeated with muon beams, which also ignore the strong force. Hofstadter was awarded a Nobel Prize in 1961 for his "photograph" of the proton.

About 1968, physicists at the Stanford Linear Accelerator Center (SLAC) bombarded protons with electrons at a much higher energy — 8 to 15 GeV — and got a vastly different set of scattering patterns. In this hard light, the proton presented quite a different picture. The relatively low-energy electrons that Hofstadter used were able to see only a "blurry" proton, a smooth distribution of charge that made the proton look like a mushy little ball. The SLAC electrons probed harder and found little guys running around inside the proton. This was the earliest indication of the reality of quarks. The new data and the old data were consistent — like morning and evening paintings by Monet — but the low-energy electrons could reveal only average charge distributions. The visualization provided by the higher-energy electrons showed that our proton contains three rapidly moving, pointlike constituents. Why did the SLAC experiment show this detail, while the Hofstadter study did not? A collision with high enough energy (determined by what goes in and what comes out) freezes the quarks in place and "feels" the pointlike force. It's the virtue of short wavelengths again. This force promptly induces large-angle scattering (remember Rutherford and the nucleus) and large energy changes. The formal name for this phenomenon is "deep inelastic scattering." In Hofstadter's earlier experiments, the quark motion was blurred out and the protons looked "smooth" and uniform inside because of the lower energy of the probing electrons. Think of taking a photograph of three rapidly vibrating, tiny light bulbs using a one-minute time exposure. The film would show one big blurry undifferentiated object. The SLAC experiment, in a crude sense, used a faster shutter, freezing the spots of light so that they could easily be counted.

Since the quark interpretation of the higher energy electron scattering was very far out and of tremendous importance, these experiments were repeated at Fermilab and at CERN (an acronym for the European Center for Nuclear Research), using muons of ten times the SLAC energy (150 GeV) as well as neutrinos. Muons, like electrons, test the electromagnetic structure of the proton, but neutrinos, impervious to both the electromagnetic *and* the strong forces, test what's called the weak-force distribution. The weak force is the nuclear force responsible for radioactive decay, among other things. These huge experiments, carried out in heated competition, each came to the same conclusion: the proton is made of three quarks. And we learned some details about how the quarks move about. Their motion defines what we call "proton."

Detailed analysis of all three types of experiments — electron, muon, and neutrino — also succeeded in detecting a new kind of particle, the gluon. Gluons are carriers of the strong force, and without them the data just could not be explained. The same analysis gave quantitative details on how the quarks whirl about each other in their proton prison. Twenty years of such study (the technical term is structure functions) has given us a sophisticated model that accounts for all the collision experiments in which protons, neutrons, electrons, muons, and neutrinos as well as photons, pions, and antiprotons are aimed at protons. This is Monet with a vengeance. Perhaps Wallace Stevens's poem "Thirteen Ways of Looking at a Blackbird" would be more to the point.

As you can see, we learn many things in order to account for what-goes-in-and-what-comes-out. We learn about the forces and how these forces result in complex structures such as protons (made of three quarks) and mesons (made of a quark and an antiquark). With so much complementary information, it becomes less and less important that we can't see inside the black box where the collision actually takes place.

One can't help being impressed by the sequence of "seeds within seeds." The molecule is made of atoms. The core of the atom is the nucleus. The nucleus is made of protons and neutrons. The proton and neutron are made of quarks. The quarks are made of . . . whoops, hold it. The quarks can't be broken down, we think, but of course we are not sure. How dare we say we've come to the end of the road? Nevertheless, that is the consensus — at present — and after all, Democritus can't live forever.

NEW MATTER: SOME RECIPES

We have yet to discuss an important process that can take place during a collision. We can make new particles. This happens all the time around the house. Look at the lamp that is valiantly trying to illuminate this dark page. What is the source of the light? It is electrons, agitated by the electrical energy squirting into the filament of the bulb or, if you are energy efficient, into the gas of the fluorescent lamp. The electrons *emit photons*. That's the process. In the more abstract language of the particle physicist, the electron in the process of a collision can radiate a photon. The energy is provided to the electron (via the wall plug) by an accelerating process.

Now we have to generalize. In the process of creation, we are constrained by the laws of conservation of energy, momentum, charge, and respect for all of the other quantum rules. Also, the object that is somehow responsible for creating a new particle has to be "connected" to the particle being created. Example: a proton collides with another proton, and a new particle, a pion, is made. We write it like this:

$$p^+ + p^+ \rightarrow p^+ + \pi^+ + n$$

That is, protons collide and produce another proton, a positive pion (π^+), and a neutron. These particles are all connected via the strong force, and this is a typical creation process. Alternatively, one can view this as a proton, "under the influence" of another proton, dissolving into a "pi plus" and a neutron.

Another kind of creation, a rare and exciting process called annihilation, takes place when matter and antimatter collide. The term *annihilation* is used in its strictest dictionary sense of putting something out of existence. When an electron collides with its antiparticle, the positron, the particle and antiparticle disappear, and in their place energy, in the form of a photon, appears momentarily. The conservation laws don't like this process, so the photon is temporary and must soon create two particles in its place — for example, another electron and a positron. Less frequently the photon may dissolve into a muon and an antimuon, or even a positive proton and a negative antiproton. Annihilation is the only phenomenon that is fully efficient in converting mass to energy in accordance with Einstein's law, $E = mc^2$. When a nuclear bomb explodes, for instance, only a fraction of 1 percent of the atomic mass is converted into energy.

When matter and antimatter collide, 100 percent of the mass disappears.

When we're making new particles, the primary requirement is that there be enough energy, and $E = mc^2$ is our accounting tool. For example, we mentioned that a collision between an electron and a positron can result in a proton and an antiproton, or a p and a p-bar, as we call them. Since the rest mass energy of a proton is about 1 GeV, the particles in the original collision must bring in at least 2 GeV to produce a p/p-bar pair. More energy increases the probability of this result and gives the newly produced objects some kinetic energy, making them easier to detect.

The glamorous nature of antimatter has given rise to the science fiction notion that it may solve the energy crisis. Indeed, a kilogram (2.2 pounds) of antimatter would provide enough energy to keep the United States going for a day. This is because the entire mass of antiproton (plus the proton it takes with it to total annihilation) is converted to energy via $E = mc^2$. In the burning of coal or oil, only one billionth of the mass is converted to energy. In fission reactors this number is 0.1 percent, and in the long-awaited fusion energy supply (don't hold your breath!) it is about 0.5 percent.

PARTICLES FROM THE VOID

Another way of thinking about these things is to imagine that all space, even empty space, is awash with particles, all that nature in her infinite wisdom can provide. This is not a metaphor. One of the implications of quantum theory is that these particles do in fact pop in and out of existence in the void. The particles, in all sizes and shapes, are all temporary. They are created and then quickly disappear — a bazaar of seething activity. As long as they occur in empty space, vacuum, nothing really happens. This is quantum spookiness, but perhaps it can help to explain what happens in a collision. Here a pair of charmed quarks (a certain kind of quark and its antiquark) appears and disappears; there a bottom quark and its anti-bottom mate. And wait, over there, what's that? Well, whatever: an X and an anti-X appear, something we have no knowledge of in 1993.

There are rules in this chaotic madness. The quantum numbers must add to zero, the zero of the void. Another rule: the heavier the objects, the less frequent their evanescent appearance. They "borrow" energy from the void to appear for the minutest fraction of a second, then disappear because they must pay it back in a time specified by Heisenberg's uncertainty relations. Now here is the key: if energy can be provided from the outside, then the transient virtual appearance of these vacuum-originated particles can be converted to real existence, existence that can be detected by bubble chambers or counters. How provided? Well, if an energetic particle, fresh out of the accelerator and shopping for new particles, can afford to pay the price — that is, at least the rest mass of the pair of quarks or X's — then the vacuum is reimbursed, and we say that our accelerated particle has created a quark-antiquark pair. Obviously, the heavier the particles we want to create, the more energy we need from the machine. In Chapters 7 and 8 you'll meet many new particles that came into being in just such a fashion. Incidentally, this quantum fantasy of an all-pervading vacuum filled with "virtual particles" has other experimental implications, modifying the mass and magnetism of electrons and muons, for example. We'll explain further when we get to the "g minus 2" experiment.

THE RACE

Beginning in the Rutherford era, the race was on to make devices that could reach very high energies. The effort was helped along in the 1920s by the electric utility companies, because electrical power is transmitted most efficiently when the voltage is high. Another motivation was the creation of energetic x-rays for cancer therapy. Radium was already being used to destroy tumors, but it was enormously expensive and higher energy radiation was thought to be a great advantage. Thus the electric utilities and medical research institutes supported the development of high voltage generators. Rutherford characteristically took the lead when he issued a challenge to England's Metropolitan-Vickers Electrical Company to "give us a potential on the order of ten million volts which can be accommodated in a reasonably sized room . . . and an evacuated tube capable of withstanding this voltage."

German physicists tried to harness the huge voltage of Alpine lightning storms. They hung an insulated cable between two mountain peaks, siphoning off charges as high as 15 million volts and inducing huge sparks that jumped 18 feet between two metal spheres — spectacular, but not too useful. This approach was abandoned when a scientist was killed while adjusting the apparatus.

The failure of the German team illustrated that one needed more than power. The terminals of the gap had to be housed in a beam tube or vacuum chamber that was a very good insulator. (High voltages love to arc across insulators unless the design is very precise.) The tube also had to be strong enough to withstand having its air pumped out. A high-quality vacuum was essential; if there were too many residual molecules floating around inside the tube they would interfere with the beam. And the high voltage had to be steady enough to accelerate lots of particles. These and other technical problems were worked on from 1926 to 1933 before they were solved.

Competition was intense throughout Europe, and American institutions and scientists joined the fray. An impulse generator built by Allgemeine Elektrizität Gesellschaft in Berlin reached 2.4 million volts but produced no particles. The idea was transported to General Electric in Schenectady, which improved the energy to 6 million volts. At the Carnegie Institution in Washington, D.C., physicist Merle Tuve drove an induction coil to several million volts in 1928 but didn't have an appropriate beam tube. Charles Lauritsen at Cal Tech succeeded in building a vacuum tube that would hold 750,000 volts. Tuve adopted Lauritsen's tube and produced a beam of 10^{13} (10 trillion) protons per second at 500,000 volts, theoretically enough particles and energy to probe the nucleus. Tuve did in fact achieve nuclear collisions, but not until 1933, by which time two other efforts had beaten him to the punch.

Another runner-up was Robert Van de Graaff, of Yale and then MIT, who built a machine that carried electric charge along an endless silk belt up to a large metal sphere, gradually increasing the voltage of the sphere until, at a few million volts, he drew a tremendous arc to the wall of the building. This was the now famous Van de Graaff generator, familiar to high school physics students across the land. Enlarging the radius of the sphere postponed the discharge. Encasing the entire sphere in dry nitrogen gas helped increase the voltage. Ultimately, Van de Graaff generators would be the machines of choice in the under-10-million-volt category, but it took years to perfect the idea.

The race continued through the late 1920s and early '30s. It was a couple of Rutherford's Cavendish gang, John Cockcroft and Ernest Walton, who won, though by a whisker. And (here I have to groan) they were given invaluable help by a theorist. Cockcroft and Walton, after numerous failures, were attempting to reach the one million volts that was perceived to be necessary to probe the nucleus. A Russian theorist, George Gamow, had been visiting Niels Bohr in

Copenhagen and decided to hop over to Cambridge before heading home. There he got into an argument with Cockcroft and Walton, telling the experimenters that they didn't need all the voltage they were playing with. He argued that the new quantum theory permitted successful nuclear penetrations even if the energy was not high enough to overcome the electrical repulsion of the nucleus. He explained that the quantum theory gave the protons wave properties, which can tunnel through the nuclear charge "barrier," as we discussed in Chapter 5. Cockcroft and Walton finally took note and redesigned their device for 500,000 volts. Using a transformer and a voltage multiplier circuit, they accelerated protons obtained from a discharge tube of the type that J. J. Thomson used to generate cathode rays.

In Cockcroft and Walton's machine, bursts of protons, about a trillion per second, accelerated down the evacuated tube and smashed into targets of lead, lithium, and beryllium. The year was 1930, and nuclear reactions had finally been induced by accelerated particles. Lithium was disintegrated by protons of only 400,000 eV, far below the millions of electron volts that had been thought necessary. It was a historic event. A new style of "knife" was now available, although still in its most primitive form.

A MOVER AND SHAKER IN CALIFORNIA

The action now switches to Berkeley, California, where Ernest Orlando Lawrence, a native of South Dakota, had arrived in 1928 after a brilliant beginning in physics research at Yale. E. O. Lawrence invented a radically different technique of accelerating particles in a machine called a cyclotron, for which he was awarded the Nobel Prize in 1939. Lawrence was familiar with the clumsy electrostatic machines, with their huge voltages and frustrating electrical breakdowns, and he figured there had to be a better way. Searching through the literature for ways to achieve high energy without high voltages, he came across the papers of a Norwegian engineer, Rolf Wideröe. Wideröe noted that one could double the energy of a particle without doubling the voltage by passing it through two gaps in a row. Wideröe's idea is the basis for what is now called the linear accelerator. One gap is positioned after another down a line, the particles picking up energy at each gap.

Wideröe's paper, however, gave Lawrence an even better idea. Why not use a single gap with modest voltage but use it over and over again? Lawrence reasoned that when a charged particle moves in a magnetic field, its path is curved into a circle. The circle's radius is determined by the strength of the magnet (strong magnet, small radius) and the momentum of the charged particle (high momentum, large radius). Momentum is simply the particle's mass times its speed. What this means is that a strong magnet will guide a particle to move in a tiny circle, but if the particle gains energy and therefore also momentum, the radius of the circle will increase.

Picture a hatbox sandwiched between the north and south poles of a large magnet. Make the box out of brass or stainless steel, something strong but nonmagnetic. Pump the air out of the box. Inside it are two hollow D-shaped copper structures that almost fill the box: the straight sides of the D open and facing each other across a small gap, the round sides closed. Suppose one D is positively charged, the other negatively charged, with the difference of potential being, say, 1,000 volts. A stream of protons generated (never mind how) near the center of the circle is aimed across the gap from the positive D to the negative D. The protons gain 1,000 volts and their radius now increases since the momentum is higher. The protons sweep around inside the D, and when they return to the gap, thanks to clever switching, they again see a negative voltage across the gap. Again they are accelerated, and now they have 2,000 eV. The process continues. Every time the protons cross the gap, they gain 1000 eV. As they gain momentum they fight against the constricting power of the magnet, and the radius of their path continues to increase. The result is that the protons spiral out from the center of the box toward the perimeter. There they strike a target, a collision takes place, and the research begins.

The key to acceleration in the cyclotron is to make sure that the protons always see a negative D on the other side of the gap. The polarity has to flip-flop rapidly from D to D in exact synchronization with the rotation of the protons. But, you may be asking yourself, isn't it difficult to synchronize the alternating voltage with the protons, whose path continues to describe larger and larger circles as the acceleration continues? The answer is no. Lawrence discovered that by God's cleverness, the spiraling protons compensate for their longer path by speeding up. They complete each half circle in the same time, a process known as resonant acceleration. To match the proton orbits, one needs a fixed-frequency alternating voltage, a technology

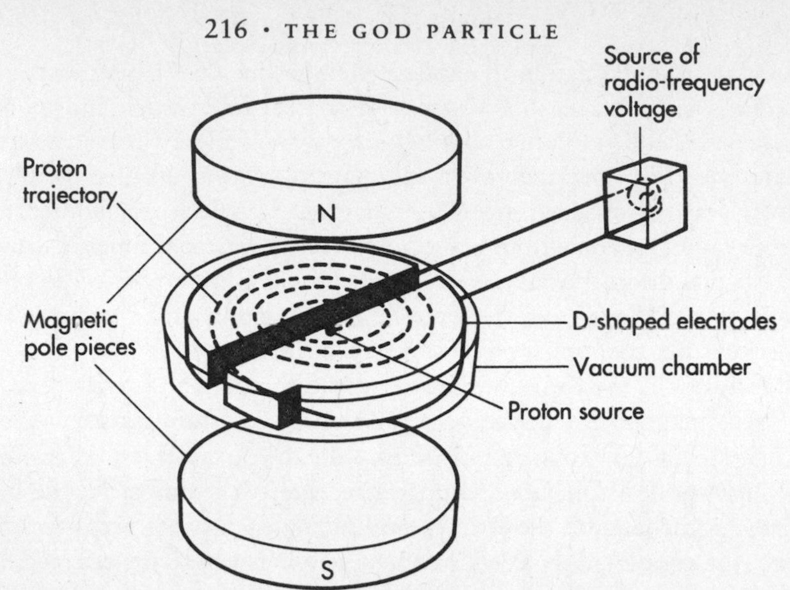

that was well known in radio broadcasting. Hence the name of the switching acceleration mechanism: radio-frequency generator. In this system the protons arrive at the edge of the gap just as the opposite D has maximum negative voltage.

Lawrence worked out the theory of the cyclotron in 1929 and 1930. Later he designed, on paper, a machine in which the protons made a hundred turns with a generation of 10,000 volts across the D-gap. That would give him a beam of 1 MeV protons (10,000 volts × 100 turns = 1 MeV). Such a beam would be "useful for the studies of atomic nuclei." The first model, actually constructed by Stanley Livingston, one of Lawrence's students, came up considerably short, reaching 80 KeV (80,000 volts). Lawrence then went big-time. He obtained a huge grant ($1,000!) to build a machine that could produce nuclear disintegrations. The pole pieces (the north and south pole pieces of the magnet) were ten inches in diameter, and in 1932 the machine accelerated protons to an energy of 1.2 MeV. These were used to produce nuclear collisions in lithium and other elements only a few months after Cockcroft and Walton's group at Cambridge. Second place, but Lawrence still lit a cigar.

BIG SCIENCE AND THE CALIFORNIA MYSTIQUE

Lawrence was a mover and shaker of enormous energy and ability. He was the father of Big Science. The term refers to huge, centralized facilities of great complexity and expense that are shared by a large number of scientists. In its evolution, Big Science created new ways of carrying out research with *teams* of scientists. It also created exquisite sociological problems, about which more later. The likes of Lawrence had not been seen since Tycho Brahe, the Lord of Uraniborg, the laboratory on Hven. In the experimental arena, Lawrence established the United States as a serious player in world physics. He contributed to the California mystique, the love of technological extravaganzas, complex and expensive undertakings. These were alluring challenges for young California and, indeed, for the young United States.

By 1934 Lawrence was producing beams of 5 MeV deuterons with a thirty-seven-inch cyclotron. The deuteron, a nucleus consisting of one proton and one neutron, had been discovered in 1931, and had proved to be a more efficient projectile than the proton for producing nuclear reactions. In 1936 he had an 8 MeV deuteron beam. In 1939 a sixty-inch machine operated at 20 MeV. A monster started in 1940 and completed after the war had a magnet that weighed 10,000 tons! Cyclotrons were built all over the world because of their ability to unravel the mysteries of the nucleus. In medicine they were used to treat tumors. The beam of particles, directed at a tumor, deposits enough energy in the malignancy to destroy it. In the 1990s over a thousand cyclotrons are in use in hospitals across the United States. Basic research in particle physics, however, has abandoned the cyclotron in favor of a new type of machine.

THE SYNCHROTRON: AS MANY TURNS AS YOU WANT

The drive to create ever higher energies intensified and spread worldwide. At each new energy domain new discoveries were made. New puzzles were also created, increasing the desire to attain even higher energies. Nature's richness seemed to be hidden in the nuclear and subnuclear microworld.

The cyclotron is limited by its design. Because the particles spiral outward, the number of orbits is obviously limited by the circumference of the device. To get more orbits and more energy, you need a bigger cyclotron. The magnetic field must be applied to the entire spiral area, so the magnets must be large . . . and expensive. Enter the synchrotron. If the particles' orbit, instead of spiraling out, could be kept to a constant radius, then the magnet would be needed only along the narrow path of the orbit. As the particles gained energy, the magnetic field could be increased synchronously to keep them imprisoned in an orbit of constant radius. Clever! Tons and tons of iron could be saved, because the magnetic pole pieces, transverse to the path of the beam, could be reduced to inches instead of feet.

Two important details must be mentioned before we proceed to the 1990s. In a cyclotron the charged particles (protons or deuterons) travel through what became thousands of turns in a vacuum chamber clamped between the poles of a magnet. To keep the particles from spreading out and striking the chamber walls, it was absolutely essential to have some kind of focusing process. Just as a lens focuses the light from a flashlight into a (nearly) parallel beam, magnetic force is used to squeeze the particles into a tight beam.

In the cyclotron this focusing action is provided by the way the magnetic field changes as the protons move toward the outer edge of the magnet. Robert R. Wilson, a young student of Lawrence's and later the builder of the Fermilab accelerator, was the first to understand the subtle but crucial effect the magnetic forces had in keeping the protons from spraying out. In the early synchrotrons, the pole pieces were shaped to provide these forces. Later, specially designed quadrupole magnets (with two north poles and two south poles) were used to focus the particles, while separate dipole magnets steered them in a fixed orbit.

Fermilab's Tevatron, a trillion-electron-volt machine completed in 1983, is a good example. The particles are steered into a circular orbit by powerful superconducting magnets, much as tracks guide a train around a turn. The highly evacuated beam pipe is a stainless steel (nonmagnetic) oval-shaped tube about 3 inches wide and 2 inches high, centered between the north and south poles of the magnets. Each dipole (steering) magnet is 21 feet long. The "quads" are 5 feet long. More than a thousand magnets are needed to cover the length of the tube. The beam pipe and magnet combination complete a circle that has a radius of 1 kilometer, or 0.6 miles — quite a change from Lawrence's first 4-inch model. You can see the advantage of the synchrotron design here. One needs a lot of magnets, but they're relatively skinny, just wide enough to cover the vacuum pipe. If the Tevatron were a cyclotron, we would need a magnet with pole pieces 1.2 miles in diameter to cover the 4-mile-around machine!

Particles make 50,000 orbits in one second around this 4-mile track. In 10 seconds the particles have traveled 2 million miles. Each time they pass a gap — actually a series of specially constructed cavities — a radio-frequency voltage kicks up the energy by about 1

MeV. The magnets that keep the particles focused allow them to deviate from their appointed rounds by less than one eighth of an inch over the entire trip. It's not perfect, but it's good enough. Like aiming a rifle at a mosquito sitting on the moon but hitting it in the wrong eye. To keep the protons in the same orbit while they are being accelerated, the strength of the magnets must increase in precise synchronism with the proton's gain in energy.

The second important detail has to do with the theory of relativity: protons become detectably heavier as their energy rises above 20 or so MeV. This increase in mass destroys the "cyclotron resonance" that Lawrence had discovered, in which the spiraling protons exactly compensate for their longer path by speeding up. This allows the rotation to be synchronized with a fixed frequency of the accelerating voltage across the gap. At higher energy the rotation time increases, and one can no longer apply a constant radio-frequency voltage. To counter the slowdown, the applied frequency has to decrease, so frequency-modulated (FM) accelerating voltages are used to track the increasing mass of the protons. The synchrocyclotron, a frequency-modulated cyclotron, was the earliest example of the influence of relativity on accelerators.

The proton synchrotron solves the problem in an even more elegant way. It is a little complicated but depends on the fact that the speed of the particle (99 point whatever percent of the speed of light) is essentially constant. Suppose the particle crosses the gap at that part of the radio-frequency cycle when the accelerating voltage is zero. No acceleration. We now increase the magnetic field a bit. The particle bends in a tighter circle and arrives a bit early at the gap, and now the radio frequency is in a phase to accelerate. Thus the mass grows, the orbit radius increases, and we are back to where we started but with higher energy. The system is self-correcting. If the particle gains too much energy (mass), its radius will increase and it will arrive later at the gap and see a decelerating voltage, which will correct the error. Raising the magnetic field has the effect of increasing the mass energy of our hero-particle. This method depends on "phase stability," which is discussed later in this chapter.

IKE AND THE PIONS

One early accelerator was near and dear to me — Columbia University's 400 MeV synchrocyclotron, built on an estate in Irvington-on-Hudson, New York, within commuting distance of Manhattan.

The estate, named after the ancestral Scottish mountain Ben Nevis, was established in colonial times by Alexander Hamilton. Later it was owned by a branch of the Du Pont family and then by Columbia University. The Nevis cyclotron, built between 1947 and 1949, was one of the most productive particle accelerators in the world during its twenty-some years of operation (1950–1972). It also produced more than a hundred and fifty Ph.D.'s, about half of whom stayed in the field of high-energy physics and became professors at Berkeley, Stanford, Cal Tech, Princeton, and many other such fly-by-night institutions. The other half went everywhere: small teaching institutions, government labs, science administration, industrial research, investment banking . . .

I was a graduate student when President (of Columbia) Dwight Eisenhower dedicated the new facility in June of 1950, in a small ceremony on the lawn of the lovely estate — magnificent trees, shrubbery, a few red brick outbuildings — sloping down to the stately Hudson River. After appropriate speechifying, Ike threw a switch and out of the loudspeakers came the amplified "cheeps" of a Geiger counter, indicating radiation. The cheeps were produced by a radioactive source I held near a particle counter because the machine had chosen that moment to crash. Ike never found out.

Why 400 MeV? The hot particle of 1950 was the pion, or pi meson, as it's also called. The pion had been predicted in 1936 by a Japanese theoretical physicist, Hideki Yukawa. It was thought to be the key to the strong force, which in those days was the big mystery. Today we think of the strong force in terms of gluons. But back then pions, which fly back and forth between the protons and neutrons to hold them together tightly in the nucleus, were the key, and we needed to make and study them. To produce pions in nuclear collisions, the particle coming in from the accelerator must have an energy greater than $m(\text{pion})c^2$, that is, greater than the pion's rest mass energy. Multiplying the pion's rest mass by the speed of light squared, we get 140 MeV, its rest mass energy. Since only a fraction of the collision energy goes into the production of new particles, we needed extra energy, and we settled on 400 MeV. The Nevis machine became a pion factory.

BEPPO'S LADIES

But wait. First a word on how we found out about pions in the first place. In the late 1940s, scientists at the University of Bristol in England noticed that when an alpha particle passes through a photographic emulsion coated on a glass plate, it "activates" the molecules in its path. After developing the film, you can see a track defined by grains of silver bromide. The track is easily discerned through a low-power microscope. The Bristol group sent batches of very thick emulsion up in balloons almost to the top of the atmosphere, where the intensity of cosmic rays is much higher than at sea level. This source of "naturally" occurring radiation far exceeded in energy Rutherford's puny 5 MeV alphas. It was in these emulsions exposed to cosmic rays in 1947 that the pion was first discovered by Cesare Lattes, a Brazilian, Giuseppe Occiallini, an Italian, and C. F. Powell, the resident professor in Bristol.

The most colorful of the above trio was Occiallini, known as Beppo to his friends. An amateur speleologist and compulsive practical joker, Beppo was the driving force of the group. He trained a bevy of young women to do the painstaking work of studying the emulsions under a microscope. My thesis supervisor, Gilberto Bernardini, a close friend of Beppo's, visited him one day in Bristol. Following directions given to him in unbroken English, a language he found very difficult, Bernardini quickly got lost. Finally he stumbled into a lab where several very proper English ladies were staring into microscopes and cursing in Italian argot that would be outlawed on the docks of Genoa. "Ecco!" cried Bernardini. "Dissa is Beppo's lab!"

What the tracks in those emulsions showed was a particle, the pion, entering at high speed, gradually slowing down (the density of the grains of silver bromide increases as the particle slows), and coming to rest. At the end of the track a new, energetic particle appears and races off. A pion is unstable, decaying within one hundredth of a microsecond into a muon (the new particle at the end of the track) and something else. The something else turned out to be a neutrino, which doesn't leave a track in the emulsion. The reaction is written $\pi \rightarrow \mu + \nu$. That is, a pion (eventually) gives rise to a muon and a neutrino. Since the emulsion provides no time sequence information, it took careful analysis of the tracks on half a dozen of these rare occurrences to understand what the particle was and how it decayed. The new particle had to be studied, but using cosmic rays yields only a handful of such events *per year.* As with nuclear disintegrations, accelerators with high enough energy were required.

At Berkeley, Lawrence's 184-inch cyclotron began to produce pions, as did the Nevis machine. Soon synchrocyclotrons in Rochester, Liverpool, Pittsburgh, Chicago, Tokyo, Paris, and Dubna (near

Moscow) were studying the pion in its strong interactions with neutrons and protons as well as the weak force in the pion's radioactive decay. Other machines at Cornell, Cal Tech, Berkeley, and the University of Illinois used electrons to produce pions, but the most successful machines were the proton synchrocyclotrons.

THE FIRST EXTERNAL BEAM: PLACE YOUR BETS!

So there I was in the summer of 1950 with a machine going through birth pains and me needing data so I could get a Ph.D. and earn a living. Pions were the name of the game. Hit a piece of something — carbon, copper, *anything* containing nuclei — with the 400 MeV protons from the Nevis machine and you'd generate pions. Berkeley had hired Lattes, who showed the physicists how to expose and develop the very sensitive emulsions used so successfully in Bristol. They inserted a stack of emulsions into the beam vacuum tank and allowed the protons to hit a target near the stack. Remove the emulsions through an air lock, develop them (a week of effort), and then subject them to microscopic study (months!). All this effort had given the Berkeley team but a few dozen pion events. There had to be an easier way. The trouble was that the particle detectors had to be installed inside the machine, in the region of the strong accelerator magnet, to record the pions, and the only device that was practical was the stack of emulsions. In fact, Bernardini was planning an emulsion experiment on the Nevis machine similar to what the Berkeley folks had done. The large, elegant cloud chamber I had built for my Ph.D. project was a much better detector, but it would never fit between the poles of a magnet inside an accelerator. Nor would it survive as a particle detector in the intense radiation inside the accelerator. Between the cyclotron magnet and the experimental area was a ten-foot-thick concrete wall to confine the stray radiation.

A new postdoc, John Tinlot, had arrived at Columbia from Bruno Rossi's famed cosmic ray group at MIT. Tinlot was the quintessential physicist. In his late teens he had been a violinist of concert quality, but he had put his violin away after an agonizing decision to study physics. He was the first young Ph.D. I had ever worked with, and I learned enormously from him. Not only physics. John was a genetically infected horse player and gambler: long shots, blackjack, craps, roulette, poker — lots of poker. We played during experiments while the data were being collected. We played on vacation, on trains and airplanes. It was a moderately expensive way to learn physics, my losses being moderated by the other players — students, technicians, and security guards whom John would recruit. He had no pity.

John and I sat on the floor of the not-yet-really-working accelerator, drinking beer and discussing the world. "What really happens to the pions that come flying off the target?" he asked suddenly. I had learned to be cautious. John was a gambler in physics as well as in horses. "Well, if the target is inside the machine [and it had to be, we didn't know how to get the accelerated protons out of the cyclotron], the powerful magnet will spray them in all directions," I answered cautiously.

JOHN: *Some* will come out of the machine and hit the shielding wall?
ME: Sure, but all over the place.
JOHN: Why don't we find out?
ME: How?
JOHN: We do magnetic tracing.
ME: That's work. [It was 8 P.M. on a Friday.]
JOHN: Do we have the table of measured magnetic fields?
ME: I'm supposed to go home.
JOHN: We'll use those huge rolls of brown wrapping paper and draw the paths of the pions on a scale of one to one . . .
ME: Monday?
JOHN: You do the slide-rule work [this was 1950] and I'll draw the paths.

Well, by 4 A.M. Saturday we had made a fundamental discovery that would change the way cyclotrons were used. We had traced eighty or so fictional particles emerging from a target in the accelerator with plausible directions and energies — we used 40, 60, 80, and 100 MeV. To our amazement, the particles didn't "just go everywhere." Instead, because of the properties of the magnetic field near and beyond the rim of the cyclotron magnet, they curved around the machine in a tight beam. We had discovered what became known as "fringe field focusing." By rotating the large sheets of paper — that is, by picking a specific target location — we could get the beam of pions in a generous energy band around 60 MeV to head right for my brand-new cloud chamber. The only catch was the wall of concrete between the machine and the experimental area where my princess chamber sat.

No one had anticipated our discovery. Monday morning we were camped outside the director's office waiting to pounce on him with it. We had three simple requests: (1) a new target location in the machine; (2) a much thinner window between the beam vacuum chamber of the cyclotron and the outside world to minimize the influence of a one-inch-thick stainless steel plate on the emerging pions; and (3) a new hole about four inches high by ten inches wide, we guessed, through the ten-foot-thick concrete wall. All this from a lowly graduate student and a postdoc!

Our director, Professor Eugene Booth, was a Georgia gentleman and a Rhodes scholar who rarely said "gosh darn." He made an exception for us. We argued, we explained, cajoled. We painted visions of glory. He would be famous! Imagine an external pion beam, the first ever!

Booth threw us out, but after lunch he called us in again. (We had been weighing the advantages of strychnine versus arsenic.) Bernardini had dropped in, and Booth had tried out our idea on this eminent visiting professor. My guess is that the details, expressed in Booth's Georgian lilt, were too much for Gilberto, who once confided in me, "Booos, Boosth, who can pronounce dese American names?" However, Bernardini supported us with typical Latin exaggeration, and we were in.

A month later it all worked — just like the wrapping paper sketches. In a few days my cloud chamber had registered more pions than all the other labs in the world put together. Each photograph (we took one each minute) had six to ten beautiful tracks of pions. Every three or four photographs would show a kink in a pion track as it disintegrated into a muon and "something else." I used the pion decays as my thesis. Within six months we had constructed four beams, and Nevis was in full production as a data factory for the properties of pions. At the earliest opportunity, John and I went to the racetrack in Saratoga where, continuing his roll, he hit a 28-to-1 shot in the eighth race, on which he had wagered our dinner and return-home gas money. I really loved that guy.

John Tinlot must have had extraordinary insight to suspect the fringe field focusing that everyone else in the cyclotron business had missed. He went on to a distinguished career as a professor at the University of Rochester but died of cancer at the age of forty-three.

A SOCIAL SCIENCE DIVERSION: THE ORIGIN OF BIG SCIENCE

World War II marked a crucial watershed between pre-WWII and post-WWII scientific research. (How's that for a controversial statement?) But it also marked a new phase in the search for the a-tom. Let's count some of the ways. The war generated a leap forward in technology, much of this centered in the United States, which was unencumbered by the loud noises of nearby explosions that Europe was experiencing. The wartime development of radar, electronics, the nuclear bomb (to use its proper name) all provided examples of what a collaboration between science and engineering could do — as long as it was unconstrained by budget considerations.

Vannevar Bush, the scientist who led U.S. science policy during the war, spelled out a new relationship between science and government in an eloquent report to President Franklin D. Roosevelt. From that time on, the U.S. government was committed to supporting basic research in the sciences. Support for research, basic and applied, climbed so rapidly that we can laugh at the $1,000 grant E. O. Lawrence worked so hard to get in the early 1930s. Even adjusting for inflation, that amount pales in contrast to federal support of basic research in 1990 — some $12 billion! World War II also saw a flood of scientist refugees from Europe become a crucial part of the research boom in the United States.

In the early 1950s some twenty universities had accelerators capable of carrying out research in nuclear physics at the cutting edge. As we came to understand the nucleus better, the frontier shifted to the subnuclear domain, where larger — more expensive — machines were required. The era became one of consolidation — scientific mergers and acquisitions. Nine universities banded together to build and manage the accelerator laboratory at Brookhaven, Long Island. They commissioned a 3 GeV machine in 1952 and a 30 GeV machine in 1960. Princeton University and the University of Pennsylvania banded together to build a proton machine near Princeton. MIT and Harvard built the Cambridge Electron Accelerator, a 6 GeV electron machine.

Over the years, as the consortia grew in size, the number of front-line machines diminished. We needed ever higher energy to address the question of "What's inside?" and to search for the true a-toms — or the zero and one of our library metaphor. As new machines were proposed, older ones were phased out to free up funds, and Big Science (a term often used as an expletive by ignorant commentators) grew bigger. In the 1950s, one could do maybe two or three experiments a year with groups of two to four scientists. In the following decades, the collaborations got larger and larger, and the experiments took longer and longer, driven in part by the necessity to build ever more complex detectors. By the 1990s the Collider Detector Facility alone at Fermilab comprised 360 scientists and students from twelve universities, two national labs, and institutions in Japan and Italy. Scheduled runs stretched to a continuous year or more of data taking — with time off for Christmas, the Fourth of July, or whenever something broke down.

Supervising the evolution from a tabletop science to one based on accelerators measured in miles around was the U.S. government. The World War II bomb program gave rise to the Atomic Energy Commission (AEC), a civilian agency that oversaw nuclear weapons research, production, and stockpiling. It was also given the mission, as a national trust, of funding and overseeing basic research in nuclear and what was later to become particle physics.

The case for Democritus's a-tom even reached the halls of Congress, which created the Joint (House and Senate) Committee on Atomic Energy to provide oversight. The committee's hearings, published in dense, government-green booklets, are a Fort Knox of information for science historians. Here one reads the testimony of E. O. Lawrence, Robert Wilson, I. I. Rabi, J. Robert Oppenheimer, Hans Bethe, Enrico Fermi, Murray Gell-Mann, and many others patiently responding to questions about how the search for the ultimate particle was going — and why did it require yet another machine? The interchange at the beginning of this chapter between Fermilab's flamboyant founding director, Robert Wilson, and Senator John Pastore was taken from one of those green books.

To complete the alphabet soup, the AEC dissolved into the ERDA (Energy Research and Development Agency), which soon gave way to the DOE (U.S. Department of Energy), which at this writing oversees the national laboratories where atom smashers operate. Presently there are five such high-energy labs in the U.S.: SLAC, Brookhaven, Cornell, Fermilab, and the Superconducting Super Collider lab, now under construction.

Accelerator labs are generally owned by the government, but operated by a contractor, which can be a university, such as Stanford in SLAC's case, or a consortium of universities and institutions, as is the case with Fermilab. The contractors appoint a director, and then they pray. The director runs the lab, makes all the important decisions, and often stays on the job too long. As Fermilab director from 1979 to 1989, my major task was to implement Robert R. Wilson's vision: the construction of the Tevatron, the first superconducting accelerator.

We also had to create a proton-antiproton collider and humongous detectors that would observe head-on collisions near 2 TeV.

I worried a lot about the process of research when I was director of Fermilab. How could students and young postdocs experience the joy, the learning, the exercise of creativity experienced by Rutherford's students, by the founders of quantum theory, by my own small group of colleagues as we sweated out the problems on the floor of the Nevis cyclotron? But the more I looked into what was happening at the lab, the better I felt. The nights I visited the CDF (when old Democritus wasn't there), I found students enormously excited as they ran their experiments. On a giant screen events were flashing, reconstructed by the computer to make sense to the dozen or so physicists on shift. Occasionally, an event would be so suggestive of "new physics" that an audible gasp would be heard.

Each large research collaboration consists of many groups of five or ten people: a professor or two, several postdocs, and several graduate students. The professor looks after his brood, making sure they are not lost in the crowd. Early on they are wrapped up in the design, building, and testing of equipment. Later on comes the data analysis. There is so much data in one of these collider experiments that much of it must wait for some group to complete one analysis before getting around to tackling the next problem. The individual young scientist, perhaps advised by her professor, selects a specific problem that receives the consensual agreement of the council of group leaders. And problems abound. For example, when W^+ and W^- particles are produced in proton-antiproton collisions, what is the precise form of the process? How much energy do the W's take away? At what angles are they emitted? And so on. This could be an interesting detail, or it could be a clue to a crucial mechanism in the strong and weak forces. The most exciting task for the 1990s is to find the top quark and measure its properties. Up to mid-1992 this search was carried out by four subgroups of the CDF collaboration at Fermilab doing four independent analyses.

Here the young physicists are on their own, fighting complex computer programs and the inevitable distortions introduced by an imperfect apparatus. Their problem is to extract a valid conclusion about how nature works, to establish one more piece of the jigsaw puzzle of the microworld. They have the benefit of a huge support group: experts in software, in theoretical analysis, in the art of seeking confirming evidence for tentative conclusions. If there is an interesting glitch in the way W's are thrown out of collisions, is it an artifact of the apparatus (metaphorically, a small crack in the microscope lens)? Is it a bug in the software? Or is it real? And if it is real, wouldn't colleague Harry see a similar effect in his analysis of Z particles — or perhaps Marjorie in her analysis of recoil jets?

Big Science is not the sole province of particle physicists. Astronomers share giant telescopes, pooling their observations in order to draw valid conclusions about the cosmos. Oceanographers share research ships elaborately equipped with sonar, diving vessels, and special cameras. Genome research is the microbiologists' Big Science program. Even chemists require mass spectrometers, expensive dye lasers, and huge computers. Inevitably, in one discipline after another, scientists are sharing the expensive facilities that are necessary to make progress.

Having said all this, I must emphasize that it is also extremely important for young scientists to be able to work in more traditional modes, clustered around a tabletop experiment with their peers and a professor. There they have the splendid option of pulling a switch, turning out the lights, and going home to think, perchance to sleep. "Small science" has also been a source of discovery, variety, and innovation, which contribute enormously to the advancement of knowledge. We must strike the proper balance in our science policy and be prayerfully grateful that both options exist. As for high-energy practitioners, one can tsk, tsk, and wish for the good old days when the lonely scientist sat in his folksy laboratory, mixing colorful elixirs. It's a charming vision, but it will never get us to the God Particle.

BACK TO THE MACHINES: THREE TECHNICAL BREAKTHROUGHS

Of the many technical breakthroughs that permitted acceleration to essentially unlimited energy (unlimited, that is, except by budgets) we'll look at three up close.

The first was the concept of *phase stability*, discovered by V. I. Veksler, a Soviet genius, and independently and simultaneously by Edwin McMillan, a Berkeley physicist. Our ubiquitous Norwegian engineer, Rolf Wideröe, independently patented the idea. Phase stability is important enough to call in a metaphor. Think of two identical hemispherical bowls with very small flat bottoms. Turn one bowl upside down, and place a ball on the small flat bottom, which is now the top. Place a second ball at the bottom of the noninverted bowl. Both balls are at rest. Are both stable? No. The test is to give each ball a nudge. Ball No. 1 rolls down the outside of the bowl, changing its condition radically. That's unstable. Ball No. 2 rolls up the side a bit, returns to the bottom, overshoots, and oscillates around its equilibrium position. That's stable.

The mathematics of particles in accelerators has much in common with the two conditions. If a small disturbance — for example, a particle's gentle collision with a residual gas atom or with a fellow accelerated particle — results in large changes in motion, there is no basic stability, and sooner or later the particle will be lost. On the other hand, if these perturbations result in small oscillatory excursions around the ideal orbit, we have stability.

Progress in the design of accelerators was an exquisite mixture of analytic (now highly computerized) study and the invention of ingenious devices, many of them building on the radar technology developed during World War II. The concept of phase stability was implemented in a variety of machines by applying radio frequency (rf) electrical forces. Phase stability in an accelerator happens when we organize the accelerating radio frequency so that a particle arrives at a gap at slightly the wrong time, resulting in a slight change in the particle's trajectory; the next time the particle hits the gap, the error is corrected. An example was given earlier with the synchrotron. What actually happens is that the error is overcorrected, and the particle's phase, relative to the radio frequency, oscillates around an ideal phase in which good acceleration is achieved, like a ball at the bottom of the bowl.

The second breakthrough occurred in 1952, when Brookhaven Laboratory was completing its 3 GeV Cosmotron accelerator. The accelerator group was expecting a visit from colleagues at the CERN lab in Geneva, where a 10 GeV machine was being designed. Three physicists preparing for the meeting made an important discovery. Stanley Livingston (a student of Lawrence's), Ernest Courant, and Hartland Snyder were a new breed of cat: accelerator theorists. They hit on a principle known as *strong focusing.* Before I describe this second breakthrough, I should make the point that particle accelerators had become a sophisticated and scholarly discipline. It pays to review the key ideas. We have a gap, or radio-frequency cavity, which is what gives the particle its increase in energy at each crossing. To use it over and over, we guide the particles into an approximate circle, using magnets. The maximum energy of particles that can be achieved in an accelerator is determined by two factors: (1) the largest radius that the magnet can provide and (2) the strongest magnetic field possible at that radius. We can build higher-energy machines by making the radius bigger, by making the maximum magnetic field stronger, or by doing both.

Once these parameters are set, giving the particles too much energy would drive them outside of the magnet. Cyclotrons in 1952 could accelerate particles to no more than 1,000 MeV. Synchrotrons provided magnetic fields to guide the particles at a fixed radius. Recall that the synchrotron magnet strength starts out very low (to match the low energy of the injected particles) at the beginning of the acceleration cycle and increases gradually to its maximum value. The machine is doughnut-shaped, and the radius of the doughnut in the various machines constructed during this era varied from 10 to 50 feet. The energies achieved were up to 10 GeV.

The problem that occupied the clever theorists at Brookhaven was how to keep the particles tightly bunched and stable relative to an idealized particle moving without disturbances in magnetic fields of mathematical perfection. Since the transits are so long, extremely small disturbances and magnetic imperfections can drive the particle away from the ideal orbit. Soon we have no beam. So we must provide conditions for stable acceleration. The mathematics was complicated enough, one wag said, "to curl a rabbi's eyebrows."

Strong focusing involves shaping the magnetic fields that guide the particles so that they are held much closer to an ideal orbit. The key idea is to machine the pole pieces into appropriate curves so that the magnetic forces on the particle generate rapid oscillations with tiny amplitudes around the ideal orbit. That is stability. Before strong focusing, the doughnut-shaped vacuum chambers had to be 20 to 40 inches wide, requiring magnet poles of similar sizes. The Brookhaven breakthrough permitted reduction in the size of the magnet's vacuum chamber to 3 to 5 inches. The result? A huge savings in cost per MeV of accelerated energy.

Strong focusing changed the economics and, early on, made it thinkable to build a synchrotron with a radius of almost 200 feet. Later we'll talk about the other parameter, the strength of the magnetic field. As long as iron is used for guiding the particles, this is limited to 2 tesla, the strongest magnetic field that iron can support without turning purple. Breakthrough is a correct description of strong focusing. Its first application was a 1 GeV electron machine built by Robert Wilson the Quick at Cornell. Brookhaven's proposal to the AEC to build a strong-focusing proton machine was said to have been a two-page letter! (Here we can lament the growth of bureaucracy but it would do no good.) This was approved, and the result was the 30 GeV machine known as AGS, completed at Brookhaven in 1960. CERN scrapped its plans for a 10 GeV weak-focusing machine and used the Brookhaven strong-focusing idea to build a 25 GeV strong-focusing accelerator for the same price. They turned it on in 1959.

By the late 1960s, the idea of using tortured pole pieces to achieve strong focusing had given way to a separated function concept. One installs a "perfect" dipole guide magnet and segregates the focusing function in a quadrupole magnet symmetrically arrayed around the beam pipe.

Using mathematics, physicists learned how complex magnetic fields direct and focus particles; magnets with larger numbers of north and south poles — sextupoles, octupoles, decapoles — became components of sophisticated accelerator systems designed to exercise precise control over the particle orbits. From the 1960s on, computers were more and more important in operating and controlling the currents, voltages, pressures, and temperatures in the machines. Strong focusing magnets and computer automation made possible the remarkable machines that were built in the 1960s and '70s.

The first GeV (billion-electron-volt) machine was the modestly named Cosmotron, which began operation at Brookhaven in 1952. Cornell followed with a 1.2 GeV machine. Here are the other stars of that era . . .

ACCELERATOR	ENERGY	LOCATION	YEAR
Bevatron	6 GeV	Berkeley	1954
AGS	30 GeV	Brookhaven	1960
ZGS	12.5 GeV	Argonne (Chicago)	1964
The "200"	200 GeV	Fermilab	1972 (upgraded to 400 GeV in 1974)
Tevatron	900 GeV	Fermilab	1983

Elsewhere in the world there were the Saturne (France, 3 GeV), Nimrod (England, 10 GeV), Dubna (USSR, 10 GeV), KEK PS (Japan, 13 GeV), PS (CERN/Geneva, 25 GeV), Serpuhkov (USSR, 70 GeV), SPS (CERN/Geneva, 400 GeV).

The third breakthrough was *cascade acceleration,* a concept attributed to Cal Tech physicist Matt Sands. Sands decided that, when one is going for high energy, it is inefficient to do it all in one machine. He envisioned a sequence of different accelerators, each optimized for a particular energy interval, say 0 to 1 MeV, 1 to 100 MeV, and so on. The various stages can be compared to gears on a sports car, with each gear designed to raise the speed to the next level in the optimal manner. As the energy increases, the accelerated beam gets tighter. At the higher energy stages, the smaller transverse dimensions thus require smaller and cheaper magnets. The cascade idea has dominated all machines since the 1960s. Its highest exemplars are the Tevatron (five stages) and the Super Collider under construction in Texas (six stages).

IS BIGGER BETTER?

A point that may have been lost in the preceding discussion of technical considerations is why it helps to make cyclotrons and synchrotrons big. Wideröe and Lawrence demonstrated that one doesn't have to produce enormous voltages, as earlier pioneers believed, to accelerate particles to high energies. One just sends the particles through a series of gaps, or designs a circular orbit so that one gap can be reused. Thus in circular machines there are but two parameters: magnet strength and the radius of the orbiting particles. Accelerator builders adjust these two factors to get the energy they want. The radius is limited by money, mostly. Magnet strength is limited by technology. If we can't boost the magnetic field, we make the circle bigger to increase the energy. In the Super Collider we know that we want to produce 20 TeV in each beam. And we know (or we think we know) how strong a magnet we can build. From that we can extrapolate how big around the tube must be: 53 miles.

A FOURTH BREAKTHROUGH: SUPERCONDUCTIVITY

Back in 1911 a Dutch physicist discovered that certain metals, when cooled to extremely low temperatures — just a few degrees above absolute zero on the Kelvin scale (−273 degrees centigrade) — lose all their resistance to electricity. A loop of wire at that temperature would carry a current forever with no use of energy.

In your house, electrical power is supplied via copper wires from the friendly power company. The wires get warm because of the frictional resistance they offer to the flow of current. This waste heat uses power and adds to your bill. In conventional electromagnets for motors, generators, and accelerators, copper wires carry currents that produce magnetic fields. In a motor the magnetic field turns bundles of current-carrying wires. Feel the warm motor. In an accelerator the magnetic field steers and focuses the particles. The magnet's copper wires get hot and are cooled by a powerful flow of water, usually through holes in the thick copper windings. To give you some idea of where the money goes, the 1975 electric bill for the Fermilab accelerator was about $15 million, some 90 percent of which was for the power used in running the magnets for the 400 GeV main ring.

Early in the 1960s a technical breakthrough took place. New alloys of exotic metals were able to maintain the fragile state of superconductivity while conducting huge currents and producing high magnetic fields. All of this at the more civilized temperatures of 5 to 10 degrees above absolute zero rather than the very difficult 1 to 2 degrees required for common metals. Helium is a true liquid at 5 degrees (everything else solidifies at this temperature), so the possibility of practical superconductivity emerged. Most of the large laboratories began working with wire made of such alloys as niobium-titanium or niobium 3-tin in place of copper and surrounding the wires with liquid helium to cool them to superconducting temperatures.

Large magnets using the new alloys were built for particle detectors — for example, to surround a bubble chamber — but not for accelerators, which required that magnetic fields increase in strength as the particles gain energy. The changing currents in the magnets generate frictional effects (eddy currents) that normally destroy the superconducting state. Much research was addressed to this problem in the 1960s and '70s, with Fermilab, under Robert Wilson, serving as a leader in the field. Wilson's team began R&D in superconducting magnets in 1973, shortly after the original "200" accelerator began operating. One motivation was the exploding costs of electrical power due to the oil crisis of that era. The other was competition from the European consortium, CERN, based in Geneva.

The 1970s were lean years for research funds in the United States. After World War II the world leadership in research had been solidly in this country, as the rest of the world labored to rebuild war-shattered economies and scientific infrastructures. By the late 1970s, balance had begun to be restored. The Europeans were building a 400 GeV machine, the Super Proton Synchrotron (SPS), which was better funded and better supplied with the expensive detectors that determine the quality of the research. (This machine marked the beginning of another cycle in international collaboration and competition. In the 1990s Europe and Japan remain ahead of the United States in some research fields and not far behind in most others.)

Wilson's idea was that if one could solve the problem of varying magnetic fields, a superconducting ring would save an enormous amount of electrical power while producing more powerful magnetic fields, which for a given radius would translate to higher energy. Aided by Alvin Tollestrup, a Cal Tech professor spending a sabbatical year at Fermilab (he eventually extended this to permanence), Wilson studied in great detail how changing currents and fields create local heating. Research going on in other labs, especially the Rutherford Lab in England, helped the Fermilab group build hundreds of models. They worked with metallurgists and materials scientists and, between 1973 and 1977, succeeded in solving the problem. One could ramp the model magnets from zero current to 5,000 amperes in 10 seconds, and the superconductivity persisted. In 1978–79 a production line began producing twenty-one-foot magnets with excellent properties, and in 1983 the Tevatron began operating as a superconducting "afterburner" at the Fermilab complex. The energy went from 400 GeV to 900 GeV, and the power consumption was reduced from 60 megawatts to 20 megawatts, with most of that used to produce liquid helium.

When Wilson began his R&D program in 1973, the annual production of superconducting material in the United States was a few hundred pounds. Fermilab's consumption of 125,000 pounds of superconducting material stimulated producers and radically changed the posture of the industry. Today the biggest customers are firms that make magnetic resonance imaging (MRI) devices, for medical diagnosis. Fermilab can take a modicum of credit for this $500-million-a-year industry.

THE COWBOY LAB DIRECTOR

The man who deserves much of the credit for Fermilab itself is our first director, artist/cowboy/machine designer Robert Rathbun Wilson. Talk about charisma! Wilson grew up in Wyoming, where he rode horseback and studied hard at school, winning a scholarship to Berkeley. There he was a student of E. O. Lawrence's.

I have already described the architectural feats of this Renaissance man in building Fermilab, but he was technologically sophisticated as well. Wilson became the founding director of Fermilab in 1967 and received an allocation of $250 million to build (so said the specifications) a 200 GeV machine with seven beam lines. Construction, started in 1968, was to take five years, but Wilson completed the machine ahead of schedule in 1972. By 1974 it was working steadily at 400 GeV with fourteen beam lines and with $10 million left over from the original allocation — all this with the most splendid architecture ever seen in a U.S. government installation. I recently calculated that had Wilson been in charge of our defense budget over the past fifteen years with the same skills, the United States would now be enjoying a tidy annual budget surplus and our tanks would be the talk of the art world.

One story has it that Fermilab first sprang into Wilson's mind in the early 1960s in Paris, where he was an exchange professor. One day he found himself sketching a beautiful, curvaceous nude model with a group of other artists in a public drawing session at the Grande Chaumière. The "200" was being discussed in the United States, and Wilson didn't like what he read in his mail. While others drew breasts, Wilson drew circles for beam tubes and adorned them with calculations. This is dedication.

Wilson wasn't perfect. He took short cuts when building Fermilab, and not all were successful. He complained bitterly that one blooper cost him a year (he would have finished in 1971), and an extra $10 million. He also gets mad, and in 1978, disgusted with the slow pace of federal funding for his superconducting work, he quit. When I was asked to become his successor I went to see him. He threatened to haunt me if I didn't take the job, and that did it. The prospect of being haunted by Wilson on his horse was too much. So I took the job and prepared three envelopes.

A DAY IN THE LIFE OF A PROTON

We can illustrate everything that has been explained in this chapter by describing Fermilab's cascade accelerator, which has five sequential machines (seven if you want to count the two rings in which we make antimatter). Fermilab is a complex choreography of five different accelerators, each a step up in energy and sophistication, like ontogeny recapitulating phylogeny (or whatever it recapitulates).

First we need something to accelerate. We run over to Ace Hardware and buy a pressurized bottle of hydrogen gas. The hydrogen atom consists of an electron and a simple nucleus of one proton. There are enough protons in this bottle to run Fermilab for a year. Cost: about twenty dollars if you return the bottle. The first machine in the cascade is nothing less than a Cockcroft-Walton electrostatic accelerator, 1930s design. Although it is the most ancient of the Fermilab series of accelerators, it is the most futuristic looking, adorned with very large and shiny balls and doughnutlike rings that photographers like to shoot. In the Cockcroft-Walton a spark strips the electron away from the atom, leaving a positively charged proton essentially at rest. The machine then accelerates the protons, creating a 750 KeV beam aimed at the entrance to the next machine, which is a linear accelerator, or linac. The linac sends the protons down a 500-foot-long series of radio-frequency cavities (gaps) to bring them to 200 MeV.

At this respectable energy they are transferred via magnetic steering and focusing to the "booster," a synchrotron, which whirls the protons around and raises their energy to 8 GeV. Just think: at this point we've produced higher energies than the Berkeley Bevatron, the first GeV accelerator, and we have two rings yet to go. This load of protons is then injected into the main ring, the almost-four-mile-around "200" machine, which in the years 1974–1982 worked at 400 GeV, twice the official energy it was designed for. The main ring was the workhorse of the Fermilab complex.

After the Tevatron came on-line in 1983, the main ring began taking life a little easier. Now it takes the protons up to only 150 GeV and then transfers them to the superconducting Tevatron ring, which is exactly the same size as the main ring and is just a few feet beneath it. In the conventional application of the Tevatron, the superconducting magnets carry the 150 GeV particle around and around, 50,000 circuits per second, gaining about 700 KeV per turn until, after about 25 seconds, they reach 900 GeV. By this time the magnets, powered by currents of 5,000 amperes, have increased their field strength to 4.1 tesla, more than twice the field that the old iron magnets could provide. And the energy required to maintain the 5,000 amperes is approximately zero! The technology of superconducting alloys is continually improving. By 1990 the 1980 Tevatron technology had been improved so that the Super Collider will use fields of 6.5 tesla, and CERN is working hard to push the technology to what may be a limit for niobium alloys — to 10 tesla. In 1987 a new kind of superconductor was discovered based on ceramic materials that require only liquid nitrogen cooling. Hopes were raised that a cost breakthrough was imminent, but the requisite strong magnetic fields are not there yet, and no one can estimate when and if these new materials will ever replace niobium titanium.

At the Tevatron, 4.1 tesla is the limit, and now the protons are kicked by electromagnetic forces into an orbit that brings them out of the machine into a tunnel, where they are divided up among some fourteen beam lines. Here experimental teams provide targets and detectors to do their experiments. Some thousand physicists work in the fixed-target program. The machine operates in cycles. It takes about 30 seconds to do all the acceleration. The beam is spilled out over another 20 seconds so as not to crowd the experimenters with too high a rate of particles for their experiments. This cycle is repeated every minute.

The external beam line is very tightly focused. My colleagues and I set up an experiment in "Proton Center," where a beam of protons is extracted, focused, and steered for about 8,000 feet onto a target 0.01 inches wide, the width of a razor blade. The protons collide with the thin edge. Every minute, day after day for weeks, a burst of protons strikes this target, never shifting by more than a small fraction of its width.

The other mode of using the Tevatron, the collider mode, is quite different, and we will discuss it in detail. In this mode, the injected protons coast around in the Tevatron at 150 GeV waiting for antiprotons, which in due course are delivered from the p-bar source and sent around the ring in the opposite direction. When both beams are in the Tevatron, we begin ramping up the magnets and accelerating both beams. (More about how this works in a moment.)

At every phase of the sequence, computers control the magnets and radio-frequency systems, keeping the protons tightly bunched and under control. Sensors give information on currents, voltages, pressures, temperatures, the location of the protons, and the latest Dow Jones averages. A malfunction could send the beam careening out of its vacuum pipe and through the enveloping magnet structure, boring a very neat and very expensive hole. This has never happened — at least not yet.

DECISIONS, DECISIONS: PROTONS VS. ELECTRONS

We've been talking a lot about proton machines here, but protons aren't the only way to go. The nice thing about protons is that they are relatively inexpensive to accelerate. We can accelerate them to thousands of billions of electron volts. The Super Collider will accelerate protons to 20 trillion electron volts. In fact, there may be no theoretical limit to what we can do. On the other hand, protons are full of other particles — quarks and gluons. This makes the collisions messy and complicated. That's why some physicists prefer to accelerate electrons, which are pointlike, a-tomlike. Because they are points, their collisions are cleaner than with protons. The downside is that they are low in mass, so they are difficult and expensive to accelerate. Their low mass results in a large amount of electromagnetic radiation when steered around a circle. Much more power must be put in to make up for the radiation loss. While this radiation is a waste from the point of view of acceleration, it's a spinoff boon to some researchers because it is very intense and of very short wavelength. Many circular electron accelerators are actually devoted to producing this synchrotron radiation. Customers include biologists who use the intense photon beams to study the structure of huge molecules, electronic chip makers who do x-ray lithography, condensed-matter scientists, who study the structure of materials, and many other practical types.

One way around this energy loss is to use a linear accelerator, such as the 2-mile-long linac at Stanford, built back in the early 1960s. The Stanford machine was originally called "M," for monster, and it was an outrageous machine for its time. It begins on the Stanford campus, about a quarter mile from the San Andreas Fault, and works its way toward San Francisco Bay. The Stanford Linear Accelerator Center owes its existence to the drive and verve of its founder and first director, Wolfgang Panofsky. J. Robert Oppenheimer told the story that the brilliant Panofsky and his equally brilliant twin brother, Hans, attended Princeton together, both achieving stellar academic records, but one doing just a hair better than the other. From that time on, claimed Oppenheimer, they became "Smart" Panofsky and "Dumb" Panofsky. Which is which? "That's a secret!" says Wolfgang. If truth be told, most of us call him Pief.

The differences between Fermilab and SLAC are obvious. One does protons; one does electrons. One is circular, the other straight. And when we say a linear accelerator is straight, we mean straight. For example, let's say we build a two-mile stretch of road. The surveyors guarantee us that it's straight, but it isn't. It follows the very slight curve of the earth. To a surveyor standing on the surface of the planet, it looks straight, but if viewed from space it's an arc. The beam tube in SLAC, on the other hand, is *straight*. If the earth were a perfect sphere, the linac would be a two-mile tangent to the earth's surface. Electron machines proliferated around the world, but SLAC remained the most spectacular, accelerating electrons to 20 GeV in 1966 and to 50 GeV in 1987. Then the Europeans took over.

COLLIDERS VERSUS TARGETS

Okay, here are our choices so far. You can accelerate protons or electrons, and you can accelerate them in circles or in a straight line. But there's one more decision to make.

Conventionally, one extracts beams from the confines of the magnetic prison and transports the beams, always in vacuum pipes, up to a target where collisions take place. We've explained how analyzing the collisions provides information about the subnuclear world. The accelerated particle brings in a certain amount of energy, but only a fraction of it is available to explore nature at small distances or to manufacture new particles via $E = mc^2$. The law of conservation of momentum says that some of the input energy will be preserved and given to the final products of the collisions. For instance, if a moving bus hits a stationary truck, much of the energy from the accelerating bus will go into knocking the various bits of sheet metal, glass, and rubber forward. This subtracts from the energy that could demolish the truck more thoroughly.

If a 1,000 GeV proton strikes a proton at rest, nature insists that whatever particles come off must have enough forward motion to equal the forward momentum of the incident proton. It turns out that this leaves a maximum of only 42 GeV for making new particles.

We came to realize in the mid-1960s that if one could get two particles, each having the full energy of the accelerator beam, to collide head on, we'd have an extraordinarily more violent collision. Twice the energy of the accelerator would be brought into the collision, and *all of it* would be available, since the total initial momentum is zero (equal and opposite momenta for the colliding objects). Ergo, in a 1,000 GeV accelerator, a head-on collision of two particles, each having 1,000 GeV, releases 2,000 GeV for the creation of new particles, compared to the 42 GeV when the accelerator is in stationary target mode. There's a penalty, however. A machine gun can easily hit the side of a barn; it is more difficult to have two machine guns shoot at each other and have the bullets collide in midair. This gives you some idea of the challenge of operating a colliding-beam accelerator.

MAKING ANTIMATTER

Stanford followed up its original collider with a very productive accelerator called SPEAR, for Stanford Positron-Electron Accelerator Ring, in 1973. Here beams of electrons are accelerated in the two-mile-long linear accelerator to an energy between 1 and 2 GeV and then injected into a small magnetic storage ring. Positrons, Carl Anderson's particles, are produced by a sequence of reactions. First, the intense electron beam impinges on a target to produce, among other things, an intense beam of photons. The debris of charged particles is swept away with magnets, which do not affect the neutral photons. Thus a clean beam of photons is allowed to strike a thin target, for example platinum. The most common result is that the pure energy of the photon converts to an electron and a positron, each sharing the original energy of the photon, minus the rest mass of the electron and positron.

A magnet system collects some fraction of the positrons, and these are injected into a storage ring in which the electrons have been patiently going around and around. The streams of positrons and electrons, having opposite electrical charges, curve in opposite directions in a magnet. If one stream goes clockwise, the other goes counterclockwise. The result is obvious: head-on collisions. SPEAR made several important discoveries, colliders became very popular, and a plethora of poetic(?) acronyms was unleashed upon the world. Before SPEAR there was ADONE (Italy, 2 GeV); after SPEAR (3 GeV), there was DORIS (Germany, 6 GeV), then PEP (Stanford again, 30 GeV), PETRA (Germany, 30 GeV), CESR (Cornell, 8 GeV), VEPP (USSR), TRISTAN (Japan, 60 to 70 GeV), LEP (CERN, 100 GeV), and SLC (Stanford, 100 GeV). Note that colliders are rated by the sum of the two beam energies. LEP, for example, has 50 GeV in each beam; ergo it's a 100 GeV machine.

In 1972, proton-proton head-on collisions were made available at the pioneering CERN Intersecting Storage Ring (ISR) facility in Geneva. Here two independent rings entwine around one another, the protons going in opposite directions in each ring and colliding head on at eight different intersection points. Matter and antimatter such as electrons and positrons can be circulated in the same ring because the magnets make them circulate in opposite directions, but two separate rings are needed to slam protons into each other.

In the ISR each ring is filled with 30 GeV protons from the more conventional CERN accelerator, the PS. The ISR was ultimately very successful. But when it was fired up in 1972, it attained only a few thousand collisions per second in the "high luminosity" collision points. "Luminosity" is the term used to describe the number of collisions per second, and ISR's early troubles demonstrated the difficulty of getting two machine gun bullets (the two beams) to collide. Eventually the machine improved to over 5 million collisions per second. As for physics, some important measurements were made, but, in general, the ISR mostly provided a valuable learning experience about colliders and detection techniques. The ISR was an elegant machine both technically and in appearance — a typical Swiss production. I worked there during my 1972 sabbatical and returned frequently over the next decade. Early on I took I. I. Rabi, who was visiting Geneva for an "Atoms for Peace" conference, on a tour. As we entered the elegant tunnel of the accelerator, Rabi's jaw dropped, and he exclaimed, "Ah, Patek Philippe!"

The most difficult collider of all, one that pits protons against antiprotons, was made possible by an invention of a fabulous Russian, Gershon Budker, working in the Novosibirsk Soviet Science City. Budker had been building electron machines in Russia, competing with his American friend Wolfgang Panofsky. Then his operation was transferred to Novosibirsk, a new university research complex in Siberia. As he put it, since Panofsky was not similarly transferred to Alaska, the competition was unfair and he was forced to innovate.

In Novosibirsk in the 1950s and '60s, Budker ran a thriving capitalistic system of selling small accelerators to Soviet industry in exchange for materials and money to keep his research going. He had been fascinated by the prospects for using antiprotons, or p-bars, as one of the colliding elements in accelerators, but realized that they are a scarce commodity. The only place to find them is in high-energy collisions, where they are produced, yes, via $E = mc^2$. A machine with many tens of GeV will have a few p-bars among the debris of collisions. To garner enough for useful collision rates, they would have to be accumulated over many hours. But as the p-bars emerge from a struck target, they are moving every which way. Accelerator scientists like to state these motions in terms of their principal direction and energy (just right!) and the superfluous sideways motions that tend to fill up the available space in the vacuum chamber. What Budker saw was the possibility of "cooling" the sideways components of their motions and compressing the p-bars into a much denser beam as they are stored. This is a complicated business. New levels of beam control, magnet stability, and ultra-high vacuum must be achieved. The antiprotons must be stored, cooled, and accumulated for upward of ten hours before there are enough to inject into the collider for acceleration. It was a lyrical idea, but the program was far too complex for Budker's limited resources in Siberia.

Enter Simon Van der Meer, a Dutch engineer at CERN who advanced this cooling technique in the late 1970s and helped to build the first p-bar source for use with the first proton-antiproton collider. He used CERN's 400 GeV ring as both the storage and collision device, and the first p/p-bar collisions went on-line in 1981. Van der Meer shared the 1985 Nobel Prize with Carlo Rubbia for his contribution of "stochastic cooling" to the program that Rubbia had designed and that resulted in the discoveries of the W^+, W^-, and Z^0 particles, which we'll discuss later.

Carlo Rubbia is so colorful that he deserves a whole book, and he has at least one. (*Nobel Dreams,* by Gary Taubes, is about him.) One of the more brilliant graduates of the awesome Scuola Normale in Pisa, where Enrico Fermi was a student, Carlo is a dynamo that can never slow down. He worked at Nevis, at CERN, at Harvard, at Fermilab, at CERN again, and then Fermilab again. Traveling so much, he invented a complex cost-minimizing scheme of interchanging his "to" and "fro" ticket halves. I once briefly convinced him he'd retire with eight tickets left over, all west to east. In 1989 he became director of CERN, by which time the European consortium's lab had held the lead in proton-antiproton collisions for about six years. However, the lead was recaptured by the Tevatron in 1987–88, when Fermilab made significant improvements in the CERN scheme and put into operation its own antiproton source.

P-bars don't grow on trees, and you can't buy them at Ace Hardware. In the 1990s Fermilab is the world's largest repository of antiprotons, which are stored in a magnetic ring. A futuristic study by the U.S. Air Force and the Rand Corporation has determined that one milligram (one thousandth of a gram) of antiprotons would be an ideal rocket fuel, for it would contain the energy equivalent of about two tons of oil. Since Fermilab is the world leader in antiproton production (10^{10} per hour), how long would it take to make a milligram? At the present rate, a few million years of twenty-four-hour operation. Some incredibly optimistic extrapolations of technology might reduce this to a few thousand years. So my advice is not to invest in Fidelity's P-Bar Mutual Fund.

The Fermilab collider scheme works as follows. The old 400 GeV accelerator (the main ring), operating at 120 GeV, throws protons against a target every two seconds. Each collision of about 10^{12} protons makes some 10 million antiprotons heading in the right direction with the right energy. With each p-bar there are thousands of unwanted pions, kaons, and other debris, but these are all unstable and they go away sooner or later. The p-bars are focused into a magnetic ring called the debuncher ring, where they are processed, organized, and compressed, then transferred to the accumulator ring. Both rings are about 500 feet around and store p-bars at 8 GeV, the same energy as the booster accelerator. It takes five to ten hours to accumulate enough p-bars to inject back into the accelerator complex. Storage is a delicate affair, as all of our equipment is made out of matter (what else?), and the p-bars are antimatter. If they come into contact with matter — annihilation. So we must be fastidious in keeping the p-bars orbiting near the center of the vacuum tube. And the quality of the vacuum must be extraordinary — the best "nothing" that technology can buy.

After accumulation and continued compression for about ten hours, we are ready to inject the p-bars back into the accelerator from whence they came. In a procedure reminiscent of a NASA launch, a tense countdown ensues to make sure that every voltage, every current, every magnet, and every switch is correct. The p-bars are zapped into the main ring, where they circulate counterclockwise because of their negative charge. They are accelerated to 150 GeV and transferred, again by magnetic legerdemain, to the Tevatron superconducting ring. Here the protons, recently injected from the booster via the main ring, have been patiently waiting, circulating tirelessly in the customary clockwise direction. Now we have two beams, running in opposite directions around the four-mile ring. Each beam is composed of six bunches of particles, with about 10^{12} protons and a somewhat smaller number of p-bars per bunch.

Both beams are accelerated from 150 GeV, the energy imparted to them from the main ring, to the full Tevatron energy of 900 GeV. The final step is "squeeze." Because the beams are counter-circulating in the same small tube, they inevitably have been crossing each other during the acceleration phase. However, their density is so low that very few collisions between particles occur. "Squeeze" energizes special superconducting quadrupole magnets that compress beam diameters from soda straws (a few millimeters) to human hairs (microns). This increases the density of particles enormously. Now, when the beams cross, there is at least one collision per crossing. Magnets are tweaked to make sure the collisions take place at the center of the detectors. The rest is up to them.

Once we have established stable operation, the detectors turn on and begin collecting data. Typically this continues for ten to twenty hours while more p-bars are being accumulated with the help of the old main ring. In time the proton and antiproton bunches become depleted and more diffuse, cutting the event rate. When the luminosity (the number of collisions per second) has gone down to about 30 percent, and if there are enough new p-bars stored in the accumulator ring, the beams are dumped and another NASA countdown ensues. It takes about a half hour to refill the Tevatron collider. About 200 billion antiprotons is considered an okay number to inject. More is better. These face some 500 billion protons, far easier to come by, to produce about 100,000 collisions per second. Improvements to all of this, designed for installation in the 1990s, will increase these numbers by about a factor of ten.

In 1990 the CERN p/p-bar collider retired, leaving the field to the Fermilab facility with its two powerful detectors.

WATCHING THE BLACK BOX: THE DETECTORS

We learn about the subnuclear domain by observing, measuring, and analyzing the collisions induced by high-energy particles. Ernest Rutherford locked his team up in a dark room so they could see and count the scintillations generated by alpha particles hitting zinc sulfide screens. Our techniques of particle counting have evolved considerably since then, especially in the post–World War II period.

Prior to World War II the cloud chamber was a major tool. Anderson used it to discover the positron, and it was found in cosmic ray laboratories around the world. My assignment at Columbia was to build a cloud chamber to operate with the Nevis cyclotron. As an absolutely green graduate student, I was unaware of the subtleties of cloud chambers and was competing with experts at Berkeley, Cal Tech, Rochester, and other such places. Cloud chambers are finicky devices, susceptible to "poisoning" — impurities that create unwanted droplets, which compete with those that delineate the particle tracks. No one at Columbia had any experience with these loathsome detectors. I read all the literature and adopted all the superstitions: clean the glass with sodium hydroxide and wash with triple distilled water; boil the rubber diaphragm in 100 percent methyl alcohol; mutter the right incantations . . . A little prayer can't hurt.

In desperation, I tried to get a rabbi to bless my cloud chamber. Unfortunately, I picked the wrong rabbi. He was Orthodox, very religious, and when I asked him to say a *brucha* (Hebrew: "blessing") for my cloud chamber, he demanded to know what a cloud chamber was. I showed him a photograph, and he was furious at my suggested sacrilege. The next guy I tried, a Conservative rabbi, upon seeing the picture, asked how the cloud chamber worked. I explained. He listened, nodded, stroked his beard, and finally said sadly that he just couldn't do it. "The law . . ." So I went to the Reform rabbi. He was just getting out of his Jaguar XKE when I came to his house. "Rabbi, can you say a brucha for my cloud chamber?" I pleaded. "Brucha?" he responded. "What's a brucha?" So I was worried.

Finally I was ready for the big test. At this point everything should have worked, but each time the chamber was operated, I got dense, white smoke. At this stage Gilberto Bernardini, a true expert, arrived at Columbia and began looking over my shoulder.

"Whatsa de brass rod, poking into de chamber?" he asked.

"That's my radioactive source," I said, "to give tracks. But all I get it white smoke."

"Tay-ka id oud."

"Take it out?"

"Si, si, oud!"

So take it out I did, and a few minutes later . . . tracks! Beautiful wavy threads of tiny droplets tearing through the chamber. The most beautiful sight I'd ever seen. What happened was that my millicurie source was so strong it was filling the chamber with ions, and each grew its own drop. The result: dense, white smoke. I didn't need a radioactive source. Cosmic rays, omnipresent in the space around us, kindly provide enough radiation. Ecco!

The cloud chamber turned out to be a very productive instrument because one could photograph the trail of tiny droplets formed along the track of particles passing through it. Equipping it with a magnetic field caused the tracks to curve, and measuring the radius of this curvature gave us the momentum of the particles. The closer the tracks were to being straight (less curvature) the more energetic the particles. (Remember the protons in Lawrence's cyclotron, which gained momentum and then described larger circles.) We took thousands of pictures that revealed a variety of data on the properties of pions and muons. The cloud chamber — looked at as an instrument rather than as a source of my Ph.D. and tenure — allowed us to observe some dozen tracks in each photograph. The pions take about a billionth of a second to pass through the chamber. We can provide a dense plate of material in which a collision can take place, which happens perhaps once in every hundred photographs. Because pictures can be taken only about one per minute, the data accumulation rate is further limited.

BUBBLE, BUBBLE, TOIL AND TROUBLE

The next advance was the bubble chamber, invented in the mid-1950s by Donald Glaser, then at the University of Michigan. The first bubble chamber was a little thimble of liquid ether. The evolution of liquid hydrogen chambers up to the 15-foot monster, retired from Fermilab in 1987, was led by the famed Luis Alvarez at the University of California.

In a chamber filled with liquid, often liquefied hydrogen, tiny bubbles form along the trail of particles passing through. The bubbles indicate the onset of boiling due to a sudden deliberate lowering of the pressure in the liquid. What this does is put the liquid above the boiling point, which depends on both temperature and pressure. (You may have experienced the difficulty of cooking an egg in your mountain chalet. At the low pressure of mountaintops, water boils well below 100 degrees C.) A clean liquid, no matter how hot, will resist boiling. For example, if you heat some oil in a deep pot above its normal boiling temperature, and if everything is really clean, it won't boil. But toss in a single piece of potato, and explosive boiling takes place. So to produce bubbles, two things are required: temperature above the boiling point and some kind of impurity to encourage the formation of a bubble. In the bubble chamber, the liquid is superheated by the sudden decrease in pressure. The charged particle, in its numerous gentle collisions with atoms of the liquid, leaves a trail of excited atoms that, after the pressure is lowered, are ideal for nucleating the bubbles. If a collision occurs between the incident particle and a proton (nucleus of hydrogen) in the vat, all the emerging charged products are also rendered visible. Since the medium is a liquid, dense plates are not necessary, and the collision point can be seen clearly. Researchers around the world took millions of photographs of collisions in bubble chambers, their analysis aided by automated scanners.

So here is how it works. The accelerator shoots a beam of particles toward the bubble chamber. If this is a charged particle beam, ten or twenty tracks begin to crowd the chamber. Within a millisecond or so after the passage of the particles, a piston is rapidly moved, lowering the pressure and thereby beginning the formation of bubbles. After another millisecond or so of growth time, a light is flashed, film is moved, and we are ready for another cycle.

It is said that Glaser (who won the Nobel Prize for his bubble chamber and promptly became a biologist) got his idea for nucleation of bubbles by studying the trick of increasing the head on a glass of beer by adding salt. The bars of Ann Arbor, Michigan, thus spawned one of the more successful instruments used to track the God Particle.

There are two keys to collision analysis: space and time. We would like to record a particle's trajectory in space and its precise time of passage. For example, a particle comes into the detector, stops, decays, and gives rise to a secondary particle. A good example of a stopping particle is a muon, which can decay into an electron, separated in time by a millionth or so of a second from the stopping event. The more precise your detector, the more information. Bubble chambers are excellent for space analysis of the event. The particles leave tracks, and in bubble chambers we can locate points on those tracks to an accuracy of about 1 millimeter. But they provide no time information.

Scintillation counters can locate particles in both space and time. Made of special plastics, they produce a flash of light when struck by a charged particle. The counters are wrapped in light-tight black plastic, and each tiny light flash is funneled to an electronic photomultiplier that converts the signal, indicating passage of a particle, into a sharply defined electronic pulse. When this pulse is superimposed on an electronic train of clock pulses, the arrival of a particle can be recorded to a precision of a few billionths of a second. If a number of scintillation strips are used, a particle will strike several in succession, leaving a series of pulses that describe its path in space. The space location depends on the size of the counter, which typically establishes the location to a precision of a few inches.

A major breakthrough was the proportional wire chamber (PWC), the invention of a prolific Frenchman working at CERN, Georges Charpak. A World War II hero of the Resistance and a concentration camp prisoner, Charpak became the preeminent inventor of particle detector devices. In his PWC, an ingenious, "simple" device, a number of fine wires, only a few tenths of an inch apart, are stretched across a frame. Typically the frame is two feet by four feet, with a few hundred two-foot-long wires strung across the four-foot span. Voltages are organized so that when a particle passes near a wire, it generates an electrical pulse in the wire, and the pulse is recorded. The accurately surveyed location of the struck wire locates one point on the trajectory. The time of the pulse is obtained by comparison with an electronic clock. By further refinements, the space and time definition can be pinpointed to approximately 0.1 millimeters and 10^{-8} seconds. With many such planes stacked in an airtight box filled with an appropriate gas, one can precisely define the trajectories of particles. Because the chamber is active for only a short interval of time, random background events are suppressed and very intense beams can be used. Charpak's PWCs have been a part of every major particle physics experiment since about 1970. In 1992 Charpak was awarded the Nobel Prize (alone!) for his invention.

All of these different particle sensors and more were incorporated into the sophisticated detectors of the 1980s. The CDF detector at Fermilab is typical of one of the most complex systems. Three stories high, weighing 5,000 tons, and built at a cost of $60 million, it is designed to observe the head-on collisions of protons and antiprotons in the Tevatron. Here some 100,000 sensors, which include scintillation counters and wires in exquisitely designed configurations, feed streams of information in the form of electronic pulses to a system that organizes, filters, and finally records data for future analysis.

As in all such detectors, there is too much information to handle in real time — that is, immediately — so the data are encoded in digital form and organized for recording on magnetic tape. The computer must decide which collisions are "interesting" and which are not, since there are over 100,000 collisions per second in the Tevatron, and this is expected to increase in the early 1990s to one million collisions per second. Now, most of these collisions are of no interest. The jewels are those in which a quark in one proton really smacks an antiquark or even a gluon in the p-bar. These hard collisions are rare.

The information-handling system has less than a millionth of a second to examine a particular collision and make a fateful decision: is this event interesting? To a human this is mind-boggling speed, but not to a computer. It is all relative. In one of the big cities, a turtle was attacked and robbed by a gang of snails. When later questioned by the police, the turtle said: "I don't know. Everything happened so fast!"

To alleviate the electronic decision making, a system of sequential levels of event selection has evolved. The experimenters program the computers with various "triggers," indicators that tell the system which events to record. For example, a common trigger would be an event that discharges a large amount of energy into the detector, for new phenomena are most likely to occur at high rather than low energies. The setting of triggers is a sweaty-palm business. Make them too loose, and you overwhelm the capability and logic of the recording technology. Set them too tight, and you may miss some new physics, or you may have done the entire experiment for nothing. Some triggers will flip "on" when an energetic electron is detected emerging from the collision. Another trigger will be convinced by the narrowness of a jet of particles, and so on. Typically there are ten to twenty different configurations of collision events that are allowed to set off a trigger. The total number of events passed by these triggers may be 5,000 to 10,000 in a second, but now the event rate is low enough (one every ten-thousandth of a second) to "think" and examine — er, have the computer examine — the candidates more carefully. Do you really want to record this event? The screening goes on through four or five levels until it gets down to about ten events per second.

Each of these events is recorded on magnetic tape in full detail. Often, at the stages where we are rejecting events, a sampling of, say, one in a hundred is recorded for future study to determine if important information is being lost. The entire data acquisition system (DAQ) is made possible by an unholy alliance of physicists who think they know what they want, clever electronic engineers who try hard to please, and, oh yes, a revolution in commercial microelectronics based on the semiconductor.

The geniuses in all of this technology are too numerous to list, but in my subjective view, one of the leading innovators was a shy electronics engineer who functioned in a garret at Columbia's Nevis

Lab, where I grew up. William Sippach was way ahead of his physicist controllers. We specified; he designed and built the DAQ. Time and again I would telephone him at three in the morning crying that we'd come up against a serious limitation in his (it was always *his* when we had trouble) electronics. He would listen quietly and ask a question: "Do you see a microswitch inside the cover plate of rack sixteen? Activate it and your problem will be solved. Good night." Sippach's fame spread, and in a typical week, visitors from New Haven, Palo Alto, Geneva, and Novosibirsk would drop in to talk to Bill.

Sippach and the many others who helped develop these complex systems continue a great tradition that began in the 1930s and '40s when the circuits for the early particle detectors were invented. These in turn become the key ingredients in the first generation of digital computers. These, in turn, begat better accelerators and detectors, which begat . . .

The detectors are the bottom line in this whole business.

WHAT WE FOUND OUT: ACCELERATORS AND PHYSICS PROGRESS

You now know everything you need to know about accelerators — perhaps more. You may in fact know more than most theorists. This is not a criticism, just a fact. More important is what these new machines told us about the world.

As I've mentioned, the synchrocyclotrons of the 1950s enabled us to learn much about pions. Hideki Yukawa's theory suggested that by exchanging a particle with a particular mass, one could create a strong attractive counterforce that would bind protons to protons, protons to neutrons, and neutrons to neutrons. Yukawa predicted the mass and lifetime of this particle being exchanged: the pion.

The pion has a rest mass energy of 140 MeV, and it was produced prolifically in the 400 to 800 MeV machines on university campuses around the world in the 1950s. Pions decay into muons and neutrinos. The muon, which was the great puzzle of the 1950s, seemed to be a heavier version of the electron. Richard Feynman was one of the prominent physicists who agonized over two objects that behave in all respects identically, except that one weighs two hundred times as much as the other. The unraveling of this mystery is one of the keys to our entire thrust, a clue to the God Particle itself.

The next generation of machines produced a generational surprise: hitting the nucleus with *billion*-volt particles was doing "something different." Let's review what you can do with an accelerator, especially since the final exam is coming soon. Essentially, the vast investment in human ingenuity described in this chapter — the development of the modern accelerator and particle detector — allows us to do two kinds of things: to *scatter* objects or — and this is the "something different" — to *produce new objects.*

1. *Scattering.* In scattering experiments we look at how incident particles after collision fly off in various directions. The technical term for the end product of a scattering experiment is angular distribution. When analyzed according to the rules of quantum physics, these experiments tell us a good deal about the nucleus that is scattering the particles. As the energy of the incoming particle from the accelerator increases, the structure comes into better focus. So we learned about the composition of nuclei — neutrons and protons and how they are arranged and how they jiggle around to maintain their arrangement. As we further increase the energy of our protons, we can "see" into the protons and the neutrons. Boxes inside boxes.

To make things simple, we can use single protons (hydrogen nuclei) as targets. Scattering experiments told us about the proton's size and about how the positive electric charge is distributed. A clever reader will ask whether the probe — the particle hitting the target — itself contributes to the confusion, and the answer is yes. So we use a variety of probes. Alpha particles from radiation gave way to protons and electrons fired from accelerators, and later we used secondary particles: photons derived from electrons, pions derived from proton-nucleus collisions. As we got better at doing this in the 1960s and '70s, we began using tertiary particles as the bombarding particles; muons from decays of pions became probes, as did neutrinos from the same source, and lots more.

The accelerator laboratory became a service center with a variety of products. By the late 1980s, Fermilab's sales force advertised to potential customers that the following hot and cold running beams were available: protons, neutrons, pions, kaons, muons, neutrinos, antiprotons, hyperons, polarized protons (all spinning in the same direction), tagged photons (we know their energy), and if you don't see it, ask!

2. *Producing new particles.* Here the object is to see if a new energy domain results in the creation of new, never-before-seen particles. If

there is a new particle, we want to know everything about it — its mass, spin, charge, family, and so on. We also need to know its lifetime and what other particles it decays into. Of course, we have to know its name and what role it plays in the great architecture of the particle world. The pion was discovered in cosmic rays, but soon we found that it doesn't spring fully grown from the forehead of cloud chambers. What happens is that cosmic-ray protons from outer space enter the earth's atmosphere where they collide with nuclei of nitrogen and oxygen (today we also have more pollutants), and out of these collisions pions are created. A few other weird objects were also identified in cosmic-ray studies, such as particles called K^+ and K^- and objects called lambda (the Greek letter Λ). When more powerful accelerators took over, starting in the mid-1950s and then with a vengeance in the 1960s, various exotic particles were created. The trickle of new objects soon became a flood. The huge energies available in collisions uncovered the existence of not one or five or ten but hundreds of new particles, undreamt of in most of our philosophies, Horatio. These discoveries were group efforts, the fruits of Big Science and a mushrooming of technologies and techniques in experimental particle physics.

Each new object was given a name, usually a Greek letter. The discoverers, typically a collaboration of sixty-three and a half scientists, would announce the new object and give as many of its properties — mass, charge, spin, lifetime, and a long list of additional quantum properties — as were known. They would then pass Go, collect two hundred dollars, write up a thesis or two, and wait to be invited to give seminars, conference papers, be promoted, all of that. Most of all, they were eager to follow up and to make sure others confirmed their results, preferably using some other technique so as to minimize instrumental biases. That is, any particular accelerator and its detectors tend to "see" events in a particular way. One needs to have the event confirmed by a different set of eyes.

The bubble chamber served as a powerful technique for discovering particles since many of the details of a close encounter could be seen and measured. Experiments using electronic detectors were generally aimed at more specific processes. Once a particle had made it to the list of confirmed objects, one could design specific collisions and specific devices to provide data on other properties, such as its lifetime — all the new particles were unstable — and decay modes. Into what does it disintegrate? A lambda decays into a proton and a pion; a sigma decays into a lambda and a pion; and so on. Tabulate, organize, try not to be overwhelmed by the data. These were the guidelines to sanity as the subnuclear world exposed deeper and deeper complexity. Collectively all the Greek-letter particles created in strong-force collisions were called *hadrons* — Greek for heavy — and there were hadrons by the hundreds, literally. This was not what we wanted. Instead of a single, tiny, uncuttable particle, the search for the Democritan a-tom had turned up hundreds of heavy, very cuttable particles. Disaster! We learned from our biology colleagues what to do when you don't know what to do: classify! And this we did with abandon. The results — and consequences — of this classification are taken up in the next chapter.

THREE FINALES: TIME MACHINE, CATHEDRALS, AND THE ORBITING ACCELERATOR

We close this chapter with a new view of what actually happens in accelerator collisions. This view comes to us courtesy of our colleagues in astrophysics. (There is a small but very funny group of astrophysicists ensconced at Fermilab.) These people assure us — and we have no reason to doubt them — that the world was created about 15 billion years ago in a cataclysmic explosion, the Big Bang. In the earliest instants after creation, the infant universe was a hot, dense soup of primordial particles colliding with one another with energies (equivalent to temperatures) vastly higher than anything we can imagine reproducing, even with acute megalomania, double time. But the universe is cooling as it expands. At some point, about 10^{-12} seconds after creation, the average energy of the particles in the hot universe soup was reduced to 1 trillion electron volts, or 1 TeV, about the same energy that Fermilab's Tevatron produces in each beam. Thus we can look at accelerators as time machines. The Tevatron replicates, for a brief instant during head-on collisions of protons, the behavior of the entire universe at age "a millionth of a millionth of a second." We can calculate the evolution of the universe if we know the physics of each epoch and the conditions handed to it by the previous epoch.

This time-machine application is really a problem for the astros. Under normal circumstances, we particle physicists would be amused and flattered but unconcerned about how accelerators mimic the early universe. In recent years, however, we've begun to see the link. Farther back in time, where the energies are considerably higher than 1 TeV — the limit of our present accelerator inventory — lies a secret that we need. This earlier, hotter universe contains a vital clue to the lair of the God Particle.

Accelerator as time machine — the astrophysics connection — is one view to consider. Another connection comes from Robert Wilson, the cowboy accelerator-builder, who wrote:

> Familiarly enough, both aesthetic and technical considerations were inextricably combined [in the design of Fermilab]. I even found, emphatically, a strange similarity between the cathedral and the accelerator: The one structure was intended to reach a soaring height in space; the other is intended to reach a comparable height in energy. Certainly the aesthetic appeal of both structures is primarily technical. In the cathedral we see it in the functionality of the ogival arch construction, the thrust and then the counterthrust so vividly and beautifully expressed, so dramatically used. There is a technological aesthetic in the accelerator, too. There is a spirality of the orbits. There is an electrical thrust and a magnetic counterthrust. Both work in an ever upward surge of focus and function until the ultimate expression is achieved, but this time in the energy of a shining beam of particles.
>
> Thus carried away, I looked into cathedral building a bit further. I found a striking similarity between the tight community of cathedral builders and the community of accelerator builders: Both of them were daring innovators, both were fiercely competitive on national lines, but yet both were basically internationalists. I like to compare the great Maître d'Oeuvre, Suger of St. Denis, with Cockcroft of Cambridge; or Sully of Notre-Dame with Lawrence of Berkeley; and Villard de Honnecourt with Budker of Novosibirsk.

To which I can only add that there is this deeper connection: both cathedrals and accelerators are built at great expense as a matter of faith. Both provide spiritual uplift, transcendence, and, prayerfully, revelation. Of course, not all cathedrals worked.

One of the glorious moments in our business is the scene in a crowded control room, where the bosses, on this special day, are at the console, staring at the screens. Everything is in place. The labor of so many scientists and engineers for so many years is now about to hatch as the beam is traced from the hydrogen bottle through the intricate viscera . . . It works! Beam! In less time than you can say hooray, the champagne is poured into Styrofoam cups, jubilation and ecstasy written on all faces. In our holy metaphor I see the workmen lowering the last gargoyle into place as priests, bishops, cardinals, and the requisite hunchback stand tensely around the altar to see if it works.

One must consider the aesthetic qualities of an accelerator as well as its GeVs and other technical attributes. Thousands of years hence, archaeologists and anthropologists may judge our culture by our accelerators. After all, they are the largest machines our civilization has ever built. Today we visit Stonehenge or the Great Pyramids, and we marvel first at their beauty and at the technological achievement of building them. But they had a scientific purpose as well; they were crude "observatories" for tracking astronomical bodies. So we must also stand in awe of how ancient cultures were driven to erect grand structures in order to measure the movements of the heavens in an attempt to understand and to live in harmony with the universe. Form and function combined in the pyramids and Stonehenge to allow their creators to seek scientific truths. Accelerators are our pyramids, our Stonehenge.

The third finale has to do with the man Fermilab is named for, Enrico Fermi, one of the most famous physicists of the 1930s, '40s, and '50s. He was Italian by birth, and his work in Rome was marked by brilliant advances in both experiment and theory and by a crowd of exceptional students gathered around him. He was a dedicated and gifted teacher. Awarded the Nobel Prize in 1938, he used the occasion to escape from fascist Italy and settle in the U.S.

His popular fame stems from heading up the team that built the first chain-reacting nuclear pile in Chicago during World War II. At the University of Chicago after the war he again gathered a brilliant group of both theoretical and experimental students. Fermi's students from both his Rome period and his Chicago period dispersed around the world, winning top positions and prizes everywhere. "You can tell a good teacher by how many of his students win Nobel Prizes," goes an ancient Aztec saying.

In 1954 Fermi gave his retiring address as president of the American Physical Society. With a mixture of respect and satire, he predicted that in the near future we would build an accelerator in orbit around the earth, making use of the natural vacuum of space. He also cheerfully noted that it could be built with the combined military budgets of the United States and the USSR. Using supermagnets and my pocket cost estimator, I get 50,000 TeV for a cost of $10 trillion, not including quantity discounts. What better way to return the world to sanity than by beating swords into accelerators?

· Interlude C ·

HOW WE VIOLATED PARITY IN A WEEKEND . . . AND DISCOVERED GOD

I cannot believe God is a weak left-hander.

— Wolfgang Pauli

LOOK AT YOURSELF in a mirror. Not too bad, hey? Suppose you raise your right hand, and your image in the mirror also raises its right hand! What? Can't be. You mean left! You'd clearly be in a state of shock if the wrong hand went up. This has never happened with people, as far as we know. But an equivalent act did occur with a fundamental particle called a muon.

Mirror symmetry had been tested in the laboratory over and over again. The scientific name for mirror symmetry is parity conservation. This is the story of an important discovery, and also of how progress oftentimes involves the killing of an exquisite theory by an ugly fact. It all started at lunch on Friday and was over by about 4 A.M. the following Tuesday morning. A very profound conception of how nature behaved turned out to be a (weak) misconception. In a few intense hours of data taking, our understanding of the way the universe is constructed was changed forever. When elegant theories are disproven, disappointment sets in. It appears that nature is clumsier, more ponderous, than we had expected. But our depression is tempered by the faith that when all is known, a deeper beauty will be revealed. And so it was with the downfall of parity in a few days of January 1957 in Irvington-on-Hudson, twenty miles north of New York City.

Physicists love symmetry because it has a mathematical and intuitive beauty. Symmetry in art is exemplified by the Taj Mahal or a

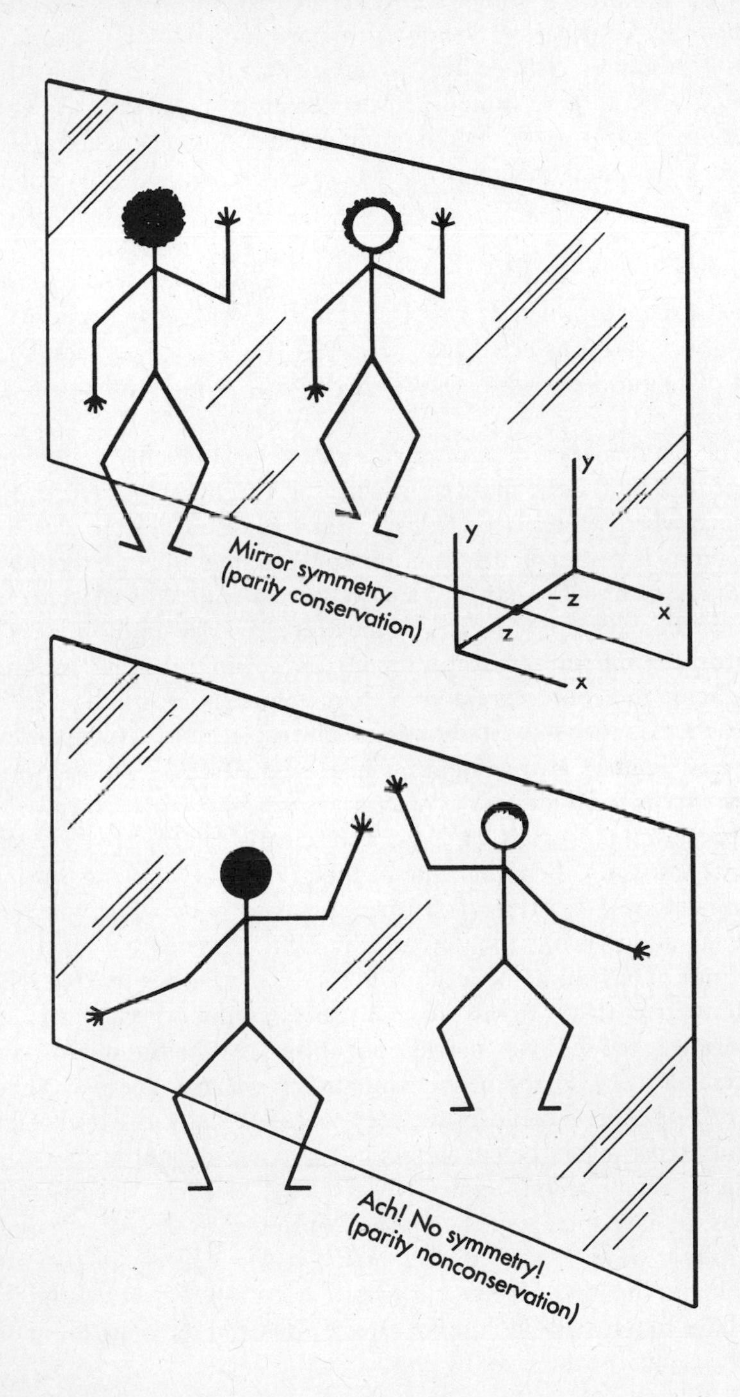
Mirror symmetry
(parity conservation)
y
y
−z
z
x
x
Ach! No symmetry!
(parity nonconservation)

Greek temple. In nature, shells, simple animals, and crystals of various kinds exhibit symmetrical patterns of great beauty, as does the almost perfect bilateral symmetry of the human body. The laws of nature contain a rich set of symmetries that for years, at least before January 1957, were thought to be absolute and perfect. They have been immensely useful in our understanding of crystals, large molecules, atoms, and particles.

THE EXPERIMENT IN THE MIRROR

One of these symmetries was called mirror symmetry, or parity conservation, and it asserted that nature — the laws of physics — could not distinguish between events in the real world and those in the mirror.

The mathematically appropriate statement, which I'll give for the record, is that the equations describing the laws of nature do not change when we replace the z-coordinates of all objects with $-z$. If the z-axis is perpendicular to a mirror, defining a plane, this replacement is exactly what happens to any system when it is reflected in the mirror. For example, if you, or an atom, are 16 units in front of a mirror, the mirror shows the image as 16 units behind the mirror. Replacing the coordinate z with $-z$ creates a mirror image. If, however, the equations are invariant to this replacement (for example, if the coordinate z always appears in the equation as z^2), then mirror symmetry is valid and parity is conserved.

If one wall of a lab is a mirror, and scientists in the lab are carrying out experiments, then their mirror images will be carrying out mirror images of these experiments. Is there any way of deciding which is the true lab and which is the mirror lab? Could Alice know where she is (in front of or behind the looking glass) by some objective test? Could a committee of distinguished scientists examining a videotape of an experiment tell if it was carried out in the real or the mirror lab? In December of 1956 the unequivocal answer was no. There was no way a panel of experts could prove they were watching the mirror image of the experiments being conducted in the real laboratory. At this point a perceptive innocent might say, "But look, the scientists in this movie all have their buttons on the left side of their coats. It must be the mirror view." "No," the scientists answer, "that is just a custom; nothing in the laws of nature insists that buttons be on the right side. We have to put aside all human affectations and see if anything in our movie is against the laws of physics."

So before January 1957 no such violations had been seen in the mirror-image world. The world and its mirror image were equally valid descriptions of nature. Anything that was happening in the mirror space could in principle and practice be replicated in the laboratory space. Parity was useful. It helped us classify molecular, atomic, and nuclear states. It also saves work. If a perfect human stands, disrobed and half concealed by a vertical screen, by studying the half that you do see, you can pretty much know what is behind the screen. Such is the poetry of parity.

The "downfall of parity," as the events of January 1957 were later described, is a quintessential example of how physicists think, how they adapt to shock, how theory and mathematics bend to the winds of measurement and observation. What is far from typical about this story is the speed and relative simplicity of the discovery.

THE SHANGHAI CAFE

Friday, January 4, 12 noon. Friday was our traditional Chinese-lunch day, and the faculty of the Columbia University Physics Department gathered outside the office of Professor Tsung Dao Lee. Between ten and fifteen physicists trooped down the hill from the 120th Street Pupin Physics Building to the Shanghai Café on 125th and Broadway. The lunches started in 1953, when Lee arrived at Columbia from the University of Chicago with a fairly new Ph.D. and a towering reputation as a theoretical superstar.

What characterized the Friday lunches was uninhibited noisy conversations, sometimes three or four simultaneously, punctuated by the very satisfactory slurping of winter melon soup and the sharing out of the dragon meat phoenix, shrimp balls, sea cucumbers, and other spicy exotica of northern Chinese cuisine, not yet trendy in 1957. Already on the walk down, it was clear that this Friday the theme would be parity and the hot news from our Columbia colleague C. S. Wu, who was conducting an experiment at the Bureau of Standards in Washington.

Before entering into the serious business of lunch discussion, T. D. Lee carried out his weekly chore of composing the lunch menu on a small pad offered by the respectful waiter-manager. T. D. composes a Chinese menu in the grand manner. It is an art form. He glances at the menu, at his pad, fires a question in Mandarin at the waiter, frowns, poises his pencil over the pad, carefully calligraphs a few symbols. Another question, a change in one symbol, a glance at the embossed tin ceiling for divine guidance, and then a flurry of rapid writing. A final review: both hands are poised over the pad, one with fingers outstretched, conveying the blessings of the pope on the assembled throng, the other holding the stub of a pencil. Is it all there? The yin and the yang, the color, texture, and flavor in proper balance? Pad and pencil are handed to the waiter, and T. D. plunges into the conversation.

"Wu telephoned and said her preliminary data indicated a huge effect!" he said excitedly.

•

Let's return to the laboratory (the real world as She made it) with one wall a mirror. Our normal experience is that whatever we hold up to the mirror, whatever experiments we do in the lab — scattering, production of particles, gravity experiments like Galileo's — all the mirror-lab reflections will conform to the same laws of nature that govern in the lab. Let's see how a violation of parity would show up. The simplest objective test of handedness, one we could communicate to inhabitants of the planet Twilo, employs a right-handed machine screw. Facing the slotted end, turn the screw "clockwise." If the screw advances into a block of wood, it is defined as right-handed. Obviously the mirror view shows a left-handed screw because the mirror guy is turning it counterclockwise, but it still advances. Now suppose we live in a world so curious (some Star Trek universe) that it is impossible — against the laws of physics — to make a left-handed screw. Mirror symmetry would break down; the mirror image of a right-handed screw could not exist; and parity would be violated.

This is the lead-in to how Lee and his Princeton colleague Chen Ning Yang proposed to examine the validity of the law for weak-force processes. We need the equivalent of a right-handed (or left-handed) particle. Like the machine screw, we need to combine a *rotation* and a direction of *motion*. Consider a spinning particle — call it a muon. Picture it as a cylinder spinning around its axis. We have *rotation*. Since the ends of the cylinder-muon are identical, we cannot say whether it is spinning clockwise or counterclockwise. To see this, place the cylinder between you and your favorite antagonist. While you swear it is rotating to the right, clockwise, she insists that it is rotating to the left. And there is no way to resolve the dispute. This is a parity-conserving situation.

The genius of Lee and Yang was to bring in the weak force (which they wanted to examine) by watching the spinning particle decay. One decay product of the muon is an electron. Suppose nature dictates that the electron comes off only one end of the cylinder. This gives us a *direction.* And we can now determine the *sense* of rotation — clockwise or counterclockwise — because one end is defined (the electron comes off here). This end plays the role of the point of the machine screw. If the sense of spin rotation relative to the electron is right-handed, like the sense of the machine screw relative to its point, we have defined a right-handed muon. Now *if* these particles always decay in such a way as to define right-handedness, we have a particle process that violates mirror symmetry. This is seen if we align the spin axis of the muon parallel to our mirror. The mirror image is a left-handed muon — *which doesn't exist.*

•

The rumors about Wu had begun over the Christmas break, but the Friday after New Year was the first gathering of the Physics Department since the holidays. In 1957 Chien Shiung Wu, like me a professor of physics at Columbia, was quite a well-established experimental scientist. Her specialty was the radioactive decay of nuclei. She was tough on her students and postdocs, exceedingly energetic, careful in evaluating her results, and much appreciated for the high quality of the data she published. Her students (behind her back) called her Generalissimo Mme. Chiang Kai-shek.

When Lee and Yang challenged the validity of parity conservation in the summer of 1956, Wu went into action almost immediately. She selected as the object of her study the radioactive nucleus of cobalt-60, which is unstable. The cobalt-60 nucleus changes spontaneously into a nucleus of nickel, a neutrino, and a positive electron (a positron). What one "sees" is that the cobalt nucleus suddenly shoots off a positive electron. This form of radioactivity is known as beta decay, because the electrons, whether negative or positive, emitted during the process were originally called beta particles. Why does this happen? Physicists call it a weak interaction, and think of a force operating in nature that generates these reactions. Forces not only push and pull, attract and repel, but are also capable of generating changes of species, such as the process of cobalt changing to nickel and emitting leptons. Since the 1930s a large number of reactions have been attributed to the weak force. The great Italian-American Enrico Fermi was the first to put the weak force into a mathematical form, enabling

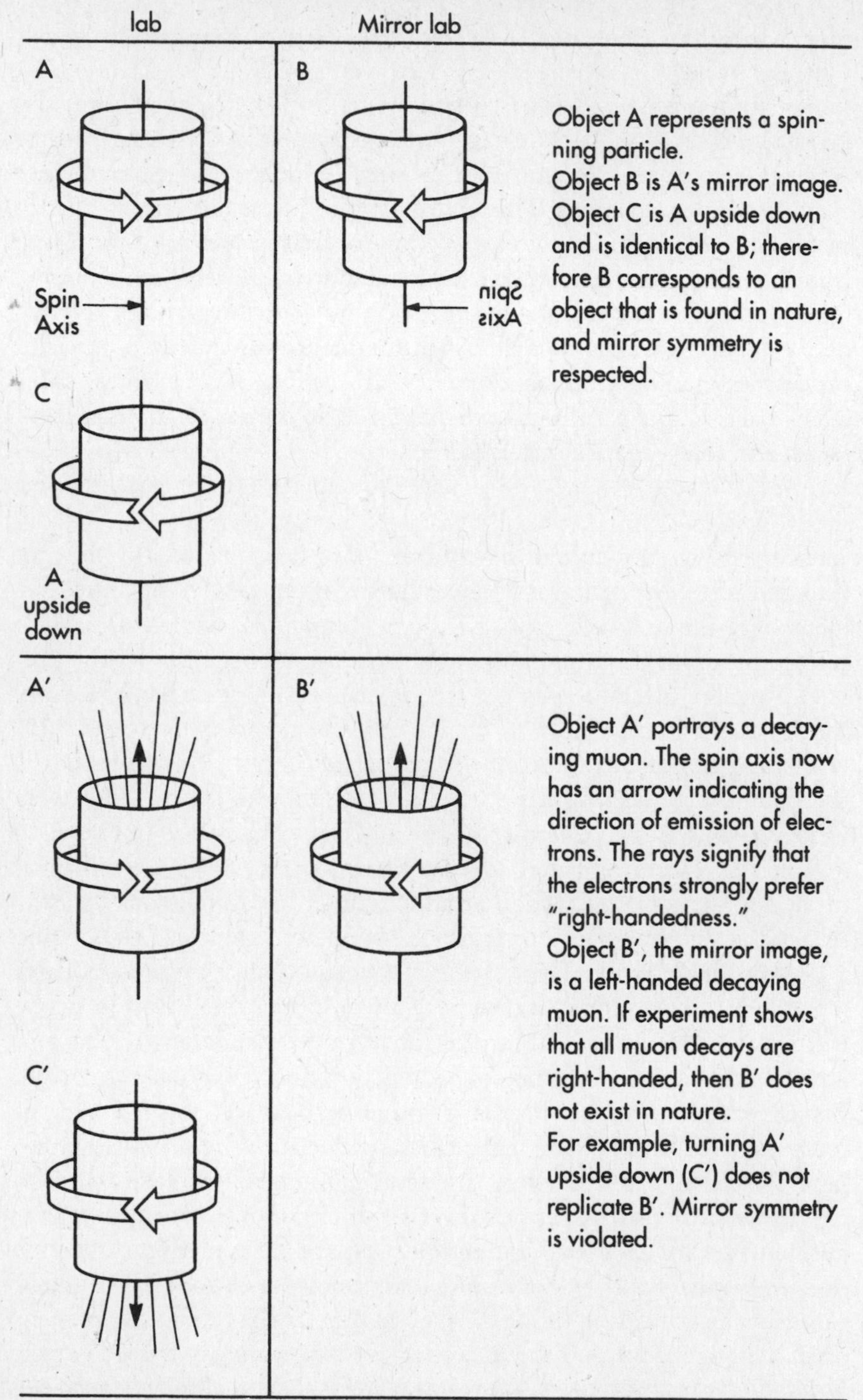
lab
Mirror lab
A
B
Spin Axis
Spin Axis
C
A upside down
A′
B′
C′
Object A represents a spinning particle.
Object B is A's mirror image.
Object C is A upside down and is identical to B; therefore B corresponds to an object that is found in nature, and mirror symmetry is respected.
Object A′ portrays a decaying muon. The spin axis now has an arrow indicating the direction of emission of electrons. The rays signify that the electrons strongly prefer "right-handedness."
Object B′, the mirror image, is a left-handed decaying muon. If experiment shows that all muon decays are right-handed, then B′ does not exist in nature.
For example, turning A′ upside down (C′) does not replicate B′. Mirror symmetry is violated.

him to predict many details of reactions such as that which occurs with cobalt-60.

Lee and Yang, in their 1956 paper called "The Question of Parity Conservation in the Weak Force," selected a number of reactions and examined the experimental implications of the possibility that parity — mirror symmetry — was not respected by the weak force. They were interested in the directions in which the emerging electron is ejected from a spinning nucleus. If the electron favored one direction over another, that would be like dressing the cobalt nuclei in buttoned shirts. One could tell which was the real experiment, which was a mirror image.

What is it that differentiates a great idea from a routine piece of scientific work? Analogous questions can be asked about a poem, a painting, a piece of music — in fact, gasp and choke, even a legal brief. In the case of the arts, it is the test of time that ultimately decides. In science, experiment determines whether an idea is "right." If it is brilliant, a new area of research is opened, a host of new questions are generated, and a large number of old questions are put to bed.

T. D. Lee's mind worked in subtle ways. In ordering a lunch or in commenting on some old Chinese pottery or on the abilities of a student, his remarks all had hard edges, like a cut precious stone. In Lee and Yang's parity paper (I didn't know Yang that well), this crystalline idea had many sharp sides. To question a well-established law of nature takes a lot of Chinese chutzpah. Lee and Yang realized that all of the vast amount of data that had led to the "well-established" parity law was irrelevant to that piece of nature that caused radioactive decay, the weak force. This was another brilliant, sharp edge: here, for the first time to my knowledge, the different forces of nature were permitted to have different conservation laws.

Lee and Yang rolled up their sleeves, poured perspiration on their inspiration, and examined a large number of radioactive decay reactions that represented likely candidates for a test of mirror symmetry. Their paper provided laboriously detailed analyses of likely reactions so dumb experimentalists could test the validity of mirror symmetry. Wu devised a version of one of these, using the cobalt reaction. The key to her approach was to make sure that the cobalt nuclei — or at least a very good fraction of them — were spinning in the same sense. This, Wu argued, could be ensured by running the cobalt-60 source at very low temperatures. Wu's experiment was extremely elaborate, requiring hard-to-find cryogenic apparatus. This led her to the Bureau of Standards, where the technique of spin alignment was well developed.

•

The next to last course that Friday was a large carp braised in black bean sauce with scallions and leeks. It was during this serving that Lee reiterated the key information: the effect Wu was observing was large, more than ten times larger than expected. The data were rumored, tentative, and therefore very preliminary but (T. D. served me the fish head, knowing I liked the cheeks) if the effect was that large, it was just what we would expect if neutrinos were two-component . . . I lost the rest of his excitement because an idea had started growing in my own mind.

After lunch there was a seminar, some departmental meetings, a social tea, and a colloquium. In all of these activities I was distracted, bugged by the notion that Wu was seeing a "big effect." From Lee's talk at Brookhaven in August I remembered that the effects produced by the suggested violations of parity when pions and muons decayed were assumed to be minuscule.

Big effect? I had looked briefly in August at the "pi-mu" (pion-muon) chain of decays and had realized that to design a reasonable experiment one would need to have parity violation in *two* sequential reactions. I kept recalling the calculations we had done in August before deciding that the experiment was borderline or less in chances of success. However, if the effect was large . . .

By 6 P.M. I was in my car heading north to dinner at home in Dobbs Ferry and then to a quiet evening shift with my graduate student at the nearby Nevis Lab in Irvington-on-Hudson. The Nevis 400 MeV accelerator was a workhorse for producing and studying the properties of mesons, relatively new particles in the 1950s. In those happy days, there were very few mesons to worry about, and Nevis worried about pions as well as muons.

At Nevis we had intense beams of pions coming off a target bombarded by protons. The pions were unstable, and during their flight from the target, out of the accelerator, through the shielding wall, and out into the experimental hall, some 20 percent would undergo the weak decay into a muon and a neutrino.

$$\pi \rightarrow \mu + \nu \quad \text{(in flight)}$$

The muons generally traveled in the same direction as the parent pion. If the parity law was violated, there would be an excess of muons with spin axis aligned in the direction of the motion of the muon over muons with spin axes pointing, say, opposite to the flight. If the effect was *large,* nature could be providing us with a sample of particles all spinning in the same sense. This is the situation Wu had to organize by cooling cobalt-60 to extremely low temperatures in a magnetic field. The key was to watch those muons whose direction of spin axis was known to decay into an electron plus some neutrinos.

THE EXPERIMENT

The heavy traffic on the drive north on the Saw Mill River Parkway on Friday evening tends to obscure the lovely forested hills that line this road, which winds along the Hudson River, past Riverdale, Yonkers, and points north. It was somewhere on this road that the implications of the "big effect" possibility dawned upon me. In the case of a spinning object, if any direction of the spin axis is favored in the decay, that is the effect. A small effect might be 1,030 electrons emitted in one direction relative to the spin axis versus 970 in the other, and this would be very difficult to determine. But a big effect, say 1,500 versus 500, would be much easier to find, and the same fortunate bigness would help in organizing the spins of the muons. To do the experiment, we need a sample of muons all spinning in the same direction. Since they will be moving from the cyclotron to our apparatus, the direction of motion of the muons becomes a reference for the muon spin. We need most of the muons to be right-handed (or left, it doesn't matter), now using the direction of motion as a "thumb." Muons will arrive, pass through a few counters, and stop in a carbon block. Then we count how many electrons are emerging in the direction in which the muons were moving against how many are emerging in the opposite direction. A significant difference would be proof of parity violation. Fame and fortune!

Suddenly, my usual Friday night calm was destroyed by the thought that we could trivially do the experiment. My graduate student, Marcel Weinrich, had been working on an experiment involving muons. His setup, with simple modifications, could be used to look for a big effect. I reviewed the way muons were created in the Columbia accelerator. In this I was a sort of expert, having worked with John Tinlot on the design of external pion and muon beams some years ago when I was a brash graduate student and the machine was brand new.

In my mind I visualized the entire process: the accelerator, a 4,000-ton magnet with circular pole pieces about twenty feet in diameter, sandwiches a large stainless steel evacuated box, the vacuum chamber. A stream of protons is injected via a tiny tube in the center of the magnet. The protons spiral outward as strong radio-frequency voltages kick them on each turn. Near the end of their spiral trip, the particles have an energy of 400 MeV. Near the edge of the chamber, almost at the place where we would run out of magnet, a small rod carrying a piece of graphite waits to be bombarded by the energetic protons. Their 400 million volts is enough to create new particles — pions — as they collide with a carbon nucleus in the graphite target.

In my mind's eye I could see the pions spewing forward from the momentum of the proton's impact. Born between the poles of the powerful cyclotron magnet, they sweep in a gradual arc toward the outside of the accelerator and do their dance of disappearance; muons appear in their place, sharing the original motion of the pions. The rapidly vanishing magnetic field outside the pole pieces helps to sweep the muons through a channel in a ten-foot-thick concrete shield and into the experimental hall where *we* are waiting.

In the experiment Marcel had been setting up, muons would be slowed down in a three-inch-thick filter and then be brought to rest in one-inch-thick blocks of various elements. The muons would lose their energy via gentle collisions with the atoms in the material and, being negative, would finally be captured by the positive nuclei. Since we did not want anything to influence the muons' direction of spin, capture into orbits could be fatal, so we switched to positive muons. What would positively charged muons do? Probably just sit there in the block spinning quietly until their time came to decay. The material of the block would have to be chosen carefully, and carbon seemed appropriate.

Now comes the key thought of the driver heading north on a Friday in January. If all (or almost all) of the muons, born in the decay of pions, could somehow have their spins aligned in the same direction, it would mean that parity is violated in the pion-to-muon reaction and violated strongly. A big effect! Now suppose the axis of spin remained parallel to the direction of motion of the muon as it swept through its graceful arc to the outside of the machine, through the channel. (If g is close to 2, this is exactly what happens.) Suppose further that the innumerable gentle collisions with carbon atoms, which gradually slowed the muon, did not disturb this relationship of spin and direction. If all this were indeed to happen, *mirabile dictu!* I would have a sample of muons coming to rest in a block all spinning in the same direction!

The muon's lifetime of two microseconds was convenient. Our experiment was already set up to detect the electrons that emerge from the decaying muons. We could try to see if equal numbers of electrons emerged in the two directions defined by the spin axis. The mirror symmetry test. If the numbers are not equal, parity is dead! And I killed it! Argggghh!

It looked as if a confluence of miracles would be needed for a successful experiment. Indeed, it was just this sequence that had discouraged us in August when Lee and Yang read their paper, which implied small effects. One small effect can be overcome with patience, but two sequential small effects — say, one percent of one percent — would make the experiment hopeless. Why two sequential small effects? Remember, nature has to provide pions that decay into muons, mostly spinning in the same sense (miracle number one). And the muons have to decay into electrons with an observable asymmetry relative to the muon spin axis (miracle number two).

By the Yonkers toll booth (1957, toll five cents) I was quite excited. I felt pretty sure that if the parity violation was large, the muons would be polarized (spins all pointing in the same direction). I also knew that the magnetic properties of the muon's spin were such as to "clamp" the spin in the direction of the particle's motion under the influence of the magnetic field. I was less certain of what happens when the muon enters the energy-absorbing graphite. If I was wrong, the muon spin axis would be twisted in a wide assortment of directions. If that happened there would be no way to observe the emission of electrons relative to the spin axis.

Let's go over that again. The decay of pions generates muons that spin in the direction in which they are moving. This is part of the miracle. Now we have to stop the muons so we can observe the direction of the electrons they emit upon decay. Since we know the direction of motion just before they hit the block of carbon, if nothing screws them up we know the spin direction when they stop and when they decay. Now all we have to do is rotate our electron detection arm about the block where the muons are at rest to check for mirror symmetry.

My palms started to sweat as I reviewed what we had to do. The counters all existed. The electronics that signaled the arrival of the high-energy muon and the entrance into the graphite block of the now slowed muon were already in place and well tested. A "telescope" of four counters for detecting the electron that emerged after muon decay also existed. All we had to do was mount these on a board of some sort that we could pivot around the center of the stopping block. One or two hours' work. Wow! I decided that it would be a long night.

When I stopped at home for a quick dinner and some bantering with the kids, a telephone call came from Richard Garwin, a physicist with IBM. Garwin was doing research in atomic processes at the IBM research labs, which were then just off the Columbia campus. Dick hung around the Physics Department a lot, but he had missed the Chinese lunch and wanted to know the latest on Wu's experiment.

"Hey, Dick, I've got a great idea on how we can test for parity violation in the simplest way you can imagine." I explained hastily and said, "Why don't you drive over to the lab and give us a hand?" Dick lived nearby in Scarsdale. By 8 P.M. we were disassembling the apparatus of one very confused and upset graduate student. Marcel saw his Ph.D. thesis experiment being taken apart! Dick was assigned the job of thinking through the problem of rotating the electron telescope so we could determine the distribution of electrons around the assumed spin axis. This wasn't a trivial problem, since wrestling the telescope around could change the distance to the muons and thus alter the yield of detected electrons.

It was then that the second key idea was invented, by Dick Garwin. Look, he said, instead of moving this heavy platform of counters around, let's leave it in place and turn the muons in a magnet. I gasped as the simplicity and elegance of the idea penetrated. Of course! A spinning charged particle is a tiny magnet and will turn like a compass needle in a magnetic field, except that the mechanical forces acting on the muon-magnet make it rotate continuously. The idea was so simple it was profound.

It was a piece of cake to calculate the value of the magnetic field needed to turn the muons through 360 degrees in a reasonable time. What is a reasonable time to a muon? Well, the muons are decaying into electrons and neutrinos with a half-life of 1.5 microseconds. That is, half of the muons have given their all in 1.5 microseconds. If we turned the muons too slowly, say 1 degree per microsecond, most of the muons would have disappeared after being rotated through a few degrees and we wouldn't be able to compare the zero-degree and 180-degree yield — that is, the number of electrons emitted from the "top" of the muon as opposed to the "bottom," the whole point of our experiment. If we increased the turning rate to, say, 1,000 degrees per microsecond by applying a strong magnetic field, the distribution would whiz past the detector so fast we would have a blurred-out result. We decided that the ideal rate of turning would be about 45 degrees per microsecond.

We were able to obtain the required magnetic field by winding a few hundred turns of copper wire on a cylinder and running a current of a few amperes through the wire. We found a Lucite tube, sent Marcel to the stockroom for wire, cut the graphite stopping block down so it could be wedged inside the cylinder, and hooked the wires to a power supply that could be controlled remotely (there was one on the shelf). In a blur of late-night activity, we had everything ready by midnight. We were in a hurry because the accelerator was always turned off at 8 A.M. Saturday for maintenance and repairs.

By 1 A.M. the counters were recording data; accumulation registers recorded the number of electrons emitted at various directions. But remember, with Garwin's scheme, we didn't measure these angles directly. The electron telescope remained stationary while the muons or, rather, their spin axis vectors, were rotated in a magnetic field. So the electrons' *time* of arrival now corresponded to their direction. By recording the time, we were recording the direction. Of course, we had lots of problems. We badgered the accelerator operators to give us as many protons hitting the target as possible. All the counters registering the muons coming in and stopping had to be adjusted. The control of the small magnetic field applied to the muons had to be checked.

After a few hours of data taking, we saw a remarkable difference in the counts of electrons emitted at zero degrees and those emitted at 180 degrees relative to the spin. The data were very crude, and we mixed excited optimism with skepticism. When we examined the data at eight the next morning, our skepticism was confirmed. The data now were much less convincing, not really inconsistent with the hypothesis that all directions of emission were equivalent — a predictor of mirror symmetry. We had pleaded with the accelerator operators to give us an additional four hours, but to no avail. Schedules are schedules. Discouraged, we walked down to the accelerator room, where the apparatus was set up. There we were greeted by a small catastrophe. The Lucite cylinder on which we had wound the wire had become warped due to the heat produced by the current in the wires. This warping had permitted the stopping block to fall. Obviously, the muons were no longer in the magnetic field we had designed for them. After some recriminations (blame the graduate student!) we cheered up. Our original impression might still be correct!

We made a plan for the weekend. Design a proper magnetic field. Think about increasing the data rate by increasing the number of muons stopping and the fraction of the decay electrons counted. Think about what happens to the positively charged muons in their collisions on the way down to rest and in the microseconds in which they sit in the lattice of carbon atoms. After all, if a positive muon managed to capture one of the many electrons that are free to move about in graphite, the electron could easily depolarize (mess up the spin of) the muon so that they would not all be doing the same thing in lockstep.

The three of us went home to sleep for a few hours before reassembling at 2 P.M. We worked through the weekend, each at an assigned task. I managed to recalculate the motion of the muon from birth as it is kicked forward by its decaying pion parent, through its sweep toward the channel and through the concrete wall into our apparatus. I kept track of spin and direction. I assumed maximum violation of mirror symmetry so that all the muons would be spinning precisely along in the direction of their motion. Everything indicated that if the violation was large, even half of maximum, we should see an oscillating curve. This not only would prove parity violation but would give us a numerical result as to how much parity was violated, from 100 percent down to (no! no!) zero. Anyone who tells you that scientists are dispassionate and coldly objective is crazy. We desperately yearned to see parity violated. Parity was not a young lady, and we weren't teenagers, but we lusted to make a discovery. The test of scientific objectivity is not to let the passion influence the methodology and the self-criticism.

Eschewing the Lucite cylinder, Garwin wound a coil directly on a new piece of graphite and tested the system at currents twice as high as we would need. Marcel rearranged the counters, improved the alignment, moved the electron telescope closer to the stopping block, tested, and improved the efficiency of all counters, all the while praying that something publishable would come out of this frantic activity.

The work went slowly. By Monday morning, some news of our intense activity had leaked out to the operator crew and to some of our colleagues. The accelerator maintenance gang found some serious problems in the machine, so Monday was out — no beam until Tuesday, 8 A.M. at best. Okay, more time to fume, fuss, check. Colleagues from the Columbia campus arrived at Nevis, curious as to what we were up to. One clever young man who had been at the Chinese lunch asked a few questions and, by my disingenuous answers, deduced that we were trying the parity experiment.

"It'll never work," he assured me. "The muons will depolarize as they lose energy in the graphite filter." I was easily depressed but not discouraged. I remembered my mentor, the great Columbia savant I. I. Rabi, telling us: spin is a very slippery thing.

About 6 P.M. on Monday, ahead of schedule, the machine began to show signs of life. We hastened our preparations, checking all the devices and arrangements. I noticed that the target with its elegant copper wire wrapping, positioned on a four-inch slab, looked a bit low. Some squinting through a surveying scope convinced me, and I looked for something that would raise it an inch or so. Over in the corner I saw a Maxwell House coffee can partially filled with wood screws, and I substituted it for the four-inch slab. Perfect! (When the Smithsonian Institution later wanted the coffee can in order to replicate the experiment, we couldn't find it.)

The loudspeaker announced that the machine was about to be turned on and that all experimenters must leave the accelerator room (or get fried). We scrambled up the steep iron staircase and across the parking lot to the lab building, where the cables from the detectors were connected to electronic racks containing circuitry, scalers, oscilloscopes. Garwin had gone home hours ago, and I sent Marcel to get some dinner while I started a checkout procedure on the electronic signals arriving from the detectors. A large, thick lab notebook was used to note all relevant information. It was gaily embellished with graffiti — "Oh shit!" "Who the hell forgot to turn off the coffee pot?" "Your wife called" — as well as the necessary record of things to do, things done, conditions of the circuits. ("Watch scaler No. 3. It tends to spark and miss counts.")

By 7:15 P.M. the proton intensity was up to standard and the pion-producing target was moved remotely into position. Instantly, the scalers began registering arriving particles. I looked at the crucial row of scalers that would register the number of electrons emitted at various intervals after the muons had stopped. The numbers were still very small: 6, 13, 8 . . .

Garwin arrived at about 9:30 P.M. I decided to get some sleep and relieve him at 6 the next morning. I drove home very slowly. I had been up for about twenty hours and was too tired to eat. It seemed as if I had just hit the pillow when the phone rang. The clock said 3 A.M. It was Garwin. "You'd better come in. We've done it!"

At 3:25 I parked at the lab and dashed in. Garwin had pasted paper strips of the scaler read-outs in the book. The numbers were devastatingly clear. More than twice as many electrons were emitted at zero degrees as at 180 degrees. Nature could tell the difference between a right-handed spin and a left-handed spin. By now the machine had come up to its best intensity, and the scaler registers were changing rapidly. The scaler corresponding to zero degrees was reading 2,560, the scaler corresponding to 180 degrees was reading 1,222. On a purely statistical basis this was overwhelming. The in-between scalers seemed satisfactorily in between. The implications of parity violation on this level were so vast . . . I looked at Dick. My breathing was becoming difficult, my palms were wet, my heartbeat accelerated, I felt lightheaded — many (not all!) of the symptoms of sexual arousal. This was big stuff. I began to make a checklist: what elements could fail in such a way as to simulate the result we were seeing? There were so many possibilities. We spent an hour, for example, checking the circuits used to count the electrons. No problem. How else could we test our conclusions?

Tuesday, 4:30 A.M. We asked the operator to shut down the beam. We ran down and physically rotated the electron telescope through 90 degrees. If we knew what we were doing, the pattern should shift by a time interval corresponding to 90 degrees. Bingo! The pattern shifted as we had predicted!

6 A.M. I picked up the telephone and called T. D. Lee. He answered after one ring. "T. D., we've been looking at the pi-mu-e chain and we now have a twenty-standard-deviation signal. The law of parity is dead." T. D.'s reaction squirted through the telephone. He asked rapid-fire questions: "What energy electrons? How did the asymmetry vary with electron energy? Was the muon spinning parallel to the direction of arrival?" To some questions we had answers. Others came later in the day. Garwin began drawing graphs and entering the

scaler readings. I made another list of things we had to do. At seven we started getting calls from Columbia colleagues who had heard. Garwin faded by eight. Marcel (temporarily forgotten!) arrived. By nine the room was crowded with colleagues, technicians, secretaries trying to find out what was going on.

It was hard to keep the experiment going. My breathing and sweating symptoms returned. We were the repository of new and profound information about the world. Physics was changed. And the violation of parity had given us a powerful new tool: polarized muons that were responsive to magnetic fields and whose spins could be tracked through the electron decay.

The phone calls from Chicago, California, and Europe came over the next three or four hours. People with particle accelerators in Chicago, Berkeley, Liverpool, Geneva, and Moscow swarmed to their machines like pilots rushing to their wartime battle stations. We continued the experiment and continued the process of checking our assumptions for a solid week, but we were desperately anxious to publish. We took data, in one form or another, twenty-four hours a day, six days a week, for the next six months. Data poured out. Other labs soon confirmed our results.

C. S. Wu was of course less than delighted by our clean, unequivocal result. We wanted to publish with her but, to her everlasting credit, she insisted she still needed a week to check her results.

It is difficult to express just how startling the results of this experiment were to the physics community. We had challenged — in fact, destroyed — a cherished belief, that nature exhibits mirror symmetry. In later years, as we shall see, other symmetries were also disproved. Even so, the experiment shook up many theorists, including Wolfgang Pauli, who made the famous statement "I cannot believe God is a weak left-hander." He didn't mean that God should be right-handed, but that She should be ambidextrous.

The annual meeting of the American Physical Society drew 2,000 physicists to the ballroom of the Hotel Paramount in New York on February 6, 1957. People hung from rafters. Front-page articles in all the major newspapers heralded the result. The *New York Times* published our press release verbatim, with pictures of particles and mirrors. But none of this matched the 3 A.M. feeling of mystical euphoria when two physicists came to know a new and profound truth.

· 7 ·

A-TOM!

> Yesterday three scientists won the Nobel Prize for finding the smallest object in the universe. It turns out that it's the steak at Denny's.
>
> — Jay Leno

THE 1950S AND '60S were great years for science in America. Compared to the much tougher 1990s, in the '50s anyone with a good idea and a lot of determination, it seemed, could get his idea funded. Perhaps this is as good a criterion for healthy science as any. The nation is still benefiting from the science that got done in these decades.

The flood of subnuclear structures opened up by the particle accelerator was as surprising as the heavenly objects revealed by Galileo's telescope. As in the Galilean revolution, mankind acquired new, previously unsuspected knowledge about the world. That this knowledge concerned inner rather than outer space made it no less profound. Pasteur's discovery of microbes and the invisible biological universe of microorganisms is an analogous event. The bizarre guess of our hero Democritus ("Guess?!" I hear him screeching. "Guess?!?!") was no longer even remarked upon. That there was a particle so small that it eluded the human eye was not a matter for further debate. Clearly, the search for the smallest particle called for extensions of the human eye: glasses, microscopes, now particle accelerators zooming down in quest of the true a-tom. And what we saw were hadrons, lots of hadrons, those Greek-letter particles created in the strong collisions induced by accelerator beams.

This is not to say that the proliferation of hadrons was an unalloyed pleasure. It did make for full employment, spreading the wealth so that the discoverers of new particles now made up a nonexclusive club. Want to find a brand-new hadron? Just wait for the next accelerator run. At a conference on the history of physics at Fermilab in 1986, Paul Dirac recounted how difficult it was for him to accept the consequences of his equation — the existence of a new particle, the positron, which Carl Anderson discovered a few years later. In 1927 it was counter to the ethos of physics to think so radically. When Victor Weisskopf remarked from the audience that in 1922 Einstein had speculated about the existence of a positive electron, Dirac waved his hand dismissively: "He was lucky." In 1930 Wolfgang Pauli had agonized before predicting the existence of the neutrino. He finally embraced the particle with great reluctance and only to favor a lesser evil, since nothing less was at stake than the principle of conservation of energy. Either the neutrino had to exist, or the conservation of energy had to go. This conservatism toward the introduction of new particles didn't last. As Professor Bob Dylan commented, the times they were a-changin'. Pioneer of the change in philosophy was theorist Hideki Yukawa, who began the process of freely postulating new particles to explain new phenomena.

In the 1950s and early '60s theorists were busy classifying the hundreds of hadrons, seeking patterns and meaning in this new layer of matter, and hounding their experimental colleagues for more data. These hundreds of hadrons were exciting, but they were a headache as well. Where was the simplicity we had been seeking since the days of Thales, Empedocles, and Democritus? There was an unmanageable zoo of these entities, and we were beginning to fear that their legions were infinite.

In this chapter, we shall see how the dream of Democritus, Boscovich, and others was finally realized. We will chronicle the construction of the *standard model*, which contains all the elementary particles needed to make all the matter in the universe, past or present, plus the forces that act upon these particles. In some ways it is more complex than Democritus's model, in which each form of matter had its own indivisible a-tom, and the a-toms joined together because of their complementary shapes. In the standard model, the matter particles bind to each other via three different forces carried by yet more particles. All of these particles interact with each other in an intricate kind of dance, which can be described mathematically but cannot be visualized. Yet in some ways the standard model is simpler than Democritus ever imagined. We don't need a separate a-tom for feta cheese, one for kneecaps, another for broccoli. There are only a small number of a-toms. Combine them in various ways, and you can make *anything*. We've already met three of these elementary particles, the electron, the muon, and the neutrino. Soon we'll meet the others and see how they all fit together.

This is a triumphant chapter, for we come to the end of the road in our search for a basic building block. In the fifties and early sixties, however, we were not feeling so sanguine about finally answering Democritus's riddle. Because of the hundred-hadron headache, the prospect of identifying a few elementary particles seemed pretty dim. Physicists were making much better progress in describing the forces of nature. Four were clearly recognized: gravity, the electromagnetic force, the strong force, and the weak force. Gravity was the domain of astrophysics, for it was too feeble to deal with in accelerator labs. This omission would come to haunt us later. But we were getting the other three forces under control.

The Electric Force

The 1940s had seen the triumph of a quantum theory of the electromagnetic force. The work of Paul Dirac in 1927 successfully blended quantum theory and special relativity in his theory of the electron. However, the marriage of quantum theory and electromagnetism, the electromagnetic force, was a stormy one, filled with stubborn problems.

The struggle to unite the two theories was known informally as the War Against Infinities, and by the mid-1940s it involved infinity on one side and, on the other, many of the brightest luminaries in physics: Pauli, Weisskopf, Heisenberg, Hans Bethe, and Dirac, as well as some new rising stars — Richard Feynman at Cornell, Julian Schwinger at Harvard, Freeman Dyson at Princeton, and Sin-itiro Tomonaga in Japan. The infinities came from this: simply described, when one calculated the value of certain properties of the electron, the answer, according to the new relativistic quantum theories, came out "infinite." Not just big, *infinite*.

One way to visualize the mathematical quantity called infinity is to think of the total number of integers — and then add one more. There is always one more. Another way, one that was more likely to appear in the calculations of these brilliant but deeply unhappy theorists, is to evaluate a fraction in which the denominator becomes zero. Most pocket calculators will politely inform you — usually with a series of EEEEEEs — that you have done something stupid. Earlier relay-driven mechanical calculators would go into a grinding cacophony that usually terminated in a dense puff of smoke. Theorists saw infinities as a sign that something was deeply wrong with the way the marriage between electromagnetism and quantum theory was being consummated — a metaphor we probably should not pursue, much as we are tempted. In any case, Feynman, Schwinger, and Tomonaga, working separately, achieved victory of a sort in the late 1940s. They finally overcame the inability to calculate the properties of charged particles such as the electron.

A major stimulus to this theoretical breakthrough came from an experiment carried out at Columbia by one of my teachers, Willis Lamb. In the early postwar years, Lamb taught most of the advanced courses and worked on electromagnetic theory. He also designed and carried out, using the wartime radar technology developed at Columbia, a brilliantly precise experiment on the properties of selected energy levels in the hydrogen atom. Lamb's data were to provide a test of some of the most subtle pieces of the newly minted quantum electromagnetic theory, which his experiment served to motivate. I'll skip the details of Lamb's experiment, but I want to emphasize that an experiment was seminal to the exciting creation of a workable theory of the electric force.

What emerged from the theorists was something called "renormalized quantum electrodynamics." Quantum electrodynamics, or QED, enabled theorists to calculate the properties of the electron, or its heavier brother the muon, to ten significant figures beyond the decimal point.

QED was a field theory, and thus it gave us a physical picture of how a force is transmitted between two matter particles, say, two electrons. Newton had problems with the idea of action-at-a-distance, as did Maxwell. What is the mechanism? One of the oh-so-clever ancients, a pal of Democritus's, no doubt, discovered the influence of the moon on the earth's tides and agonized over how that influence could manifest itself through the intervening void. In QED, the field is quantized, that is, broken down into quanta — more particles. These are not matter particles, however. They are particles *of the field.* They transmit the force by traveling, at the speed of light, between the two interacting matter particles. These are messenger particles, which in QED are called photons. Other forces have their own distinct messengers. Messenger particles are the way we visualize forces.

VIRTUAL PARTICLES

Before we go on, I should explain that there are two manifestations of particles: real and virtual. Real particles can travel from point A to point B. They conserve energy. They make clicks in Geiger counters. Virtual particles do none of these things, as I mentioned in Chapter 6. Messenger particles — force carriers — can be real particles, but more frequently they appear in the theory as virtual particles, so the two terms are often synonymous. It is virtual particles that carry the force message from particle to particle. If there is plenty of energy around, an electron can emit a real photon, which produces a real click in a real Geiger counter. A virtual particle is a logical construct that stems from the permissiveness of quantum physics. According to quantum rules, particles can be created by borrowing the necessary energy. The duration of the loan is governed by Heisenberg's rules, which state that the borrowed energy times the duration of the loan must be greater than Planck's constant divided by twice pi. The equation looks like this: $\Delta E \Delta t$ *is* greater than $h/2\pi$. This means that the larger the amount of energy borrowed, the shorter the time the virtual particle can exist to enjoy it.

In this view, so-called empty space can be awash with these ghostly objects: virtual photons, virtual electrons and positrons, quarks and antiquarks, even (with oh god how small a probability) virtual golf balls and anti–golf balls. In this swirling, dynamic vacuum, a real particle's properties are modified. Fortunately for sanity and progress, the modifications are very small. Nevertheless, they are measurable, and once this was understood, life became a contest between increasingly precise measurements and ever more patient and determined theoretical calculations. For example, think about a real electron. Around the electron, because of its existence, there is a cloud of transient virtual photons. These notify all and sundry that an electron is present, but they also influence the electron's properties. What's more, a virtual photon can dissolve, very transiently, into an $e^+ e^-$ pair (a positron and an electron). In a blink of a mosquito's eye, the pair is back together as a photon, but even this evanescent transformation influences the properties of our electron.

In Chapter 5, I wrote the g-value of the electron as calculated theoretically from QED and as measured by inspired experiments. As you may recall, the two figures agreed to eleven places past the decimal. Equally successful was the g-value of the muon. Because the muon is heavier than the electron, it provides an even more incisive test of the concept of messenger particles; the muon's messengers can have higher energy and cause more mischief. The effect is that the field influences the properties of the muon even more strongly. Very abstract stuff, but the agreement between theory and experiment is sensational and indicates the power of the theory.

THE PERSONAL MAGNETISM OF THE MUON

As for the verifying experiment . . . On my first sabbatical year (1958–59) I went to CERN in Geneva, using a Ford Fellowship and a Guggenheim Fellowship to supplement my half-salary. CERN was the creation of a twelve-nation European consortium to build and share the expensive facilities required to do high-energy physics. Founded in the late forties, when the rubble of World War II was still warm, this collaboration of former military adversaries became a model for international cooperation in science. There my old sponsor and friend, Gilberto Bernardini, was director of research. My main reason for going was to enjoy Europe, learn to ski, and dabble in this new laboratory nestled on the Swiss-French border just outside of Geneva. Over the next twenty years I spent about four years doing research in this magnificent multilingual facility. Although French, English, Italian, and German were common, the official language of CERN was broken Fortran. Grunts and sign language also worked. I used to contrast CERN and Fermilab as follows: "CERN is a lab of culinary splendor and architectural catastrophe and Fermilab is the other way around." Then I convinced Bob Wilson to hire Gabriel Tortella, the legendary CERN chef and cafeteria manager, as a consultant to Fermilab. CERN and Fermilab are what we like to call cooperative competitors; each loves to hate the other.

At CERN, with Gilberto's help, I organized a "g minus 2" experiment, designed to measure the g-factor of the muon with mind-boggling precision, using some tricks. One trick was made possible by the fact that muons come out of pion decay polarized; that is, the vast majority have spins that point in the same direction relative to their motion. Another clever trick is implied by the title of the experiment,

"Gee minus two" or "Jzay moins deux," as the French call it. The g-value has to do with the strength of the little magnet built into the properties of spinning charged particles like the muon and electron.

Dirac's "crude" theory, remember, predicted that the g-value was exactly 2.0. However, as QED evolved, it was found that important but tiny adjustments to Dirac's 2 were required. These small terms appear because the muon or electron "feels" quantum pulsations of the field around it. Recall that a charged particle can emit a messenger photon. This photon, as we saw, can virtually dissolve into a pair of oppositely charged particles — just fleetingly — and then restore itself before anyone can see. The electron, isolated in its void, is perturbed by the virtual photon, influenced by the virtual pair, twisted by the transient magnetic forces. These and other, even more subtle, processes in the seething broth of virtual happenings connect the electron, ever so weakly, to all the charged particles that exist. The effect is a modification of the electron's properties. In the whimsical linguistics of theoretical physics, the "naked" electron is an imaginary object cut off from the influences of the field, whereas a "dressed" electron carries the imprint of the universe, but it is all buried in extremely tiny modifications to its bare properties.

In Chapter 5, I described the electron's g-factor. Theorists were even more interested in the muon; because its mass is two hundred times greater, the muon can emit virtual photons, which reach out farther to the more exotic processes. The result of one theorist's labor of many years was the g-factor of the muon:

$$g = 2(1.001165918)$$

This result (in 1987) was the culmination of a long sequence of calculations, using the new QED formulations of Feynman and the others. The collection of terms that add up to the sum .001165918 are known as radiative corrections. Once at Columbia we were listening to theorist Abraham Pais lecture on radiative corrections when a janitor entered the hall carrying a wrench. Pais leaned over to ask the man what he wanted. "Bram," someone yelled from the audience, "I think he's here to correct the radiator."

How do we match the theory with experiment? The trick was to find a way to measure the *difference* of the muon's g-value from 2.0. By finding a way to do this, we are measuring the correction (.001165918) *directly* rather than as a tiny add-on to a large number. Imagine trying to weigh a penny by first weighing a person carrying a penny and then weighing the person without the penny, then subtracting the second weight from the first. Better to weigh the penny directly. Suppose we trap a muon in an orbit in a magnetic field. The orbiting charge is also a "magnet" with a g-value, which Maxwell's theory says is precisely 2, whereas the spin-related magnet has this minuscule excess above 2. So the muon has two different "magnets": one internal (its spin) and the other external (its orbit). By measuring the spin-magnet while the muon is in its orbital configuration, the 2.0 gets subtracted, allowing us to measure directly the deviation from 2 in the muon, no matter how small.

Picture a little arrow (the spin axis of the muon) moving in a large circle with the arrow always tangential to the orbit. That's what would happen if g = 2.000 exactly. No matter how many orbits the particle executes, the little spin arrow will always be tangent to the orbit. However, if there is ever so small a difference between the true value of g and 2, the arrow will move away from tangency perhaps about a fraction of a degree for each orbit. After, say, 250 orbits, the arrow (spin axis) may be pointing toward the center of the orbit, like a radius. Continue the orbital motion, and in 1,000 orbits the arrow will make a full turn (360 degrees) relative to its initial direction as tangent. Thanks to parity violation, we can (triumphantly) detect the direction of the arrow (the muon's spin) by the direction in which the electrons come off when the muon decays. Any angle between the spin axis and a tangent line to the orbit represents a difference between g and 2. A precise measurement of this angle yields a precise measurement of the *difference*. See? No? Oh well, believe!

The proposed experiment was complicated and ambitious, but in 1958 it was easy to collect very bright young physicists to help. I returned to the United States in mid-1959 and revisited the experiment in Europe periodically. It went through several phases, each one suggesting the next phase, and didn't really end until 1978, when the final CERN g-value of the muon was published — a triumph of experimental cleverness and determination (*sitzfleisch*, the Germans call it). The electron's g-value was more precise, but don't forget that electrons are forever and muons stay in the universe for only two millionths of a second. The result?

$$g = 2(1.001165923 \pm .00000008)$$

The error of eight parts per hundred million clearly covers the theoretical prediction.

All of this is to suggest that QED is a great theory, and it's partly why Feynman, Schwinger, and Tomonaga are considered great physicists. It does have some pockets of mystery, one of which is noteworthy and relevant to our theme. It has to do with these infinities — for example, the electron's mass. Early efforts at quantum field theory calculated a point electron as infinitely heavy. It is as if Santa, manufacturing electrons for the world, must squeeze a certain quantity of negative charge into a very small volume. This takes work! The effort should show up as a huge mass, but the electron, weighing in at 0.511 MeV, or about 10^{-30} kilograms, is a lightweight, the lowest mass of any particle whose mass is clearly not zero.

Feynman and his colleagues proposed that whenever we see this dreaded infinity appearing, we in effect bypass it by inserting the known mass of the electron. In the real world one could call this fudging. In the world of theory, the word is "renormalization," a mathematically consistent method for circumventing the embarrassing infinities that a real theory would never have. Don't worry. It worked, and allowed for the super-precise calculations we talked about. Thus, the problem of mass was bypassed — but not solved — and remained behind as a quietly ticking time bomb to be activated by the God Particle.

The Weak Force

One of the mysteries that nagged Rutherford and others was this radioactivity thing. How is it that nuclei and particles decay willy-nilly into other particles? The physicist who first elucidated this question with an explicit theory, in the 1930s, was Enrico Fermi.

There are legions of stories about Fermi's brilliance. At the first nuclear bomb test at Alamogordo, New Mexico, Fermi was lying on the ground about nine miles from the bomb tower. After the bomb went off, he stood up and dropped small pieces of paper on the ground. The pieces fell at his feet in the quiet air, but a few seconds later the shock wave arrived and knocked them a few centimeters away. Fermi calculated the yield of the explosion from the displacement of the paper bits, and his on-the-spot result agreeed closely with the official measurement, which took several days to calculate. (A friend of his, the Italian physicist Emilio Segré, pointed out, however, that Fermi was human. He had trouble figuring out his University of Chicago expense account.)

Like many physicists, Fermi loved making up math games. Alan Wattenberg tells of the time he was eating lunch with a group of physicists when Fermi noticed dirt on the windows and challenged everyone to figure out how thick the dirt could get before it would fall off the window from its own weight. Fermi helped them all get through the exercise, which required starting from fundamental constants of nature, applying the electromagnetic interaction, and proceeding to calculate the dielectric attractions that keep insulators stuck to each other. At Los Alamos during the Manhattan Project, a physicist ran over a coyote one day in his car. Fermi said it was possible to calculate the total number of coyotes in the desert by keeping track of the vehicle-coyote interactions. These were just like particle collisions, he said. A few rare events yielded clues about the entire population of such particles.

Well, he was very smart, and he has been well recognized. He has more things named after him than anyone I know. Let's see . . . there is Fermilab, the Enrico Fermi Institute, fermion particles (all the quarks and leptons), and Fermi statistics (never mind). The fermi is a unit of size equal to 10^{-13} centimeters. My ultimate fantasy is to leave behind one thing that's named after me. I begged my Columbia colleague T. D. Lee to propose a new particle that, when discovered, would be named the Lee-on. To no avail.

But over and above Fermi's work on the first nuclear reactor, beneath the football stadium at the University of Chicago, and his seminal studies of squished coyotes was a contribution more central to the understanding of the universe. Fermi described a new force in nature, the weak force.

Let's backtrack quickly to Becquerel and Rutherford. Recall that Becquerel had serendipitously discovered radioactivity in 1896 when he stored some uranium in a drawer where he kept his photographic paper. When the photographic paper came out black, he eventually traced the cause to invisible rays shooting out of the uranium. After the discovery of radioactivity and the elucidation by Rutherford of alpha, beta, and gamma radiation, many physicists the world over concentrated on the beta particles, which were soon identified as electrons.

Where did the electrons come from? Physicists very quickly figured out that the electron was emitted from the nucleus when it underwent a spontaneous change of state. In the 1930s researchers determined that nuclei consist of protons and neutrons, and traced the radioactivity of nuclei to the instability of their constituent protons and neutrons. Obviously, not all nuclei are radioactive. The conservation of energy and the weak force play important roles in whether and how readily a proton or a neutron decays in a nucleus.

In the late 1920s careful before-and-after measurements of radioactive nuclei were made. One measures the mass of the initial nucleus, the mass of the final nucleus, and the energy and mass of the emitted electron (remembering that $E = mc^2$). And here an important discovery was made: it didn't add up. Energy was missing. The input was bigger than the output. Wolfgang Pauli made his (then) daring suggestion that a small neutral object was carrying the energy away.

In 1933 Enrico Fermi put it all together. The electrons were coming from the nucleus, but not directly. What happens is that the neutron in the nucleus decays into a proton, an electron, and the small neutral object that Pauli had invented. Fermi named it the *neutrino,* meaning "little neutral one." A force is responsible for this reaction in the nucleus, said Fermi, and he called it the weak force. It is enormously feeble compared to the strong nuclear force and electromagnetism. For example, at low energy the weak force is about one thousandth the strength of electromagnetism.

The neutrino, having no charge and almost no mass, could not be directly detected in the 1930s; it can be detected today only with great effort. Though the neutrino's existence was not proven experimentally until the 1950s, most physicists accepted it as a fact because it *had* to exist to make the bookkeeping come out right. In today's more exotic reactions in accelerators, involving quarks and other weird things, we still assume that any missing energy flies out of the collision in the form of undetectable neutrinos. This artful little dodger seems to leave its invisible signature all over the universe.

But back to the weak force. The decay that Fermi described — neutron gives way to proton, electron, and neutrino (actually, an antineutrino) — occurs routinely with free neutrons. When the neutron is imprisoned in the nucleus, however, it can happen only under special circumstances. Conversely, the proton as a free particle cannot decay (as far as we know). Inside the crowded nucleus, however, the bound proton can give rise to a neutron, a positron, and a neutrino. The reason that the free neutron can undergo weak decay is simple energy conservation. The neutron is heavier than the proton, and when a free neutron changes into a proton there is enough additional rest mass energy to make the electron and the antineutrino and send each of them off with a little energy. A free proton has too little mass to do this. However, inside the nucleus the presence of all the other guys in effect alters the mass of a bound particle. If the protons and neutrons inside can, by decaying, increase the stability and lower the mass of the nucleus in which they are stuck, they do it. However, if the nucleus is already in its lowest mass-energy state, it is stable and nothing happens. It turns out that all the hadrons — the protons, neutrons, and their hundreds of cousins — are induced to decay via the weak force, with the free proton being the only apparent exception.

The theory of the weak force was gradually generalized and, in constant confrontation with new data, evolved to a quantum field theory of the weak force. A new breed of theorists emerging mostly in American universities helped to mold the theory: Feynman, Gell-Mann, Lee, Yang, Schwinger, Robert Marshak, and many others. (I keep having this nightmare in which all the theorists I've failed to cite meet in a suburb of Teheran and offer a reward of prompt admission to Theory Heaven for anyone who instantly and totally renormalizes Lederman.)

SLIGHTLY BROKEN SYMMETRY, OR WHY WE ARE ALL HERE

A crucial property of the weak force is parity violation. All the other forces respect this symmetry; that one force can violate it was a shock. Another deep symmetry, one that compares the world to the antiworld, had been demonstrated to fail by the same experiments that showed P (parity) violation. This second symmetry was called C, for charge conjugation. The failure of C symmetry also occurred only with the weak force. Before C violation was demonstrated, it was thought that a world in which all objects are made of antimatter would obey the same laws of physics as the regular old matter world. No, said the data. The weak force doesn't respect that symmetry.

What were the theorists to do? They quickly retreated to a new symmetry: CP symmetry. This says that two physical systems are essentially identical if one is related to the other by *simultaneously* reflecting all objects in a mirror (P) and also changing all particles to antiparticles (C). CP symmetry, the theorists said, is a much deeper symmetry. Even though nature does not respect C and P separately, simultaneous CP symmetry must endure. It did until 1964 when Val Fitch and James Cronin, two Princeton experimenters studying neutral kaons (a particle my group discovered in Brookhaven experiments in 1956–1958), came upon clear and compelling data that CP symmetry was, in fact, not perfect.

Not perfect? The theorists sulked, but the artist in all of us rejoiced. Artists and architects love to tweak us with canvases or architectural structures that are almost, but not exactly, symmetrical. The asymmetric towers in the otherwise symmetric cathedral at Chartres is a good example. The CP violation effect was small — a few events out of a thousand — but clear, and theorists were back to square one.

I mention CP violation for three reasons. First, it is a good example of what became recognized, in the other forces, as "slightly broken symmetry." If we believe in the intrinsic symmetry of nature, *something,* some physical agency, must enter to break that symmetry. A closely related agency doesn't actually destroy the symmetry, it just hides it so that nature appears to be asymmetrical. The God Particle is such a disguiser of symmetry. We will return to it in Chapter 8. The second reason for mentioning CP violation is that in the 1990s understanding this concept is one of the most pressing needs for clearing up the problems in our standard model.

The final reason, and the element that brought the Fitch-Cronin experiment to the respectful attention of the Royal Swedish Academy of Science, is that when applied to cosmological models of the evolution of the universe, CP violation explained a puzzle that had plagued astrophysicists for fifty years. Before 1957 a large number of experiments indicated perfect symmetry between matter and antimatter. If matter and antimatter are so symmetric, why is our planet, our solar system, our galaxy, and, evidence indicates, all other galaxies devoid of antimatter? And how could an experiment carried out on Long Island in 1965 explain it all?

Models indicated that as the universe cooled after the Big Bang, all the matter and antimatter annihilated, leaving essentially pure radiation, ultimately too cool — too low in energy — to create matter. But matter, that's us! Why are we here? The Fitch-Cronin experiment shows the way out. The symmetry isn't perfect. A slight excess of matter over antimatter (for every 100 million quark-antiquark pairs there is one extra quark) is a result of the slightly broken CP symmetry, and this tiny excess accounts for all the matter in the presently observed universe, including us. Thanks Fitch, thanks Cronin. Splendid fellows.

TRAPPING THE LITTLE NEUTRAL ONE

Much of the detailed information on the weak force was provided by neutrino beams, and herein lies another story. Pauli's 1930 hypothesis — that a small, neutral particle exists that feels only the weak force — was tested in many ways from 1930 to 1960. Precise measurements of an increasingly large number of weakly decaying nuclei and particles tended to confirm the hypothesis that a little neutral thing was escaping from the reaction carrying away energy and momentum. This was a convenient way to understand decay reactions, but could we actually detect neutrinos?

This was no easy task. Neutrinos float through vast thicknesses of matter unscathed because they obey only the weak force, whose short range reduces the probability of a collision enormously. It was estimated that to ensure a collision of a neutrino with matter would require a target of lead one light-year thick! Quite an expensive experiment. However, if we use a very large number of neutrinos, the required thickness to see a collision every once in a while is correspondingly reduced. In the mid-1950s, nuclear reactors were used as intense sources of neutrinos (so much radioactivity!), to which a huge vat of cadmium dichloride (cheaper than a light-year's worth of lead) was exposed. With so many neutrinos (actually, antineutrinos, which is mostly what you get from reactors), it was inevitable that some of them would strike protons, causing inverse beta decay; that is, a positron and a neutron were released. The positron, in its wandering, would eventually find an electron and annihilate into two oppositely moving photons. These fly outward into dry cleaning fluid, which flashes when struck by the photons. The detection of a neutron and a pair of photons represented the first experimental evidence of the neutrino, about thirty-five years after Pauli thought up the critter.

By 1959 another crisis, two in fact, arose to tweak the physicist's mind. The center of the storm was at Columbia University, but the crisis was liberally shared and appreciated around the world. All of the data on the weak force to that time were kindly provided by particles during natural decay. Greater love hath no particle than to give its all for the edification of physicists. To study the weak force we simply watched particles, such as the neutron or the pion, decay into other particles. The energies involved were provided by the rest masses of the decaying particles — typically from a few MeV to around 100 MeV or so. Even the free neutrinos shooting out of reactors and undergoing weak-force collisions involved only a few MeV. After we had modified the weak-force theory with the experimental results of parity violation, we had one zinger of an elegant theory that fit all the available data provided by zillions of nuclear decays as well as the decays of pions, muons, lambdas, and probably, though difficult to prove, Western civilization.

THE EXPLODING EQUATION

Crisis No. 1 had to do with the mathematics of the weak force. In the equations, the energy at which the force is measured appears. Depending on the data, you stick in the rest mass energy of the decaying particle — 1.65 MeV or 37.2 MeV or whatever — and out comes the right answer. You manipulate the terms, bump and grind, and, sooner or later, out come predictions as to the lifetimes, decays, spectra of electrons — things that can be compared to experiment — and they are right. But if one puts in, say, 100 GeV (*billion* electron volts), the theory goes haywire. The equation explodes in your face. In the jargon of physics, this is called "the unitarity crisis."

Here's the dilemma. The equation was okay, but it had a pathology at high energy. Little numbers worked; big ones didn't. We didn't have the ultimate truth, only a truth valid for the low-energy domain. There had to be some new physics that modified the equations at high energy.

Crisis No. 2 was the mystery of the unobserved reaction. One could calculate how often a muon decayed into an electron and a photon. Our theory of the weak processes said that this should happen. Looking for this reaction was a favorite Nevis experiment, and several new Ph.D.'s spent godknowshowmany beam hours searching with no success. Murray Gell-Mann, the pundit on all matters arcane, is often quoted as the source of something called the Totalitarian Rule of Physics: "Anything that isn't forbidden is compulsory." If our laws do not rule out an event, it not only *can* happen, it *must* happen! Since a muon decaying into an electron and a photon was not forbidden, why weren't we seeing it? What forbade this mu-e-gamma decay? (For "gamma" read "photon.")

Both crises were exciting. Both offered up the possibility of new physics. Theoretical speculations abounded, but experimental blood boiled. What to do? We experimenters must measure, hammer, saw, file, stack lead bricks — *do* something. So we did.

MURDER INC. AND THE TWO-NEUTRINO EXPERIMENT

Melvin Schwartz, an assistant professor at Columbia, after listening to a detailed review of the troubles by Columbia theorist T. D. Lee in November 1959 came up with his GREAT IDEA. Why not create a beam of neutrinos by letting a high-energy pion beam drift through enough space that some fraction, say 10 percent, of the pions decayed into a muon and a neutrino. Pions, in flight, would disappear; muons and neutrinos, sharing the pion's original energy, would appear. So here, flying through space, we have muons and neutrinos from the 10 percent of pions that decayed, plus the 90 percent of pions that didn't decay, plus a bunch of nuclear debris originating from the target that produced the pions. Now, said Schwartz, let's aim it all into a big thick wall of steel, forty feet thick, as it turned out. The wall would stop everything but the neutrinos, which would have no trouble passing through forty million miles of steel. We'd have a pure beam of neutrinos on the other side of the wall, and since the neutrino obeys only the weak force, we'd have a handy way of studying both the neutrino and the weak force via neutrino collisions.

The scheme addressed both Crisis No. 1 and Crisis No. 2. Mel's idea was that this neutrino beam would allow us to study the weak force at energies of billions rather than millions of electron volts. It would give us a view of the behavior of the weak force at high energy. It might also provide some ideas on why we don't see muons decay into electrons plus photons, based on the notion that neutrinos are somehow involved.

As happens so often in science, an almost equivalent idea was published almost simultaneously by a Soviet physicist, Bruno Pontecorvo. If the name seems more Italian than Russian, it is because Bruno is an Italian who defected to Moscow in the 1950s on ideological grounds. His physics, ideas, and imagination were nevertheless outstanding. Bruno's tragedy was in trying to carry out his imaginative ideas within a system of stultifying bureaucracy. International conferences are venues for displaying the traditional warm friendship of scientists. At one such conference in Moscow, I asked a friend, "Yevgeny, tell me, which one of you Russian physicists is really a communist?" He looked around the hall and pointed to Pontecorvo. But that was in 1960.

When I returned to Columbia from a pleasant sabbatical at CERN in late 1959, I listened to the discussions about crises in the weak force, including Schwartz's idea. Schwartz had somehow concluded that no existing accelerator was powerful enough to make a sufficiently intense neutrino beam, but I disagreed. The 30 GeV AGS (for Alternating Gradient Synchrotron) was nearing completion at Brookhaven, and I did the numbers and convinced myself and then Schwartz that the experiment was, in fact, doable. We designed what was, for 1960, a huge experiment. Jack Steinberger, a colleague at Columbia, joined us and with students and postdocs we formed a group of seven. Jack, Mel, and I were well known for our gentle and kindly demeanor. Once as we were walking across the Brookhaven accelerator floor I overheard a physicist in a group exclaim, "There goes Murder, Incorporated!"

To block all the particles except the neutrinos, we made a thick wall around a massive detector, using thousands of tons of steel from outdated naval vessels. I once made the mistake of telling a reporter that we took apart the battleship *Missouri* to make the wall. I must have gotten the name wrong, because the *Missouri* is apparently still out there someplace. But we certainly had a battleship cut up for scrap. I also made the mistake of joking that if there was a war we'd have to paste the ship back together, and that story got embellished and pretty soon there was a rumor that the navy had confiscated our experiment to fight some war (what war this could have been — it was 1960 — remains a puzzle).

What is also somewhat fabricated is my story about the cannon. We got a twelve-inch naval cannon with a suitable bore and thick walls — it made a beautiful collimator, a device for focusing and aiming a beam of particles. We wanted to fill it up with beryllium as a filter, but the bore had these deep rifling grooves. So I sent a skinny graduate student inside to stuff steel wool into the grooves. He spent about an hour in there and crawled out all hot, sweaty, and irritated and said, "I quit!" "You can't quit," I cried. "Where will I find another student of your caliber?"

Once our preparations were finished, steel from obsolete ships surrounded a detector made from ten tons of aluminum tastefully arranged so that if neutrinos collided with an aluminum nucleus, the products of the collision would be observed. The detector idea we eventually used, called a spark chamber, had been invented by a Japanese physicist, Shuji Fukui. We learned a lot by talking to Jim Cronin of Princeton who had mastered the new technique. Schwartz won the ensuing contest as to the best design that could be scaled up from a few pounds to ten tons. In this spark chamber, nicely machined one-inch-thick plates of aluminum were spaced about a half inch apart and a huge voltage difference applied between adjacent plates. If a charged particle passed through the gap, a spark would follow the trail of the particle and could be photographed. How easily this is said! The technique was not without its technical problems. But the results! Zap — and the path of a subnuclear particle was rendered visible in the red-yellow light of glowing neon gas. It was a lovely device.

We built models of spark chambers and put them in beams of electrons and pions to learn their characteristics. Most chambers of that day were about a foot square and had ten to twenty plates. The design we set about had one hundred plates, each four feet square. Each plate was one inch thick, pleading with the neutrinos to collide. Seven of us worked day and night as well as other times to assemble the apparatus and the electronics, inventing all sorts of devices — hemispherical spark gaps, automated gluing facilities, circuitry. We had help from engineers and several technicians.

We started the run late in 1960 and were immediately plagued by background "noise" created by neutrons and other debris from the target sneaking around our formidable forty feet of steel, crudding up our spark chambers, and skewing our results. Even if only one particle in a billion got through, it created problems. Leave it to background to know that one chance in a billion is the legal definition of a miracle. We struggled for weeks plugging cracks anywhere neutrons could sneak in. We searched diligently for electrical ducts under the floor. (Mel Schwartz, exploring, crawled into one, got stuck, and had to be hauled out by several strong technicians.) Every thin area was plugged with blocks of rusty steel from the ex-battleship. At one point, the director of Brookhaven's brand-new accelerator drew the line: "You'll pile those dirty blocks near my new machine over my dead body," he thundered. We didn't take him up on his offer as this would have made an unsightly lump in the shielding. So we compromised — only slightly. By late November, the background was reduced to manageable proportions.

Here is what we were doing.

The protons from the AGS smashed into a target, producing about three pions on average for each collision. We produced about 10^{11} (100 billion) collisions per second. Assorted neutrons, protons, occasional antiprotons, and other debris were also generated. The debris that headed our way crossed a space of about fifty feet before smashing into our impenetrable steel wall. In that distance some 10 percent of the pions decayed so we had something like a few tens of billions of neutrinos. A much smaller number headed in the right direction, toward our forty-foot-thick steel wall. On the other side of the wall, about a foot away, our detector, the spark chamber, lay waiting. We estimated that if we were lucky we'd see one neutrino collision in our aluminum spark chamber per week! In that week the target would spray about 500 million billion (5×10^{17}) particles in our general direction. This is why we had to reduce background so severely.

We expected two kinds of neutrino collisions: (1) a neutrino hits an aluminum nucleus, which results in a muon and an excited nucleus, or (2) a neutrino hits a nucleus, which results in an electron and an excited nucleus. Forget about the nuclei. What's important is that we expected muons and electrons to emerge from the collision in equal numbers, accompanied by occasional pions and other debris from the excited nucleus.

Virtue triumphed and in an eight-month exposure we observed fifty-six neutrino collisions, of which perhaps five were spurious. Sounds easy, but I will never, never forget that first neutrino event. We had developed a roll of film, the result of a week of data taking. Most of the frames were empty or showed some obvious cosmic ray tracks. But suddenly, there it was: a spectacular collision with a long, long muon track speeding away. That first event was the mini-Eureka moment, the flash of certainty, after so much effort, that the experiment would work.

Our first task was to prove that these were indeed neutrino events since this was the first experiment of its kind ever. We pooled all of our experience and took turns playing devil's advocate in trying to pick holes in our own conclusion. But the data were in fact rock solid, and it was time to go public. We felt secure enough to present the results to our colleagues. You should have heard Schwartz's talk to a jammed Brookhaven auditorium. Like a lawyer, he ruled out, one by one, all possible alternatives. There were smiles and tears in the audience. Mel's mother had to be helped out, sobbing uncontrollably.

There were three (always three) major consequences of the experiment. Remember that Pauli first posited the existence of the neutrino to explain the missing energy in beta decay, in which an electron is ejected from the nucleus. Pauli's neutrinos were always associated with electrons. In almost all of our events, however, the product of the neutrino collision was a muon. *Our* neutrinos refused to produce electrons. Why?

We had to conclude that the neutrinos we were using had a new specific property of "muon-ness." Since these neutrinos were born with a muon in the decay of pions, somehow "muon" was imprinted on them.

To prove this to the audience of genetically conditioned skeptics, we had to know and show that our apparatus did not more readily see muons, and that it therefore — by stupid design — was incapable of detecting electrons. Galileo's telescope problem all over again. Fortunately, we were able to demonstrate to our critics that we had built electron-detection capability into our equipment and had indeed verified this in test beams of electrons.

Another background effect came from cosmic radiation, which at sea level consists of muons. A cosmic-ray muon coming in from the back of our detector and stopping in the middle could be mistaken by lesser physicists as a muon from neutrinos going out, which is what we were looking for. We had installed a "block" against this, but how could we be sure it worked?

The key was to keep the detector going whenever the machine was shut down — which was about 50 percent of the time. When the accelerator was off, any muons that showed up would be uninvited cosmic rays. But none appeared; cosmic rays were unable to get past our block.

I mention all these technical details to show you that experimentation is not so easy and that the interpretation of an experiment is a subtle affair. Heisenberg once commented to a colleague outside the entrance to a swimming pool, "These people go in and out all very nicely dressed. Do you conclude from this that they swim dressed?"

The conclusion we — and most others — drew from the experiment was that there are (at least) two neutrinos in nature — one associated with electrons (the plain vanilla Pauli neutrinos) and one associated with muons. So we call them electron neutrinos (plain) and muon neutrinos, the kind we produced in our experiment. The distinction is now known as "flavor," in the whimsical lingo of the standard model, and people began to draw a little table:

electron neutrino	muon neutrino
electron	muon

or in physics shorthand:

ν_e	ν_μ
e	μ

The electron is placed under its cousin, the electron neutrino (indicated by the subscript), and the muon under its muon neutrino cousin. Let's recall that before this experiment we knew of three leptons — e, ν, and μ — which were not subject to the strong force. Now there were four: e, ν_e, μ, and ν_μ. The experiment was forever called the experiment of the Two Neutrinos, which ignorant people think is an Italian dance team. This turns out to be the button upon which the standard model overcoat is sewn. Note that we have two "families" of leptons, pointlike particles, arranged vertically. The electron and electron neutrino are the first family, which is found everywhere in our universe. The second family consists of the muon and the muon neutrino. Muons are not found readily today in the universe, but must be manufactured in accelerators or in other high-energy collisions, such as those produced by cosmic rays. When the universe was young and hot, these particles were abundant. When the muon, a heavy brother of the electron, was first discovered, I. I. Rabi asked, "Who ordered that?" The two-neutrino experiment provided one of the early clues to the answer.

Oh yes. The fact that two different neutrinos existed solved the crisis of the missing mu-e-gamma reaction. To review, a muon should decay into an electron and a photon, but no one was able to detect this reaction, though many tried. There should be a sequence of processes: a muon should first decay into an electron and two neutrinos — a regular neutrino and an antineutrino. These two neutrinos, being matter and antimatter, then annihilate, producing the photon. But nobody was seeing these photons. The reason why was now obvious. Clearly, the positive muon decays into a positron and two neutrinos, but these are an electron neutrino and an antimuon neutrino. These neutrinos don't annihilate each other because they're from different families. They simply stay neutrinos, and no photon is produced, thus no mu-e-gamma reaction.

The second consequence of the Murder Inc. experiment was the creation of a new tool for physics: hot and cold running neutrino beams. These appeared, in due course, at CERN, Fermilab, Brookhaven, and Serpuhkov (USSR). Remember, previous to the AGS experiment, we weren't totally sure neutrinos existed. Now we had beams of them on demand.

Some of you might have noticed that I'm avoiding an issue here. What happened to Crisis No. 1, the fact that our equation for the weak force doesn't work at high energies? Indeed, our 1961 experiment demonstrated that the collision rate *was* increasing with energy. By the 1980s, the accelerator labs mentioned above — using more intense beams at higher energies and detectors weighing hundreds of tons — were collecting millions of neutrino events at the rate of several per minute (a lot better than our 1961 yield of one or two a week). Even so, the high-energy crisis of weak interactions was not solved, though it was greatly illuminated. The rate of neutrino collisions did increase with higher energy, as the low-energy theory predicted. However, the fear that the collision rate would become impossibly large was alleviated by the discovery of the W particle in 1982. This was part of the new physics that modified the theory and led to a gentler and kinder behavior. This postponed the crisis to which, yes, we will return.

BRAZILIAN DEBT, SHORT SKIRTS, AND VICE VERSA

The third consequence of the experiment was that Schwartz, Steinberger, and Lederman were awarded the Nobel Prize in physics, but not until 1988, some twenty-seven years after the research had been done. Somewhere I heard of a reporter interviewing the young son of a new laureate: "Would you like to win a Nobel Prize like your father?" "No!" said the young man. "No? Why not?" "I want to win it alone."

The Prize. I do have some comments. The Nobel is awesome to most of us in the field, probably because of the luster of the recipients, starting with Roentgen (1901) and going through so many of our heroes including Rutherford, Einstein, Bohr, and Heisenberg. The Prize gives a colleague who wins it a certain aura. Even when your best friend, one with whom you have peed together in the woods, wins the Prize it somehow changes him in your eyes.

I had known that at various times I had been nominated. I suppose

I could have received the Prize for the "long-lived neutral kaon," which I discovered in 1956, for this was quite an unusual object, used today as a tool for studies of crucial CP symmetry. I could have gotten it for the pion-muon parity research (with C. S. Wu), but Stockholm chose to honor the theoretical instigators instead. Actually, that was a reasonable decision. Still, the byproduct discovery of polarized muons and their asymmetric decay has had extensive applications to condensed matter and atomic and molecular physics, so much so that international conferences on this subject are held regularly.

As the years passed, October was always a nervous month, and when the Nobel names were announced, I would often be called by one or another of my loving offspring with a "How come . . . ?" In fact, there are many physicists — and I'm sure this is true of candidates in chemistry and medicine as well as in the nonsciences — who will not get the Prize but whose accomplishments are equivalent to those of the people who have been recognized. Why? I don't know. It's partly luck, circumstances, the will of Allah.

But I have been lucky and have never lacked recognition. For doing what I love to do, I was promoted to full professor at Columbia in 1958 and paid reasonably well. (Being a professor in an American university is the best job in Western civilization. You can do anything you want to do, even teach!) My research was vigorous, aided by some fifty-two graduate students over the years 1956–1979 (at which time I became Fermilab director). Most of the time the rewards came when I was too busy to anticipate them: election to the National Academy of Science (1964), the President's Medal of Science (Lyndon Johnson gave it to me in 1965), and other assorted medals and citations. In 1983 Martin Perl and I shared the Wolf Prize, given by the state of Israel, for discovering the third generation of quarks and leptons (the **b** quark and the tau lepton). Honorary degrees also came in, but that's a seller's market, since hundreds of universities are each seeking four or five people to honor every year. With all that, one begins to acquire a modicum of security and a calm attitude toward the Nobel.

When the announcement finally came, in the form of a 6 A.M. phone call on October 10, 1988, it released a hidden store of uncontrolled mirth. My wife, Ellen, and I, after very respectfully acknowledging the news, laughed hysterically until the phone starting ringing and our lives started changing. When a reporter from the *New York Times* asked me what I was going to do with the prize money, I told him I couldn't decide between buying a string of racehorses or a castle in Spain, a quote he duly printed. Sure enough, a real estate agent called me the next week, telling me about a great deal on a chateau in Castille.

Winning the Nobel Prize when you are already reasonably prominent has interesting side effects. I was director of Fermilab, which has 2,200 employees, and the staff basked in the publicity, taking the occasion as a sort of early Christmas present. A lab-wide meeting had to be repeated several times so everyone could listen to the Boss, who was already pretty funny, but who was suddenly considered on a par with Johnny Carson (and was being taken seriously by really important people). The Chicago *Sun-Times* shook me up by headlining NOBEL STRIKES HOME, and the *New York Times* put a picture of me, sticking my tongue out, on the front page — above the crease!

All of this fades, but what didn't fade was the public awe at the title. At receptions all over the city I was introduced as the winner of the 1988 Nobel *Peace* Prize in physics. And when I wanted to do something rather spectacular, perhaps foolhardy, to help the Chicago public schools, the Nobel holy water worked. People listened, doors opened, and suddenly we had a program for improving science education in inner-city schools. The Prize is an incredible ticket to help one effect socially redeeming activities. The other side of the coin is that no matter what you won the Prize for, you become an instant expert in all things. Brazilian debt? Sure. Social Security? Yeah. "Tell me, Professor Lederman, what length will women's dresses be?" "As short as possible!" responds the laureate with lust in his heart. But what I do intend is to use the Prize shamelessly to help advance science education in the United States. For this task a second Prize would be helpful.

The Strong Force

The triumphs in working out the intricacies of the weak force were considerable. But there were still those hundreds of hadrons nagging us, a plethora of particles, all of which were subject to the strong force, the force that holds the nucleus together. The particles had a variety of properties: charge, mass, and spin are some we have mentioned.

Pions, for example. There are three different pions closely spaced in mass, which, after being studied in a variety of collisions, were placed together in a family — the pion family, oddly enough. Their electric charges are plus one, minus one, and zero (neutral). All the hadrons, it turned out, came in family clusters. The kaons line up like this: K^+, K^-, K^0, $\overline{K}^0$. (The signs, +, −, and 0, indicate the electric charge. The bar atop the second neutral kaon indicates that it is an antiparticle. The sigma family portrait looks like this: Σ^+, Σ^0, Σ^-. A more familiar group to you is the nucleon family: the neutron and proton, components of the atomic nucleus.

The families consist of particles of similar mass and similar behavior in strong collisions. To express this idea more specifically, the term "isotopic spin," or isospin, was invented. Isospin is useful in that it allows us to look at the concept of "nucleon" as a single object coming in two isospin states: neutron or proton. Similarly "pion" comes in three isospin states: π^+, π^-, π^0. Another useful property of isospin is that in strong collisions it is a conserved quantity, like charge. A violent collision of a proton and an antiproton may produce forty-seven pions, eight baryons, and other stuff, but the total isotopic spin number remains constant.

The point is that physicists were trying to make some sense out of these hadrons by sorting through as many properties as they could find. So there are lots of properties with whimsical names: strangeness number, baryon number, hyperon number, and so on. Why "number"? Because all these are quantum properties, hence quantum numbers. And quantum numbers obey conservation principles. This permitted theorists or out-of-experiment experimentalists to play with the hadrons, organize them, and, inspired perhaps by biologists, classify them into larger family structures. Theorists were guided by rules of mathematical symmetry, following the belief that the fundamental equations would respect such deep symmetries.

One particularly successful organization was devised in 1961 by the Cal Tech theorist Murray Gell-Mann, who called his scheme the Eightfold Way, after the teaching of the Buddha: "This is the noble Eightfold Way: namely, right views, right intention, right speech . . ." Gell-Mann correlated hadrons almost magically into coherent groups of eight and ten particles. The allusion to Buddhism was yet another excursion into whimsy, so common in physics, but various mystics seized upon the name as proof that the true order of the world is related to Eastern mysticism.

I got into trouble in the late 1970s, when I was asked to write a little biography of myself for the Fermilab newsletter on the occasion of the discovery of the bottom quark. Not expecting anyone other than my coworkers in Batavia to read the piece, I entitled the story "An Unauthorized Autobiography" by Leon Lede-rman. To my horror the story was picked up and reprinted in the CERN newsletter and then in *Science,* the official journal of the American Association for the Advancement of Science, read by hundreds of thousands of scientists in the United States. The story included the following: "His [Lederman's] period of greatest creativity came in 1956 when he heard a lecture by Gell-Mann on the possible existence of neutral K mesons. He made two decisions: First, he hyphenated his name . . ."

Anyway, by any other name, a theorist would smell as sweet, and Gell-Mann's Eightfold Way gave rise to charts of hadron particles that were reminiscent of the Mendeleev periodic table of the elements, though admittedly more arcane. Remember Mendeleev's chart with its columns of elements having similar chemical properties? This periodicity was a clue to the existence of an internal organization, to the shell structure of electrons, even before we knew about electrons. Something inside the atoms was repeating, making a pattern as the atoms increased in size. In retrospect, after the atom was understood, it should have been obvious.

THE SCREAM OF THE QUARK

The pattern of hadrons, arranged by assorted quantum numbers, also screamed for substructure. It isn't easy, however, to hear the screams of subnuclear entities. Two keen-eared physicists did, and wrote about it. Gell-Mann proposed the existence of what he referred to as mathematical structures. In 1964 he postulated that the patterns of organized hadrons could be explained if three "logical constructs" existed. He called these constructs "quarks." It is generally assumed that he lifted the word from James Joyce's diabolical novel *Finnegans Wake* ("Three quarks for Muster Mark!"). George Zweig, a colleague of Gell-Mann's, had an identical idea while working at CERN; he named his three things "aces."

We will probably never know precisely how this seminal idea came about. I know one version because I was there — at Columbia University in 1963. Gell-Mann was giving a seminar on his Eightfold Way symmetry of hadrons when a Columbia theorist, Robert Serber, pointed out that one basis for the "eight" organization would involve three subunits. Gell-Mann agreed, but if these subunits were particles they would have the unheard-of property of having *third*-integral electric charges — ⅓, ⅔, −⅓, and so on.

In the particle world, all electric charges are measured in terms of the charge on the electron. All electrons have exactly 1.602193×10^{-19} coulombs. Never mind what coulombs are. Just know that we use the previous complicated figure as a unit of charge and call it 1 because it's the charge on the electron. Conveniently, the proton's charge is also 1.0000, as is that of the charged pion, the muon (here the precision is much higher), and so on. In nature, charges come in integers — 0, 1, 2 . . . All the integers are understood to be multiples of the number of coulombs given above. Charges also come in two styles: plus and minus. We don't know why. That's the way it is. One might imagine a world in which the electron could, in a bruising collision or in a poker game, lose 12 percent of its electric charge. Not in this world. The electron, proton, pi plus, et al. always have charges of 1.0000.

So when Serber brought up the idea of particles with third-integral charges — forget it. Such things had never been seen, and the rather curious fact that all observed charges were equal to an integral multiple of a unique, unchanging standard charge became, over time, incorporated into the intuition of physicists. This "quantization" of electric charge was in fact used to seek some deeper symmetry that would account for it. However, Gell-Mann reconsidered and proposed the quark hypothesis, simultaneously blurring the issue, or so it seemed to some of us, by suggesting that quarks aren't real but are convenient mathematical constructs.

The three quarks born in 1964 are today called "up," "down," and "strange," or **u**, **d**, and **s**. There are, of course, three antiquarks: $\bar{\mathbf{u}}$, $\bar{\mathbf{d}}$, and $\bar{\mathbf{s}}$. The properties of the quarks had to be delicately chosen so that they could be used to build all of the known hadrons. The **u** quark is given a charge of +⅔; the **d** quark is −⅓ as is the **s** quark. The antiquarks have equal but opposite charges. Other quantum numbers are also selected so that they add up correctly. For example, the proton is made of three quarks — **uud** — with charges +⅔, +⅔, and −⅓, the sum being +1.0, which jibes with what we know about the proton. The neutron is a **udd** combination, with charges +⅔, −⅓, −⅓, for a sum of 0.0, which makes sense because the neutron is neutral, zero charge.

All hadrons consist of quarks, sometimes three and sometimes two, according to the quark model. There are two classes of hadrons: baryons and mesons. Baryons, which are relatives of protons and neutrons, are three-quark jobs. Mesons, which include pions and kaons, consist of two quarks — but they must be a quark combined with an antiquark. An example is the positive pion (π^+), which is $\mathbf{u\bar{d}}$. The charge is $+\frac{2}{3} + \frac{1}{3}$, which is equal to 1. (Note that the **d**-bar, the antidown quark, has a charge of $+\frac{1}{3}$.)

In fashioning this early hypothesis, the quantum numbers of the quarks, and properties such as spin, charge, isospin, and so on, were fixed in order to account for just a few of the baryons (proton, neutron, lambda, and so on) and mesons. Then these numbers and other relevant combinations were found to fit all the hundreds of known hadrons. It all worked! And all the properties of a composite — for example, a proton — are subsumed by the properties of the constituent quarks, moderated by the fact that they are in intimate interaction with one another. At least, that is the idea and the task for generations of theorists and generations of computers, given, of course, that they are handed the data.

Quark combinations raise an interesting question. It is a human trait to modify one's behavior in company. However, as we shall see, quarks are never alone, so their true unmodified properties can only be deduced from the variety of conditions under which we can observe them. In any case, here are some typical quark combinations and the hadrons they produce:

BARYONS

uud proton
udd neutron
uds lambda
uus sigma plus
dds sigma minus
uds sigma zero
dss xi minus
uss xi zero

MESONS

$\mathbf{u\bar{d}}$ positive pion
$\mathbf{d\bar{u}}$ negative pion
$\mathbf{u\bar{u}} + \mathbf{d\bar{d}}$ neutral pion
$\mathbf{u\bar{s}}$ positive kaon
$\mathbf{s\bar{u}}$ negative kaon
$\mathbf{d\bar{s}}$ neutral kaon
$\mathbf{\bar{d}s}$ neutral antikaon

Physicists gloried in the spectacular success of reducing hundreds of seemingly basic objects to composites of just three varieties of quarks.

(The term "aces" faded — no one can compete with Gell-Mann when it comes to naming.) The test of a good theory is whether it can *predict,* and the quark hypothesis, guarded or not, was a brilliant success. For example, the combination of three strange quarks, sss, was not among the record of discovered particles, but that didn't stop us from giving it a name: omega minus (Ω^-). Because particles containing the strange quark had established properties, the properties of a hadron with three strange quarks, sss, would also be predictable. The omega minus was a very strange particle with a spectacular signature. In 1964 it was discovered in a Brookhaven bubble chamber and was exactly what Dr. Gell-Mann had ordered.

Not that all issues were settled — not by a long shot. Lots of questions: for starters, how do quarks stick together? This strong force would be the subject of thousands of theoretical and experimental papers over the next three decades. The jawbreaking title "quantum chromodynamics" would propose a new breed of messenger particles, gluons, to cement(!!) quarks together. All in due course.

CONSERVATION LAWS

In classical physics there are three great conservation laws: energy, linear momentum, and angular momentum. They have been shown to be deeply related to concepts of space and time, as we will see in Chapter 8. Quantum theory introduced a great number of additional quantities that are conserved; that is, they do not change during a variety of subnuclear, nuclear, and atomic processes. Examples are electric charge, parity, and a host of new properties like isospin, strangeness, baryon number, and lepton number. We have already learned that the forces of nature differ in their respect for different conservation laws; for example, parity is respected by the strong and electromagnetic forces but not by the weak force.

To test a conservation law, one examines a huge number of reactions in which a particular property, say the electric charge, can be ascertained before and after the reaction. We recall that energy conservation and momentum conservation were so solidly established that when certain weak processes appeared to violate them, the neutrino was postulated as a saving mechanism, and it was right. Other clues to the existence of a conservation law have to do with the refusal of certain reactions to take place. For example, an electron does not decay with two neutrinos because that would violate charge conservation. Another example is proton decay. Recall that it doesn't. Protons are assigned a baryon number that is ultimately derived from its three-quark structure. So protons, neutrons, lambdas, sigmas, and so on — all three-quark fellows — have baryon number +1. The corresponding antiparticles have baryon number −1. All mesons, force carriers, and leptons have baryon number 0. If baryon number is strictly conserved, then the lightest baryon, the proton, can never decay, since all the lighter decay-product candidates have baryon number 0. Of course, a proton-antiproton collision has total baryon number 0 and can give rise to anything. So baryon number "explains" why the proton is stable. The neutron, decaying into a proton, an electron, and an antineutrino, and the proton inside the nucleus, which is able to decay into a neutron, a positron, and a neutrino, conserve baryon number.

Pity the guy who lives forever. The proton can't decay into pions because it would violate baryon number conservation. It can't decay into a neutron and a positron and a neutrino because of energy conservation. It can't decay into neutrinos or photons because of charge conservation. There are more conservation laws, and we feel that the conservation laws shape the world. As should be obvious, if the proton could decay it would threaten our existence. Of course, that does depend on the proton's lifetime. Since the universe is fifteen or so billion years old, a lifetime *much* longer than this would not influence the fate of the Republic too much.

Newer unified field theories, however, predict that baryon number will not be *strictly* conserved. This prediction has stimulated impressive efforts to detect proton decay, so far without success. But it does illustrate the existence of *approximate* conservation laws. Parity was one example. Strangeness was devised to understand why a number of baryons lived much longer than they should, given all the possible final states into which they could decay. We learned later that strangeness in a particle — lambda or kaon, for example — means the presence of the **s** quark. But lambda and kaon do decay, and the **s** quark does change into a lighter **d** quark in the process. However, this involves the weak force — the strong force will have no part of an **s** → **d** process; in other words, the strong force conserves strangeness. Since the weak force is weak, the decay of lambda, kaon, and its family members is slow, and the lifetime is long — 10^{-10} seconds instead of an allowed process that typically takes 10^{-23} seconds.

The many experimental handles on conservation laws are fortunate, because an important mathematical proof showed that conservation laws are related to symmetries that nature respects. (And symmetry, from Thales to Sheldon Glashow, is the name of the game.) This connection was discovered by Emmy Noether, a woman mathematician, about 1920.

But back to our story.

NIOBIUM BALLS

Despite the omega minus and other successes, no one had ever seen a quark. I'm speaking here in the physicist sense, not the skeptical-lady-in-the-audience sense. Zweig claimed from the beginning that aces/quarks were real entities. But when John Peoples, the current director of Fermilab, was a young experimenter in search of quarks, Gell-Mann told him not to worry about them, that quarks were merely "an accounting device."

Saying this to an experimenter is like throwing down a gauntlet. Searches for quarks began everywhere. Of course, any time you put up a "Wanted" sign, false sightings appear. People looked in cosmic rays, in deep ocean sediment, in old, fine wine ('Shno quarks here, hic!) for a funny electric charge trapped in matter. All the accelerators were used in attempts to smash quarks out of their prisons. A charge of ⅓ or ⅔ would have been relatively easy to find, but still most searches came up empty. One Stanford University experimenter, using tiny, precisely engineered balls made of pure niobium, reported trapping a quark. The experiment languished when it couldn't be repeated, and disrespectful undergrads wore T-shirts inscribed "You have to have niobium balls if you want to trap quarks."

Quarks were spooky; the failure to find free quarks and the ambivalence of the original concept slowed the acceptance of the concept until the late sixties, when a different class of experiments demanded quarks, or at least quarklike things. Quarks were invented to explain the existence and classification of the huge number of hadrons. But if a proton had three quarks, why didn't they show up? Well, we gave it away earlier. They can be "seen." It's Rutherford all over again.

"RUTHERFORD" RETURNS

A series of scattering experiments was undertaken using new electron beams at SLAC in 1967. The objective was a more incisive study of the structure of the proton. The electron at high energy goes in, hits a proton in a hydrogen target, and an electron of much lower energy comes out, but at a large angle to its initial path. The pointlike structures inside the proton act in some sense as the nucleus did for Rutherford's alpha particles. The issue here, however, was more subtle.

The Stanford team, led by SLAC physicist Richard Taylor, a Canadian, and two MIT physicists, Jerome Friedman and Henry Kendall, were enormously aided by the theoretical kibitzing of Richard Feynman and James Bjorken. Feynman had been lending his energy and imagination to the strong interactions and in particular to "what's inside the proton?" He was a frequent visitor to Stanford from his base at Cal Tech in Pasadena. Bjorken (everyone calls him "Bj"), a Stanford theorist, was intensely interested in the experimental process and in the rules underlying seemingly inchoate data. These rules, Bjorken reasoned, would be indicators of the basic laws (inside the black box) controlling the structure of the hadrons.

Here we have to go back to our good friends Democritus and Boscovich, both of whom shed light on the subject. Democritus's test for an a-tom is that it must be indivisible. In the quark model the proton is actually a gooey agglomerate of three quickly moving quarks. But because those quarks are always inextricably tethered to one another, experimentally the proton appears indivisible. Boscovich added a second test. An elementary particle, or a-tom, must be pointlike. This test the proton fails decidedly. The MIT-SLAC team, with assists from Feynman and Bj, came to realize that the operative criterion in this instance was "points" rather than indivisibility. Translating their data into a model of pointlike constituents required much more subtlety than Rutherford's experiment did. That's why it was so convenient to have two of the world's best theorists on the team. The outcome was that the data did indeed indicate the presence of pointlike moving objects inside the proton. In 1990, Taylor, Friedman, and Kendall picked up their Nobel for establishing the reality of quarks. (They are the scientists referred to by Jay Leno at the beginning of the chapter.)

A good question: how can these guys see quarks when quarks are never free? Consider a sealed box with three steel balls inside. You shake the box, tilt it in various ways, listen, and conclude: three balls. The more subtle point is that quarks are always detected in proximity to other quarks, which may change their properties. This factor had to be dealt with but . . . *piano, piano.*

The quark theory made more converts, especially as theorists watching the data began imbuing the quarks with increasing reality, adding to their properties and converting the inability to see free quarks into a virtue. "Confinement" became the buzzword. Quarks are permanently confined because the energy required to separate quarks *increases* as the distance between quarks increases. Then, as one tries harder, the energy becomes sufficient to create a quark-antiquark pair, and now we have four quarks, or two mesons. It's like trying to take home one end of a string. One snips it and, oops, two strings.

Reading quark structure out of electron-scattering experiments was very much a West Coast monopoly. I must note, however, that very similar data were being collected at the same time by my group at Brookhaven. I've often joked that if Bjorken had been an East Coast theorist, *I* would have discovered quarks.

The two contrasting experiments at SLAC and Brookhaven demonstrate that there is more than one way to skin a quark. In both experiments the target particle was a proton. But Taylor, Friedman, and Kendall were using electrons as probes, and we were using protons. At SLAC they sent electrons into the "black box of the collision region" and measured the electrons coming out. Lots of other things, such as protons and pions, also came out, but these were ignored. At Brookhaven we were colliding protons on a piece of uranium (going after the protons therein) and concentrating on pairs of muons coming out, which we measured carefully. (For those of you who haven't been paying attention, electrons and muons are both leptons with identical properties except that the muon is two hundred times heavier.)

I said earlier that the SLAC experiment was similar to Rutherford's scattering experiment that revealed the nucleus. But Rutherford simply bounced alpha particles off the nucleus and measured the angles. At SLAC the process was more complicated. In the language of the theorist and in the mental image evoked by the mathematics, the incoming electron in the SLAC machine sends a messenger photon into the black box. If the photon has the right properties, it can be absorbed by one of the quarks. When the electron tosses a successful messenger photon (one that gets eaten), the electron alters its energy and motion. It then leaves the black box area and goes out and gets itself measured. In other words, the energy of the outgoing electron tells us something about the messenger photon it threw, and, more important, what ate it. The pattern of messenger photons could be interpreted only as being absorbed by a pointlike substructure in the proton.

In the dimuon experiment (so called because it produces two muons) at Brookhaven, we send high-energy protons into the black box region. The energy from the proton stimulates a messenger photon to be radiated from the black box. This photon, before leaving the box, converts into a muon and its antimuon, and these particles leave the box and get measured. This tells us something about the properties of the messenger photon, just as the SLAC experiment did. However, the muon-pair experiment was not theoretically understood until 1972 and, indeed, required many other subtle proofs before its unique interpretation was given.

This interpretation was first done by Sidney Drell and his student Tung Mo Yan at Stanford, not surprisingly, where quarks ran in the blood. Their conclusion: the photon that generates our muon pair is generated when a quark in the incoming proton collides with and annihilates an antiquark in the target (or the other way around). This is widely known as the Drell-Yan experiment even though we invented it and Drell "merely" found the right model.

When Richard Feynman called my dimuon experiment the "Drell-Yan experiment" in a book — surely he was joking — I phoned Drell and told him to call all the people who bought the book and ask them to cross out Drell and Yan on page 47 and write in Lederman. I didn't dare bug Feynman. Drell cheerfully agreed, and justice triumphed.

Since those days, Drell-Yan-Lederman experiments have been carried out in all the labs and have given complementary and confirmatory evidence of the detailed way in which quarks make protons and mesons. Still, the SLAC/Drell-Yan-Lederman studies did not convert all physicists into quark believers. Some skepticism remained. At Brookhaven there was a clue right in front of our eyes that would have answered the skeptics had we known what it meant.

In our 1968 experiment, the first of its kind, we were examining the smooth decrease in the yield of muon pairs as the mass of the messenger photons increased. A messenger photon can have a transitory mass of any value, but the higher the mass, the shorter the time it lives and the harder it is to generate. Heisenberg again. Remember, the higher the mass, the smaller the region of space that is being explored, so we should see fewer and fewer events (numbers of pairs of muons) as the energy increases. We chart this on a graph. Along the bottom of the graph, the x-axis, we show increasing masses. On the vertical y-axis we show numbers of muon pairs. So what we should get is a graph that looks like this:

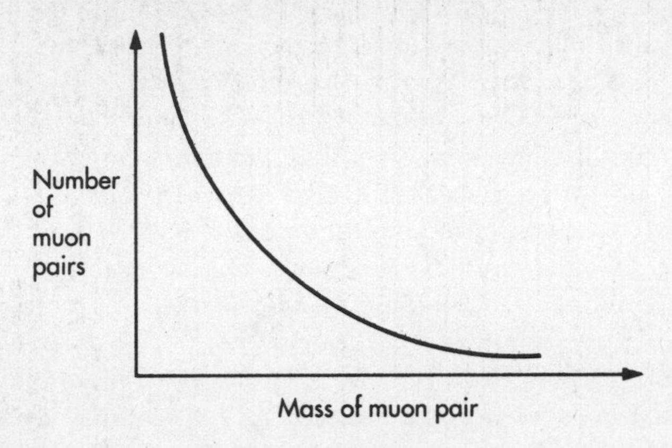

We should see a smooth descending line indicating ever-decreasing muon pairs as the energy of the photons coming out of the black box increases. But instead we got something that looked like this:

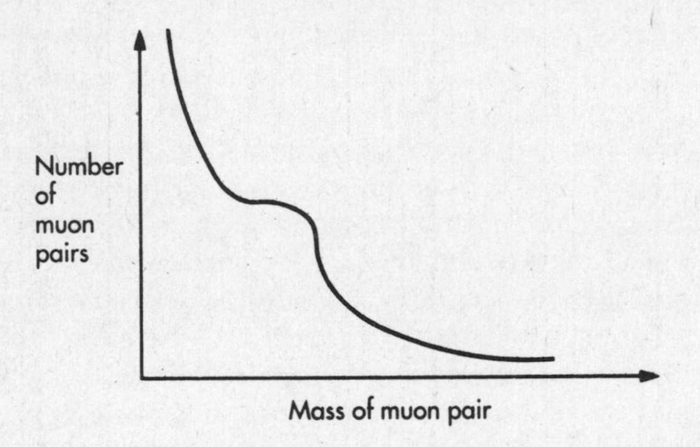

At about the 3 GeV mass level this smooth decrease was interrupted by a "shoulder," now called the Lederman Shoulder. A shoulder or a bump in the graph indicates an unexpected event, something that can't be explained by the messenger photons alone, something sitting on top of the Drell-Yan events. We did not report this shoulder as a new particle. It was the first clear miss of a discovery that would finally establish the reality of the quark hypothesis.

Incidentally, our chagrin at missing the discovery of pointlike structures in the proton, a discovery that by Swedish decree went to Friedman, Kendall, and Taylor, is mock chagrin. Even Bjorken might not have seen through the subtleties of relating the Brookhaven dimuons to quarks in 1968. The dimuon experiment, in retrospect, is my favorite. The concept was original and imaginative. Technically it was childishly simply — so simple that I missed the discovery of the decade. The data had three components — Drell-Yan proof of pointlike structures, proof of the concept of "color" in its absolute rates (discussed later), and the J/Psi discovery (directly ahead) — each of which was of Nobel quality. The Royal Swedish Academy could have saved at least two prizes had we done it right!

THE NOVEMBER REVOLUTION

Two experiments began in 1972 and 1973 that would change physics. One took place at Brookhaven, an old army camp amid the scrub pines and sand, a mere ten minutes from some of the most beautiful beaches in the world, on the south shore of Long Island, host to the Atlantic rollers coming straight from Paris. The other site was SLAC, in the brown hills above the Spanish-style campus of Stanford University. Both experiments were fishing expeditions. Neither was sharply motivated but both would come together in November of 1974 with a crash heard round the world. The events of late 1974 go down in physics history as the November Revolution. It is told around fireplaces wherever physicists gather to talk of old times and great heroes and to sip Perrier. The prehistory is the almost religious idea of theorists that nature must be pretty, symmetrical.

We should first mention that the quark hypothesis did not threaten the electron's status as an elementary particle, as an a-tom. Now there were two classes of pointlike a-toms — the quarks and the leptons. The electron, along with the muon and the neutrino, is a lepton. That would have been fine, except that Schwartz, Steinberger, and Lederman had fouled up the symmetry with the two-neutrino experiment. Now we had four leptons (electron, electron neutrino, muon, and muon neutrino) but three quarks (up, down, and strange). A chart in 1972 might have looked like this in physics shorthand:

quarks:	**u d s**	
leptons:	e	μ
	ν_e	ν_μ

Ugh. Well, you wouldn't have made such a chart because it didn't make much sense. The leptons are in a nice two-by-two pattern, but the quark sector was relatively ugly in a threesome, when theorists were already disillusioned with the number 3.

Theorists Sheldon Glashow and Bjorken had more or less noted (in 1964) that it would be simply charming if there were a fourth quark. This would restore the symmetry between quarks and leptons, which had been destroyed by our discovery of the muon neutrino, the fourth lepton. In 1970 a more cogent theoretical reason for suspecting the fourth quark appeared in a complicated but lovely argument made by Glashow and his collaborators. It converted Glashow into a passionate quark advocate. Shelly, as he is known to his admirers and his enemies, has written a number of books that establish just how passionate he can get. A major architect of our standard model, Shelly is also much appreciated for his stories, his cigars, and his critical commentaries on theoretical trends.

Glashow became an active marketer of the theoretical invention of a fourth quark, which of course he called *charm*. He traveled from seminar to workshop to conference, insisting that experimenters look for a charmed quark. His idea was that this new quark and a new symmetry in which quarks also come in matched pairs — up/down and charm/strange — would cure many pathologies (Doctor, here is where it hurts) in the theory of the weak force. It would for example serve to cancel certain reactions that had not been seen but had been predicted. Slowly he won adherents, at least among theorists. In the summer of 1974, a seminal review paper, "The Search for Charm," was written by theorists Mary Gaillard (one of the tragically few women in physics and one of the top theorists of any sex), Ben Lee, and Jon Rosner. The paper was especially instructive for experimenters because it pointed out that such a quark, call it **c**, and its antiparticle $\bar{c}$, or **c**-bar, could be made in the black collision box and emerge as a neutral meson in which **c** and $\bar{c}$ were bound together. They even proposed that the old Brookhaven data my group had taken of muon pairs may have been evidence of a $c\bar{c}$ decaying into two muons, and that this could be the interpretation of the Lederman Shoulder near 3 GeV. That is, 3 GeV was presumably the mass of the $c\bar{c}$ thing.

BUMP HUNTING

Still, these were only theorists talking. Other published accounts of the November Revolution have implied that the experimenters involved were somehow working their tails off to verify the ideas of the theorists. Dream on. They were fishing. In the case of the Brookhaven physicists, they were "bump hunting," looking for blips in the data that might indicate some new physics — something that would upset the apple cart, not steady it.

At the time that Glashow, Gaillard, and others were talking charm, experimental physics was having its own problems. By then, the competition between electron-positron ($e^- e^+$) colliders and proton accelerators was clearly recognized. The "lepton people" and the "hadron people" had a spirited debate going. Electrons hadn't done much. But you should have heard the propaganda! Because electrons are thought to be structureless points, they offer a clean initial state: an e^- (electron) and an e^+ (positron, the electron's antiparticle) heading toward each other in the black-box collision domain. Clean, simple. The initial step here, the model insisted, is that the particle-antiparticle collision generates a messenger photon of energy equal to the sum of the two particles.

Now, the messenger photon has a brief existence, then materializes into pairs of particles of appropriate mass, energy, spin, and other quantum numbers imposed by the laws of conservation. These come out of the black box and what we commonly see are (1) another $e^+ e^-$ pair, (2) a muon-antimuon pair, or (3) hadrons in a wide variety of combinations but constrained by the initiating condition — the energy and quantum properties of the messenger photon. The variety of possible final states, all derived from a simple initial state, speaks to the power of the technique.

Contrast this with the collision of two protons. Each proton has three quarks, which are exerting strong forces on one another. This means that they are rapidly exchanging gluons, the messenger particles of the strong force (we'll meet gluons later in the chapter). To add to the complexity of our unlovely proton, a gluon, on its way from, say, an up quark to a down quark, can momentarily forget its mission and materialize (like the messenger photons) into any quark and its antiquark, say s and $\bar{s}$ (s-bar). The $s\bar{s}$ appearance is very fleeting, since the gluon has to get back together again in time to be absorbed, but in the meantime it makes for a complicated object.

Physicists who were stuck with using electron accelerators sneeringly called protons "garbage cans" and portrayed a proton-proton or proton-antiproton collision, not without some justice, as a collision of two garbage cans, out of which flew eggshells, banana peels, coffee grounds, and torn parimutuel tickets.

In 1973–74, the Stanford electron-positron ($e^- e^+$) collider, called SPEAR, began taking data, and ran into an inexplicable result. It appeared that the fraction of collisions yielding hadrons was higher than theoretical estimates. The story is complicated and not too interesting until October of 1974. The SLAC physicists, led by Burton Richter, who, in the hallowed tradition of group leaders, was away at the time, began to close in on some curious effects that appeared when the sum of the energies of the two colliding particles was near 3.0 GeV, a suggestive mass, as you may recall.

What added salsa to the affair was that three thousand miles east at Brookhaven, a group from MIT was repeating our 1967 dimuon experiment. Samuel C. C. Ting was in charge. Ting, who is rumored to have been the leader of all the Boy Scouts in Taiwan, got his Ph.D. at Michigan, did a postdoc term at CERN, and in the early sixties joined my group as assistant professor at Columbia, where his rough edges were sharpened.

A meticulous, driven, precise, organized experimenter, Ting worked with me at Columbia for a few years, had several good years at the DESY lab near Hamburg, Germany, and then went to MIT as a professor. He quickly became a force (the fifth? sixth?) to be reckoned with in particle physics. My letter of recommendation deliberately played up some of his weak points — a standard ploy in getting someone hired — but I did it in order to conclude: "Ting — a hot and sour Chinese physicist." In truth, I had a hang-up about Ting, which dates back to the fact that my father operated a small laundry, and as a child I listened to many stories about the Chinese competition across the street. Since then, any Chinese physicist has made me nervous.

When Ting worked with the electron machine in the DESY lab, he became an expert in analyzing $e^+ e^-$ pairs from electron collisions, so he decided that detecting electron pairs is the better way to do the Drell-Yan, oops, I mean the Ting dilepton experiment. So here he was in 1974 at Brookhaven, and, unlike his counterparts at SLAC who were colliding electrons and positrons, Ting was using high-energy protons, directing them into a stationary target, and looking at the $e^+ e^-$ pairs that came out of the black box with the latest word in instrumentation — a vastly more precise detector than the crude instrument we had put together seven years earlier. Using Charpak wire chambers, he was able to determine precisely the mass of the messenger photon or whatever else would give rise to the observed electron-positron pair. Since muons and electrons are both leptons, which pair you chose to detect is a matter of taste. Ting was bump hunting, fishing for some new phenomenon rather than trying to verify some new hypothesis. "I am happy to eat Chinese dinners with theorists," Ting once reportedly said, "but to spend your life doing what they tell you is a waste of time." How appropriate that such a personality would be responsible for finding a quark named charm.

The Brookhaven and SLAC experiments were destined to make the same discovery, but until November 10, 1974, neither group knew much about the other's progress. Why are the two experiments connected? The SLAC experiment collides an electron against a positron, creating a virtual photon as the first step. The Brookhaven experiment has an unholy complicated mishmash initial state, but it looks at virtual photons only if and when they emerge and dissolve into an $e^+ e^-$ pair. Both deal then with the messenger photon, which can have any transitory mass/energy; it depends on the force of the collision. The well-tested model of what goes on in the SLAC collision says a messenger photon is created that can dissolve into hadrons — three pions, say, or a pion and two kaons, or a proton, antiproton, and two pions, or a pair of muons or electrons, and so on. There are many possibilities, consistent with the input energy, momentum, spin, and other factors.

So if something new exists whose mass is less than the sum of the two colliding beam energies, it also can be made in the collision. Indeed, if the new "thing" has the same popular quantum numbers as the photon, it can dominate the reaction when the sum of the two energies is precisely equal to the new thing's mass. I've been told that just the right pitch and force in a tenor's voice can shatter a glass. New particles come into being in a similar fashion.

In the Brookhaven version the accelerator sends protons into a fixed target, in this case a small piece of beryllium. When the relatively large protons hit the relatively large beryllium nuclei, all kinds of things can and do happen. A quark hits a quark. A quark hits an antiquark. A quark hits a gluon. A gluon hits a gluon. No matter what the energy of the accelerator, collisions of much lower energies occur, because the quark constituents share the total energy of the proton. Thus, the lepton pairs that Ting measured in order to interpret his experiment came out of the machine more or less randomly. The advantage of such a complex initial state is that you have some probability of producing everything that can be reached at that energy. So much is going on when two garbage cans collide. The disadvantage is that you have to find the new "thing" among a big pile of debris. To prove the existence of a new particle, you need many runs to get it to show up consistently. And you need a good detector. Fortunately, Ting had a beauty.

SLAC's SPEAR machine was the opposite. It collided electrons with positrons. Simple. Pointlike particles, matter and antimatter, colliding, annihilating one another. The matter turns into pure light, a messenger photon. This packet of energy in turn coalesces back into matter. If each beam is, say, 1.5525 GeV, you get double that, a 3.105 GeV collision, every time. And if a particle exists at that mass, you can produce this new particle instead of a photon. You're almost forced to make the discovery; that's all the machine can do. The collisions it produces have a predetermined energy. To switch to another energy, the scientists have to reset the magnets and make other adjustments. The Stanford physicists could fine-tune the machine energy to a precision far beyond what had been designed into it, a remarkable technological accomplishment. Frankly, I didn't think it could be done. The disadvantage of a SPEAR-type machine is that you must scan the energy domain, very slowly, in extremely small steps. On the other hand, when you hit the right energy — or if you're tipped off somehow, and this was to become an issue — you can discover a new particle in a day or less.

Let's return for a moment to Brookhaven. In 1967–68, when we observed the curious dimuon shoulder, our data went from 1 GeV to 6 GeV, and the number of muon pairs at 6 GeV was only one millionth of what it was at 1 GeV. At 3 GeV there was an abrupt leveling of the yield of muon pairs, and above approximately 3.5 GeV the plunge resumed. In other words, there was this plateau, this shoulder from 3 to 3.5 GeV. In 1969, when we were getting ready to publish our data, we seven authors argued about how to describe the shoulder. Was it a new particle whose effect was smeared out by the highly distorting detector? Was it a new process that produced messenger photons with a different yield? No one knew, in 1969, how the muon pairs were produced. I decided that the data were not good enough to claim a discovery.

Well, in a dramatic confrontation on November 11, 1974, it turned out that the SLAC and Brookhaven groups each had clear data on an enhancement at 3.105 GeV. At SLAC, when the machine was tuned to that energy (no mean feat!), the counters recording collisions went mad, increasing by a hundredfold and dropping back to the base value when the accelerator was tuned to 3.100 or 3.120. The sharpness of the resonance was the reason it had taken so long to find; the group had gone over that territory before and had missed the enhancement. In Ting's Brookhaven data, the outgoing pairs of leptons, precisely measured, showed a sharp bump centered near 3.10 GeV. He, too, concluded that the bump could mean only one thing: he had discovered a new state of matter.

The problem of scientific priority in the Brookhaven/SLAC discovery was a very thorny controversy. Who did it *first?* Accusations and rumors flew. One charge was that the SLAC scientists, aware of Ting's preliminary results, knew where to look. The countercharge was that Ting's initial bump was inconclusive and was massaged in the hours between SLAC's discovery and Ting's announcement. The SLAC people named the new object ψ (psi). Ting named it J. Today it is generally called the J/Ψ or J/psi. Love and harmony have been restored in the community. More or less.

WHY THE FUSS? (AND SOME SOUR GRAPES)

All very interesting, but why the tremendous fuss? Word of the November 11 joint announcement spread instantly around the world. One CERN scientist recalled: "It was indescribable. Everybody in the corridors was talking about it." The Sunday *New York Times* put the discovery on its front page: NEW AND SURPRISING TYPE OF ATOMIC PARTICLE FOUND. *Science*: TWO NEW PARTICLES DELIGHT AND PUZZLE PHYSICISTS. And the dean of science writers, Walter Sullivan, wrote later in the *New York Times*: "Hardly, if ever, has physics been in such an uproar . . . and the end is not in sight." A brief two years later, Ting and Richter shared the 1976 Nobel Prize for the J/psi.

The news came to me, hard at work on a Fermilab experiment with the exotic designation E-70. Can I now, writing in my study seventeen years later, recall my feelings? As a scientist, as a particle physicist, I was overjoyed at the breakthrough, a joy tinged, of course, with envy and even just a touch of murderous hatred for the discoverers. That's the normal reaction. But I had been there — Ting was doing my experiment! True, the kinds of chambers that made Ting's experiment sharp weren't available in 1967–68. Still, the old Brookhaven experiment had the ingredients of two Nobel Prizes — if we had had a more capable detector and if Bjorken had been at Columbia and if we had been slightly more intelligent . . . And if my grandmother had had wheels — as we used to taunt "iffers" — she would have been a trolley car.

Well, I can only blame myself. After spotting the mysterious bump in 1967, I had decided to pursue the physics of dileptons at the newer high-energy machines coming on the air. CERN, in 1971, was scheduled to inaugurate a proton-proton collider, the ISR, with an effective energy twenty times that of Brookhaven's. Abandoning my Brookhaven bird in hand, I submitted a proposal to CERN. When that experiment started taking data in 1972, I again failed to see the J/psi, this time because of a fierce background of unexpected pions and our newfangled leaded-glass particle detector, which was, unknown to us, being irradiated by the new machine. The background turned out to be a discovery in itself: we detected high-transverse-momentum hadrons, another kind of data signifying the quark structure inside protons.

Meanwhile, also in 1971, Fermilab was getting ready to start a 200 GeV machine. I gambled on this new machine too. The Fermilab experiment turned on in early 1973, and my excuse was . . . well, we really didn't get down to doing what we had proposed to do, being diverted by curious data several groups had been seeing in the brand-new Fermilab environment. It turned out to be a red herring or a blue shrimp, and by the time we got around to dileptons, the November Revolution was in the history books. So not only did I miss the J at Brookhaven, I missed it at both new machines, a new record of malpractice in particle physics.

I haven't yet answered the question, what was the big deal? The J/psi was a hadron. But we have discovered hundreds of hadrons, so why blow a gasket over one more, even if it has a fancy name like J/psi? It has to do with its high mass, three times heavier than the proton, and the "sharpness" of the mass, less than 0.05 MeV.

Sharpness? What that means is the following. An unstable particle cannot have a unique, well-defined mass. The Heisenberg uncertainty relations spell it out. The shorter the lifetime, the wider the distribution of masses. It is a quantum connection. What we mean by a distribution of masses is that a series of measurements will yield different masses, distributed in a bell-shaped probability curve. The peak of this curve, for example 3.105 GeV, is called the mass of the particle, but the spread in mass values is in fact a measurement of the particle's lifetime. Since uncertainty is reflected in measurement, we can understand this by noting that for a stable particle, we have infinite time to measure the mass and therefore the spread is infinitely narrow. A very short lived particle's mass cannot be determined precisely (even in principle), and the experimental result, even with superfine apparatus, is a broad spread in the mass measurements. As an example, a typical strong-interaction particle decays in 10^{-23} seconds and has a mass spread of about 100 MeV.

One more reminder. We noted that all hadron particles are unstable except the free proton. The higher the mass of a hadron (or any particle), the shorter its lifetime because it has more things into which it can decay. So now we find a J/psi with a huge mass (in 1974 it was the heaviest particle yet found), but the shock is that the observed mass distribution is exceedingly sharp, more than a thousand times narrower than that of a typical strong-interaction particle. Thus it has a *long* lifetime. Something is preventing it from decaying.

NAKED CHARM

What inhibits its decay?

Theorists all raise their hands: a new quantum number or, equivalently, a new conservation law is operating. What kind of conservation? What new thing is being conserved? Ah, now all the answers were different, for a time.

Data continued to pour in, but now only from the $e^+ e^-$ machines. SPEAR was eventually joined by a collider in Italy, ADONE, and later by DORIS, in Germany. Another bump showed at 3.7 GeV. Call it ψ' (psi prime), no need to mention J, since this was Stanford's baby entirely. (Ting and company had gotten out of the game; their accelerator had been barely capable of discovering the particle and not capable of examining it further.) But despite feverish effort, attempts to explain the surprising sharpness of J/psi were at first stymied.

Finally one speculation began to make sense. Maybe J/psi was the long-awaited bound "atom" of **c** and $\bar{c}$, the charm quark and its antiquark. In other words, perhaps it was a meson, that subclass of hadron consisting of quark and antiquark. Glashow, exulting, called J/psi "charmonium." As it turned out, this theory was correct, but it took another two years for the speculation to be verified. The reason for the difficulty is that when **c** and $\bar{c}$ are combined, the intrinsic properties of charm are wiped out. What **c** brings, $\bar{c}$ cancels. While all mesons consist of quark and antiquark, they don't have to consist of a quark with its own particular antiquark, as does charmonium. A pion, for example, is $\mathbf{u\bar{d}}$.

The search was on for "naked charm," a meson that was a charm quark tethered with, say, an antidown quark. The antidown quark wouldn't cancel the charm qualities of its partner, and charm would be exposed in all its naked glory, the next best thing to what is impossible: a free charm quark. Such a meson, a $\mathbf{c\bar{d}}$, was found in 1976 at the Stanford $e^+ e^-$ collider by a SLAC-Berkeley group led by Gerson Goldhaber. The meson was named D^0 (D zero), and studies of D's were to occupy the electron machines for the next fifteen years. Today, mesons like $\mathbf{c\bar{d}}$, $\mathbf{c\bar{s}}$, and $\mathbf{c\bar{d}}$ are grist for the Ph.D. mill. A complex spectroscopy of states enriches our understanding of quark properties.

Now the sharpness of J/psi was understood. Charm is a new quantum number, and the conservation laws of the strong force did not permit a **c** quark to change into a lower-mass quark. To do this, the weak and electromagnetic forces had to be invoked, and these are much slower to act — hence the long lifetime and narrow width.

The last holdouts against the idea of quarks gave up about this time. The quark idea had led to a far-out prediction, and the prediction had been verified. Probably even Gell-Mann began to give quarks elements of reality, although the confinement problem — there can be no such thing as a free quark — still differentiates quarks from other matter particles. With charm, the periodic table now was balanced again:

QUARKS

up (**u**)	charm (**c**)
down (**d**)	strange (**s**)

LEPTONS

electron neutrino (ν_e)	muon neutrino (ν_μ)
electron (e)	muon (μ)

Now there were four quarks — that is, four flavors of quarks — and four leptons. We now spoke of two generations, arranged vertically in the above table. The **u**-**d**-ν_e-e is the first generation, and since the up and down quarks make protons and neutrons, the first generation dominates our present world. The second generation, **c**-**s**-ν_μ-μ, is seen in the intense but fleeting heat of accelerator collisions. We can't ignore these particles, exotic as they may seem. Intrepid explorers that we are, we must struggle to figure out what role nature had planned for them.

I have not really given due attention to the theorists who anticipated and helped to establish the J/psi as charmonium. If SLAC was the experimental heart, Harvard was the theoretical brain. Glashow and his Bronx High School of Science classmate Steve Weinberg were aided by a gaggle of young whizzes; I'll mention only Helen Quinn because she was in the thick of the charmonium euphoria and is on my role-model team.

THE THIRD GENERATION

Let's pause and step away. It's always more difficult to describe recent events, especially when the describer is involved. There is not enough of the filter of time to be objective. But we'll give it a try anyway.

Now it was the 1970s, and thanks to the tremendous magnification of the new accelerators and the matching ingenious detectors, progress toward finding the a-tom was very rapid. Experimenters were going in all directions, learning about the various charmed objects, examining the forces from a more microscopic point of view, poking at the energy frontier, addressing the outstanding problems of the minute. Then a brake on the pace of progress was applied as research funds became increasingly difficult to find. Vietnam, with its drain on the spirit and the treasury, as well as the oil shock and general malaise resulted in a turning away from basic research. This hurt our colleagues in "small science" even more. High-energy physicists are in part protected by the pooling of efforts and sharing of facilities in large laboratories.

Theorists, who work cheap (give them a pencil, some paper, and a faculty lounge), were thriving, stimulated by the cascade of data. We still saw the same pros: Lee, Yang, Feynman, Gell-Mann, Glashow, Weinberg, and Bjorken, but other names would soon appear: Martinus Veltman, Gerard 't Hooft, Abdus Salam, Jeffrey Goldstone, Peter Higgs, among others.

Let's just quickly touch on the experimental highlights, thereby unfairly favoring the "bold salients into the unknown" over the "slow steady advance of the frontier." In 1975, Martin Perl, almost singlehandedly and while dueling, d'Artagnan-like with his own colleague-collaborators, convinced them, and ultimately everyone, that a fifth lepton lurked in the SLAC data. Called tau (τ), it, like its lighter cousins the electron and the muon, comes in two signs: τ^+ and τ^-.

A third generation was in the making. Since both the electron and the muon have neutrinos associated with them, it seemed natural to assume that a neutrino-sub-tau (ν_τ) existed.

Meanwhile, Lederman's group at Fermilab finally learned how to carry out the dimuon experiment correctly, and a new, vastly more effective organization of apparatus exploded open the mass domain from the J/psi peak at 3.1 all the way to pretty nearly 25 GeV, the limit allowed by Fermilab's 400 GeV energy. (Remember, we're talking about stationary targets here, so the effective energy is a fraction of the beam energy.) And there, at 9.4, 10.0, and 10.4 GeV sat three new bumps, as clear as the Tetons viewed on a brilliant day from Grand Targhee ski resort. The huge mass of data multiplied the world's collection of dimuons by a factor of 100. Christened the upsilon (it was the last Greek letter available, we thought), the new particle repeated the story of the J/psi, and the new thing that was conserved was the beauty quark — or, as some less artistic physicists call it, the bottom quark. The interpretation of the upsilon was that it was an "atom" made of a new **b** quark bound to an anti-**b** quark. The higher mass states were simply excited states of this new "atom." The excitement over this discovery nowhere near matched that of J/psi, but a third generation was indeed news and raised an obvious question: how many more? Also, why does nature insist on Xerox copies, one generation replicating the previous one?

Let me offer a brief description of the work that led to the upsilon. Our group of physicists from Columbia, Fermilab, and Stony Brook (Long Island) included some crackerjack young experimenters. We had constructed a state-of-the-art spectrometer with wire chambers, magnets, scintillation hodoscopes, more chambers, more magnets. Our data acquisition system was "dernier cri," based on electronics designed by genius engineer William Sippach. We had all worked in the same domain of Fermilab beams. We knew the problems. We knew one another.

John Yoh, Steve Herb, Walter Innes, and Charles Brown were four of the best postdocs I have seen. The important software was reaching the state of sophistication required for work at the frontier. Our problem was that we had to be sensitive to reactions that happened as rarely as only once in every hundred trillion collisions. Since we needed to record many of these rare dimuon events, we needed to harden the apparatus to a huge rate of irrelevant particles. Our team had developed a unique understanding of how to work in a high-radiation environment and still have survivable detectors. We had learned how to build in redundancy so that we could ruthlessly suppress false information no matter how cleverly nature tried to fool us.

Early in the learning process, we ran in the dielectron mode and obtained about twenty-five electron pairs above 4 GeV. Strangely, twelve of these were clustered around 6 GeV. A bump? We debated and decided to publish the possibility that there was a particle at 6 GeV. Six months later, after the data had increased to three hundred events, poof — no bump at 6 GeV. We had suggested the name "upsilon" for the fake bump, but when better data contradicted the earlier data, the incident became known as *oops-leon.*

Then came our new setup, with all of our experience invested in a rearrangement of target, shielding, placement of magnets, and chambers. We began taking data in May of 1977. The era of month-long runs of twenty-seven events or three hundred events were over; thousands of events per week were now coming in, essentially free of background. It isn't often in physics that a new instrument permits one to survey what amounts to a new domain. The first microscope and the first telescope are historic examples of far greater significance, but the excitement and joy when they were first used cannot have been much more intense than ours. After one week, a wide bump appeared near 9.5 GeV, and soon this enhancement became statistically solid. John Yoh had, in fact, seen a clustering near 9.5 GeV in our three-hundred-event run, but having been burned at 6 GeV, he merely labeled a bottle of Mumm's champagne "9.5" and hid it in our refrigerator.

In June we drank the champagne and broke the news (which had leaked anyway) to the laboratory. Steve Herb gave the talk to a packed and excited auditorium. This was Fermilab's first major discovery. Later that month we wrote up the discovery of a broad bump at 9.5 GeV with 770 events in the peak — statistically secure. Not that we didn't spend endless man-hours (unfortunately we had no women collaborators) looking for a malfunction of the detector that could simulate a bump. Dead regions of the detector? A software glitch? We ruthlessly tracked down dozens of possible errors. All of our built-in security measures — testing the validity of the data by asking questions to which we knew what the answers should be — checked out. By August, thanks to additional data and more sophisticated analysis, we had three narrow peaks, the upsilon family: upsilon, upsilon prime, and upsilon double prime. There was no way to account for these data on the basis of the known physics of 1977. Enter beauty (or bottom)!

There was little resistance to our conclusion that we were seeing a bound state of a new quark — call it the **b** quark — and its antiparticle twin. The J/psi was a $\mathbf{c\bar{c}}$ meson. Upsilon was a $\mathbf{b\bar{b}}$ meson. Since the mass of the upsilon bump was near 10 GeV, the **b** quark must have a mass near 5 GeV. This was the heaviest quark yet recorded, the **c** quark being near 1.5 GeV. Such "atoms" as $\mathbf{c\bar{c}}$ and $\mathbf{b\bar{b}}$ have a lowest-energy ground state and a variety of excited states. Our three peaks represented the ground state and two excited states.

One of the fun things about the upsilon was that we experimentalists could handle the equations of this curious atom, composed of a heavy quark circling a heavy antiquark. Good old Schrödinger's equation worked fine, and with only a brief look at our grad school notes, we raced the professional theorists to calculate the energy levels and other properties that we had measured. We had fun . . . but they won.

Discoveries are always quasi-sexual experiences, and when John Yoh's "bicycle-on-line" quick analysis first indicated the existence of the bump, I experienced the now (for me) familiar feeling of intense euphoria, but tinged with the anxiety that "it can't really be true." The most obvious impulse is to communicate, to tell people. Who? Wives, best friends, children, in this case Director Bob Wilson, whose lab badly needed a discovery. We telephoned our colleagues at the DORIS machine in Germany and asked them to see if they could reach the energy required to make upsilons with their $e^+ e^-$ collider. DORIS was the only other accelerator that had a chance at this energy. In a tour de force of machine magic, they succeeded. More joy! (And more than a little relief.) Later you think about rewards. Will this do it?

The discovery was made traumatic by a fire that interrupted data taking after a good week of running. In May 1977 a device that measures the current in our magnets, supplied no doubt by a low bidder, caught fire, and the fire spread to the wiring. An electrical fire creates chlorine gas, and when your friendly firemen charge in with hoses and spray water everywhere, they create an atmosphere of hydrochloric acid. The acid settles on all the transistor cards and slowly begins to eat them.

Electronic salvage is an art form. Friends at CERN had told me about a similar fire there, so I called to get advice. I was given the name and telephone numbers of a Dutch salvage expert working for a German firm and living in central Spain. The fire occurred on Saturday, and it was now 3 A.M. on Sunday. From my room at Fermilab, I called Spain and reached my man. Yes, he'd come. He'd get to Chicago Tuesday, and a cargo plane from Germany filled with special chemicals would arrive Wednesday. But he needed a U.S. visa, which usually takes ten days. I called the U.S. embassy in Madrid and spouted, "Atomic energy, national security, millions of dollars at stake . . ." I was connected to an assistant to the ambassador who was not impressed until I identified myself as a Columbia professor. "Columbia! Why didn't you say so? I'm class of fifty-six," he shouted. "Tell your fellow to ask for me."

On Tuesday, Mr. Jesse arrived and sniffed at 900 cards, each carrying about 50 transistors (1975 technology). On Wednesday the chemicals arrived. Customs gave us more heartburn, but the U.S. Department of Energy helped. By Thursday we had an assembly line: physicists, secretaries, wives, girlfriends, all dipping cards in secret solution A, then B, then drying with clean nitrogen gas, then brushing with camel's-hair brushes, then stacking. I half expected that we'd be required to accompany the ritual with a low moan of Dutch incantation, but this was not necessary.

Jesse, a horseman, lived in Spain to train with the Spanish cavalry. When he learned I had three horses, he ran off to ride with my wife and the Fermilab horse club. A real expert, he gave everybody pointers. Pretty soon the prairie riders were trading tips on flying changes, passages, lavade, corbette, and capriole maneuvers. We now have a trained Fermilab cavalry to defend the lab should the hostile forces from CERN or SLAC decide to attack on horseback.

Friday we installed all the cards, testing each one carefully. By Saturday morning we were up and running, and a few days later a quick analysis showed that the bump was still there. Jesse stayed on for two weeks, riding horses, charming everyone, advising on fire prevention. We never got a bill from him, but we did pay for the chemicals. And that was how the world acquired a third generation of quarks and leptons.

The very name "bottom" suggests that there must be a "top" quark. (Or if you prefer the name "beauty," then there is a "truth" quark.) The new periodic table now reads:

First generation	*Second generation*	*Third generation*
	QUARKS	
up (**u**)	charm (**c**)	top? (**t**)
down (**d**)	strange (**s**)	bottom (**b**)
	LEPTONS	
electron neutrino (ν_e)	muon neutrino (ν_μ)	tau neutrino (ν_τ)
electron (e)	muon (μ)	tau (τ)

At this writing, the top quark has yet to be found. The tau neutrino has also never been pinned down experimentally, but no one really doubts its existence. Various proposals for a "three-neutrino experiment," a souped-up version of our two-neutrino experiment, have been submitted over the years at Fermilab, but all have been rejected because such a project would be enormously expensive.

Note that the lower left-hand grouping (ν_e-e-ν_μ-μ) in our table was established in the 1962 two-neutrino experiment. Then the bottom quark and the tau lepton put the (almost) finishing touches on the model in the late 1970s.

The table, once the various forces are added to it, is a compact summary of all the data emerging from all of the accelerators since Galileo dropped spheres of unequal weights from the nearly vertical tower at Pisa. This table is called the *standard model* or, alternately, the standard picture or standard theory. (Memorize.)

In 1993 this model is still the ruling dogma of particle physics. The machines of the 1990s, primarily Fermilab's Tevatron and CERN's electron-positron collider (called LEP), are concentrating the efforts of thousands of experimentalists on clues to what lies beyond the standard model. The smaller machines at DESY, Cornell, Brookhaven, SLAC, and KEK (Tsukuba, Japan) are also attempting to refine our knowledge of the many parameters of the standard model and trying to find clues to a deeper reality.

There is much to do. One task is to explore the quarks. Remember, in nature only two kinds of combinations exist: (1) quark plus antiquark (**q$\bar{\text{q}}$**) — these are the mesons — and (2) three quarks (**qqq**) —

the baryons. Now we can play and compose hadrons such as **u$\bar{\text{u}}$, u$\bar{\text{c}}$, u$\bar{\text{t}}$**, and **$\bar{\text{u}}$c, $\bar{\text{u}}$t, d$\bar{\text{s}}$, d$\bar{\text{b}}$** . . . Have fun! And **uud, ccd, ttb** . . . Hundreds of combinations are possible (somebody knows how many). All are particles that either have been discovered and listed in the tables or are ready to be discovered. By measuring the mass and the lifetimes and the decay modes, one learns more and more about the strong quark force mediated by gluons and about weak-force properties. *Much* to do.

Another experimental high point is called "neutral currents," and it is crucial to our story of the God Particle.

The Weak Force Revisited

By the 1970s lots of data had been collected on the decay of unstable hadrons. This decay is really the manifestation of the constituent quarks undergoing reactions — for example, an up quark changing to a down quark or vice versa. Even more informative were the results of several decades of neutrino-scattering experiments. Together, the data insisted that the weak force had to be carried by three massive messenger particles: a W^+, a W^-, and a Z^0. These had to be massive because the weak force has a very small sphere of influence, reaching no farther than approximately 10^{-19} meters. Quantum theory enforces a rough rule that the range of a force varies inversely as the mass of the messenger particle. The electromagnetic force reaches out to infinity (although it gets weaker with distance), and its messenger particle is the zero-mass photon.

But why three force carriers? Why three messenger particles — one positively charged, one negatively charged, and one neutral — to propagate the field that induces the changes of species? To explain, we're going to have to do some physics bookkeeping, making sure that things come out equal on both sides of the arrow (→). This includes the electric-charge signs. If a neutral particle decays into charged particles, for example, the positive charges have to offset the negatives.

First, here's what happens when a neutron decays into a proton, a typical weak-force process. We write it like this:

$$n \rightarrow p^+ + e^- + \bar{\nu}_e$$

We have seen this before: a neutron decays into a proton, an electron, and an antineutrino. Note that the positive proton cancels the negative charge of the electron on the right side of the reaction, the antineutrino being neutral. Everything works out. But this is a superficial view of the reaction, like watching an egg hatch into a blue jay. You don't see what the fetus is doing inside. The neutron is really a conglomerate of three quarks — one up and two downs (**udd**); a proton is two ups and a down (**uud**). So when a neutron decays into a proton, a down quark changes into an up quark. Thus it's more instructive to look inside the neutron and describe what's happening to the quarks. And in quark language, the same reaction can be written:

$$\mathbf{d} \rightarrow \mathbf{u} + e^- + \bar{\nu}_e$$

That is, a down quark in the neutron changes to an up quark, emitting an electron and an antineutrino. However, this too is a simplified version of what really happens. The electron and antineutrino don't come directly out of the down quark. There's an intermediate reaction involving a W^-. The quantum theory of the weak force therefore writes the neutron decay process in two stages:

$$1)\ \mathbf{d}^{-1/3} \rightarrow W^- + \mathbf{u}^{+2/3}$$

and then

$$2)\ W^- \rightarrow e^- + \bar{\nu}_e$$

Note that the down quark decays first into a W^- and an up quark. The W in turn decays into the electron and antineutrino. The W is the mediator of the weak force and participates in the decay reaction. In the above reaction it must be a negative W to balance the change in electric charge when **d** goes to **u**. When you add the -1 charge of the W^- to the $+2/3$ charge of the up quark, you get $-1/3$, the charge of the down quark that started the reaction. Everything works out.

In nuclei, up quarks can also decay into down quarks, turning protons into neutrons. In quark language the process is described: $\mathbf{u} \rightarrow W^+ + \mathbf{d}$ and then $W^+ \rightarrow e^+ + \nu_e$. Here we need a positive W to balance the change of charge. Thus the observed decays of quarks, via the changes of neutrons to protons and vice versa, require both a W^+ and a W^-. But that's not the whole story.

Experiments carried out in the mid-1970s involving neutrino beams established the existence of "neutral currents," which in turn required a neutral heavy force carrier. These experiments were stimulated by theorists like Glashow who were working the unification-of-forces frontier and were frustrated by the fact that weak forces seemed to require only charged force carriers. The hunt was on for neutral currents.

A current is basically anything that flows. A current of water flows in a river or a pipe. A current of electrons flows in a wire or through a solution. The W^- and W^+ mediate the flow of particles from one state to another, and the need to keep track of the electric charge probably generated the "current" concept. The W^+ mediates a positive current; the W^- mediates the negative current. These currents are studied in spontaneous weak decays, such as those just described. But they can also be generated by neutrino collisions in accelerators, made possible by the development of neutrino beams in the Brookhaven two-neutrino experiment.

Let's look at what happens when a muon neutrino, the kind we discovered at Brookhaven, collides with a proton — or, more specifically, with an up quark in the proton. The collision of a muon antineutrino with an up quark generates a down quark and a positive muon.

$$\bar{\nu}_\mu + \mathbf{u}^{-2/3} \rightarrow \mathbf{d}^{-1/3} + \mu^{+1}$$

Or, in English, muon antineutrino plus up quark → down quark plus positive muon. Effectively, when the neutrino and up quark collide, the up turns into a down and the neutrino converts to a muon. Again, what really happens in the weak-force theory is a two-reaction sequence:

$$1)\ \bar{\nu}_\mu \rightarrow W^- + \mu^+$$
$$2)\ W^- + \mathbf{u} \rightarrow \mathbf{d}$$

The antineutrino collides with the up quark and leaves the collision as a muon. The up turns into a down, the whole reaction mediated by the negative W. So we have a negative current. Now, even as early as 1955, theorists (notably Glashow's teacher, Julian Schwinger) noted that it would be possible to have a neutral current, like so:

$$\nu_\mu + \mathbf{u} \rightarrow \mathbf{u} + \nu_\mu$$

What's happening here? We have muon neutrinos and up quarks on both sides of the reaction. The neutrino bounces off the up quark but emerges as a neutrino, not a muon as in the previous reaction. The up quark gets nudged but remains an up quark. Since the up quark is part of a proton (or a neutron), the proton, albeit jostled, remains a proton. If we were to look at this reaction superficially, we would see a muon neutrino hitting a proton and bouncing off intact. But it's more subtle than that. In the previous reactions, either a negative or a positive W was required to help facilitate the metamorphosis of an up quark into a down or vice versa. Here, the neutrino must emit a messenger particle to kick the up quark (and be swallowed by it). When we try to write this reaction, it's clear that this messenger particle must be *neutral.*

This reaction is similar to the way we understand the electrical force, say between two protons; there is an exchange of a neutral messenger, the photon, and this produces the Coulomb law of force, which allows one proton to kick another. There is no change of species. The similarity is not fortuitous. The unification crowd (not the Reverend Moon but Glashow and his friends) needed such a process if they were to have a prayer of unifying the weak and electromagnetic forces.

So the experimental challenge was: can we do reactions in which neutrinos collide with nuclei and come out as neutrinos? A crucial ingredient is that we observe the impact on the struck nucleus. There was some ambiguous evidence of such reactions in our two-neutrino experiment at Brookhaven. Mel Schwartz called them "crappers." A neutral particle goes in; a neutral particle comes out. There's no change in electric charge. The struck nucleus breaks up, but very little energy appears in the relatively low-energy neutrino beam at Brookhaven — hence Schwartz's description. Neutral currents. For reasons I forget, the neutral weak messenger particle is called Z^0 (zee zero, we say), rather than W^0. But if you want to impress your friends, use the term "neutral currents," a fancy way of expressing the idea that a neutral messenger particle is required to kick off a weak-force reaction.

TIME TO BREATHE FASTER

Let's review a bit of what the theorists were thinking.

The weak force was first recognized by Fermi in the 1930s. When he wrote down his theory, Fermi modeled it in part on the quantum field theory of the electromagnetic force, quantum electrodynamics (QED). Fermi tried to see if this new force would follow the dynamics of the older force, electromagnetism (older, that is, in terms of our knowledge of it). In QED, remember, the field idea is carried by messenger particles, the photons. So the Fermi theory of the weak force should have messenger particles, too. But what would they be like?

The photon has zero mass, and that gives rise to the famous long-range inverse-square law of the electric force. The weak force was very short range, so in effect Fermi simply gave his force carriers infinite mass. Logical. Later versions of the Fermi theory, most notably by Schwinger, introduced the heavy W^+ and W^- as weak-force carriers. So did several other theorists. Let's see: Lee, Yang, Gell-Mann . . . I hate to credit any theorists because 99 percent of them will be upset. If I occasionally neglect to cite a theorist, it's not because I've forgotten. It's probably because I hate him.

Now comes the tricky part. In program music, a recurring theme introduces an idea or person or animal — like the leitmotif in *Peter and the Wolf* that tells us Peter is about to come onstage. Perhaps more appropriate in this case is the ominous cello that signals the appearance of the great white shark in *Jaws*. I am about to slip in the first thematic notes of the denouement, the sign of the God Particle. But I don't want to reveal her too early. As in any tease show, slow is better.

In the late sixties and early seventies, several young theorists began to study quantum field theory in the hopes of extending the success of QED to the other forces. You may recall that these elegant solutions to action-at-a-distance were subject to mathematical troubles: quantities that should be small and measurable appear in the equation as infinite — and that's a lot. Feynman and friends invented the process of renormalization to hide the infinities in the measured quantities, for example, *e* and *m*, the charge and mass of the electron. QED was said to be a renormalizable theory; that is, you can get rid of the stultifying infinities. However, when quantum field theory was applied to the other three forces — the weak force, the strong force, and gravity — it met with total frustration. It couldn't have happened to nicer guys. With these forces infinities ran wild, and things got so sick that the entire usefulness of quantum field theory was questioned. Some theorists reexamined QED to try to understand why that theory worked (for electromagnetism) and the other theories did not.

QED, the super-accurate theory that gives the g-value to eleven significant places, belongs to a class of theories known as gauge theories. The term gauge in this context means scale, as in HO-gauge model railroad tracks. Gauge theory expresses an abstract symmetry in nature that is very closely tied to experimental facts. A key paper by C. N. Yang and Robert Mills in 1954 stressed the power of gauge symmetry. Rather than proposing new particles to explain observed phenomena, one sought for symmetries that would predict these phenomena. When applied to QED, gauge symmetry actually generated the electromagnetic forces, guaranteed the conservation of charge, and provided, at no extra cost, a protection against the worst infinities. Theories exhibiting gauge symmetry are renormalizable. (Repeat this sentence until it rolls trippingly from the tongue, then try it out at lunch.) But the gauge theories implied the existence of gauge particles. They were none other than our messenger particles: photons for QED, and W^+ and W^- for weak. And for strong? Gluons, of course.

Some of the best and brightest theorists were motivated to work on the weak force for two, no three, reasons. The first is that the weak force was full of infinities, and it was not clear how to make it into a gauge theory. Second was the quest for unification, extolled by Einstein and very much on the minds of this group of young theorists. Their focus was on unifying the weak and electromagnetic forces, a daunting task since the weak force is vastly weaker than the electric force, has a much, much shorter range, and violates symmetries such as parity. Otherwise, the two forces are exactly alike!

The third reason was the fame and glory that would accrue to the guy who solved the puzzle. The leading contestants were Steven Weinberg, then at Princeton; Sheldon Glashow, a fellow science fiction club member with Weinberg; Abdus Salam, the Pakistani genius at Imperial College in England; Martinus Veltman at Utrecht, Netherlands; and his student Gerard 't Hooft. The more elderly theorists (well into their thirties) had set the stage: Schwinger, Gell-Mann, Feynman. There were lots of others around; Jeffrey Goldstone and Peter Higgs were crucial piccolo players.

Eschewing a blow-by-blow account of the theoretical brouhaha from about 1960 to the mid-1970s, we find that a renormalizable theory of the weak force was finally achieved. At the same time it was found that a marriage with the electromagnetic force, QED, now seemed more natural. But to do all this, one had to assemble a common messenger family of particles for the combined "electroweak" force: W^+, W^-, Z^0, and the photon. (It looks like one of those mixed families, with stepbrothers and stepsisters from previous marriages trying to live, at all odds, in harmony while sharing a common bathroom.) The new heavy particle, Z^0, helped to satisfy the demands of gauge theory, and the foursome satisfied all the requirements of parity violation, as well as the apparent weakness of the weak force. Yet at this stage (before 1970) not only hadn't the W's and Z been seen, but neither had the reactions that Z^0 might produce. And how can we talk about a unified electroweak force, when any child in the laboratory can demonstrate huge differences in behavior between the electromagnetic and weak forces?

One problem that the experts confronted, each in his own aloneness, at office or home or airplane seat, was that the weak force, being short range, needed heavy force carriers. But heavy messengers are not what gauge symmetry predicted, and the protest came in the form of infinities, sharp steel into the intellectual guts of the theorist. Also, how do three heavies, W^+, W^-, and Z^0, coexist in a happy family with the massless photon?

Peter Higgs, of the University of Manchester (England), supplied a key — yet another particle, to be discussed soon — which was exploited by Steven Weinberg, then at Harvard, now at the University of Texas. Clearly, we plumbers in the lab see no weak-electromagnetic symmetry. The theorists know that, but they desperately want the symmetry in the basic equations. So we are faced with finding a way to install the symmetry, then break it when the equations get down to predicting the results of the experiment. The world is perfect in the abstract, see, but then it becomes imperfect when we get down to details, right? Wait! I didn't think up any of this.

But here's how it works.

Weinberg, via the work of Higgs, had discovered a mechanism by which a pristine set of zero-mass messenger particles, representing a unified electroweak force, acquired mass by feeding, in a very poetic manner of speaking, on the unwanted components of the theory. Okay? No? Using Higgs's idea to destroy the symmetry, lo! — the W's and Z's acquired mass, the photon remained the same, and in the ashes of the destroyed unified theory there appeared: the weak force and the electromagnetic force. Massive W's and Z's waddled around to create the radioactivity of particles and the reactions that occasionally interfered with neutrino transits of the universe, whereas the messenger photons gave rise to the electricity we all know, love, and pay for. There. Radioactivity (weak force) and light (electromagnetism) neatly(?) tied to one another. Actually, the Higgs idea didn't destroy the symmetry; it just hid it.

Only one question remained. Why would anyone believe any of this mathematical gobbledegook? Well, Tini Veltman (far from tiny) and Gerard 't Hooft had worked the same ground, perhaps more thoroughly, and had shown that if you did the (still mysterious) Higgs trick to break the symmetry, all the infinities that had characteristically lacerated the theory vanished, and the theory was squeaky clean. Renormalized.

Mathematically, a whole set of terms appeared in the equations with signs such as to cancel terms that were traditionally infinite. But there were so many such terms! To do this systematically, 't Hooft wrote a computer program and, on a day in July 1971, watched the output as complicated integrals were subtracted from other complicated integrals. Each of these, if evaluated separately, would give an infinite result. As the readout emerged, term by term the computer printed "0." The infinities were all gone. This was 't Hooft's thesis, and it must go down with de Broglie's as a Ph.D. thesis that made history.

FIND THE ZEE ZERO

Enough for theory. Admittedly, it's complicated stuff. But we'll return to it later, and a firm pedagogical principle acquired from forty or so years of facing students — freshmen to postdocs — says that even if the first pass is 97 percent incomprehensible, the next time you see it, it will be, somehow, hauntingly familiar.

What implications did all this theory have for the real world? The grand implications will have to wait for Chapter 8. The immediate implication in 1970 for experimenters was that a Z^0 had to exist to make everything work. And if the Z^0 was a particle, we should find it. The Z^0 was neutral, like its stepsister the photon. But unlike the massless photon, Z^0 was supposed to be very heavy like its brothers, the twin W's. So our task was clear: look for something that resembles a heavy photon.

W's had been searched for in many experiments, including several of mine. We looked in neutrino collisions, didn't see any, and asserted that failure to find the W could be understood only if the mass of the W was greater than 2 GeV. Had it been lighter, it would have shown up in our second series of neutrino experiments at Brookhaven. We looked in proton collisions. No W. So now its mass had to be greater than 5 GeV. Theorists also had opinions about the W properties and kept raising the mass until, by the late seventies, it was predicted to be about 70 GeV. Way too high for the machines of that era.

But back to Z^0. A neutrino scatters from a nucleus. If it sends out a W^+ (an antineutrino will send out a W^-), it changes to a muon. But if it can send out a Z^0, then it remains a neutrino. As mentioned, since there is no change of electric charge as we follow the leptons, we call it a neutral current.

A real experiment to detect neutral currents isn't easy. The signature is an invisible neutrino coming in, an equally invisible neutrino going out, along with a cluster of hadrons resulting from the struck nucleon. Seeing only a cluster of hadrons in your detector isn't very impressive. It's just what a background neutron would do. At CERN a giant bubble chamber called Gargamelle began operating in a neutrino beam in 1971. The accelerator was the PS, a 30 GeV machine that produced neutrinos of about 1 GeV. By 1972 the CERN group was hot on the trail of muonless events. Simultaneously the new Fermilab machine was sending 50 GeV neutrinos toward a massive electronic neutrino detector managed by David Cline (University of Wisconsin), Alfred Mann (University of Pennsylvania), and Carlo Rubbia (Harvard, CERN, northern Italy, Alitalia . . .).

We can't do full justice to the story of this discovery. It's full of sturm und drang, human interest, and the sociopolitics of science. We'll skip all that and simply say that by 1973 the Gargamelle group announced, somewhat tentatively, the observation of neutral currents. At Fermilab the Cline-Mann-Rubbia team also had so-so data. Obfuscating backgrounds were serious, and the signal was not one that knocked you on your rear. They decided they had found neutral currents. Then they withdrew. Then decided again. A wag dubbed their efforts "alternating neutral currents."

By the 1974 Rochester Conference (a biennial international meeting) in London, it was all clear: CERN had discovered neutral currents, and the Fermilab group had convincing confirmation of this signal. The evidence indicated that "something like a Z^0" had to exist. But if we go strictly by the book, although neutral currents were established in 1974, it took another nine years to prove directly the existence of Z^0. CERN got the credit, in 1983. The mass? Z^0 was indeed heavy: 91 GeV.

By mid-1992, incidentally, the LEP machine at CERN had registered more than 2 million Z^0's, collected by its four huge detectors. Studying the production and the subsequent decay of Z^0's is providing a bonanza in data and keeps some 1,400 physicists busy. Recall that when Ernest Rutherford discovered alpha particles, he then explained them and went on to use them as a tool to discover the nucleus. We did the same thing with neutrinos; and neutrino beams, as we've just seen, have become an industry also, useful for finding messenger particles, studying quarks, and a number of other things. Yesterday's fantasy is today's discovery is tomorrow's device.

The Strong Force Revisited: Gluons

We needed one more discovery in the 1970s to complete the standard model. We had the quarks, but they bind together so strongly that there's no such thing as a free quark. What is the binding mechanism? We called on quantum field theory, but the results were once again frustrating. Bjorken had elucidated the early experimental results at Stanford in which electrons were bounced off the quarks in the proton. Whatever the force was, the electron scattering indicated that it was surprisingly weak when the quarks were close together.

This was an exciting result because one wanted to apply gauge symmetry here, too. Gauge theories could predict the counterintuitive idea that the strong force gets very weak at close approach and stronger as the quarks move apart. The process, discovered by some kids, David Politzer at Harvard and David Gross and Frank Wilczek at Princeton, carried a name that would be the envy of any politician: asymptotic freedom. Asymptotic roughly means "getting closer and closer, but never touching." Quarks have asymptotic freedom. The strong force gets weaker and weaker as one quark approaches a second quark. What this means, paradoxically, is that when quarks are close together they behave almost as if they are free. But when they are farther apart, the forces get effectively stronger. Short distances imply high energies, so the strong force gets weaker at high energies. This is just the opposite of the electrical force. (Things do get curiouser, said Alice.) More important, the strong force needed a messenger particle like the other forces. Somewhere the messenger acquired the name gluon. But to name it is not to know it.

Another idea, rattling around in the theoretical literature, is relevant now. Gell-Mann named this one. It's called color — or colour in Europe — and it has nothing to do with color as you and I recognize it. Color explains certain experimental results and predicts others. For example, it explained how a proton could have two up quarks and a down quark, when the Pauli principle specifically excluded two identical objects in the same state. If one of the up quarks is blue, and the other is green, we satisfy Pauli's rule. Color gives the strong force the equivalent of electric charge.

Color must come in *three* types, said Gell-Mann and others who had worked in this garden. Remember that Faraday and Ben Franklin had determined that electric charge comes in two styles, designated plus and minus. Quarks need three. So now all quarks come in three colors. Perhaps the color idea was stolen from the palette because there are three primary colors. A better analogy might be that electric charge is one-dimensional, with plus and minus directions, and color is three-dimensional (three axes: red, blue, and green). Color explained why quark combinations are, uniquely, either quark plus antiquark (mesons) or three quarks (baryons). These combinations show no color; the quarkness vanishes when we stare at a meson or a baryon. A red quark combines with an antired antiquark to produce a colorless meson. The red and antired cancel. Likewise, the red, blue, and green quarks in a proton mix to make white (try this by spinning a color wheel). Again colorless.

Even though these are nice reasons for using the word "color," it has no literal meaning. We are describing another abstract property that the theorists gave to quarks to account for the increasing amount of data. We could have used Tom, Dick, and Harry or A, B, and C, but color was a more appropriate (colorful?) metaphor. So color, along with quarks and gluons, seemed to be forever a part of the black box, abstract entities that won't make a Geiger counter click, will never leave a track in a bubble chamber, will never tickle wires in an electronic detector.

Nevertheless, the concept that the strong force gets weaker as quarks approach one another was exciting from the point of view of further unification. As the distance between particles decreases, their relative energy increases (small distance implies high energy). This asymptotic freedom implies that the strong force gets weaker at high energy. The unification seekers were then given the hope that at sufficiently high energy, the strength of the strong force may approach that of the electroweak force.

And what about the messenger particles? How do we describe the color-force-carrying particles? What emerged was that gluons carry *two* colors — a color and another anticolor — and, in their emission or absorption by quarks, they change the quark color. For example, a red-antiblue gluon changes a red quark to an antiblue quark. This exchange is the origin of the strong force, and Murray the Great Namer dubbed the theory quantum chromodynamics (QCD) in resonance with quantum electrodynamics (QED). The color-changing task means that we need enough gluons to make all possible changes. It turns out that eight gluons will do it. If you ask a theorist, "Why eight?" he'll wisely say, "Why, eight is nine minus one."

Our uneasiness with the fact that quarks were never seen outside of hadrons was only moderately tempered by a physical picture of why quarks are permanently confined. At close distances, quarks exert relatively weak forces on one another. This is the glory domain for theorists, where they can calculate properties of the quark state and the quark's influence on collision experiments. As the quarks separate, however, the force becomes stronger, and the energy required to add distance between them rises rapidly until, long before we have actually separated the quarks, the energy input results in the creation of a new quark-antiquark pair. This curious property is a result of the fact that gluons are not simple, dumb messenger particles. They actually exert forces on each other. This is where QED differs from QCD, since photons ignore each other.

Still, QED and QCD had many close analogies, especially in the high-energy domain. QCD's successes were slow in coming, but steady. Because of the fuzzy long-distance part of the force, calculations were never very precise, and many experiments would conclude with the rather nebulous statement that "our results are consistent with the predictions of QCD."

So what kind of a theory do we have if we can never, ever see a free quark? We can do experiments that sense the presence of electrons and measure them, this way and that, even when they are all bound up in atoms. Can we do the same with quarks and gluons? Bjorken and Feynman had suggested that in very hard collisions of particles, the energized quarks would initially head out and, just before leaving the influence of their quark partners, would mask themselves into a narrow bundle of hadrons — three or four or eight pions, for example, or add some kaons and nucleons. These would be narrowly directed along the path of the parent quark. They were given the name "jets," and the search was on.

With the machines of the 1970s, these jets were not easily distinguished because all we could produce were slower quarks that gave rise to broad jets of a small number of hadrons. We wanted dense, narrow jets. The first success belonged to a young woman experimentalist named Gail Hanson, a Ph.D. from MIT working at SLAC. Her careful statistical analysis revealed that a correlation of hadrons *did* appear in the debris of a 3 GeV e^+ e^- collision at SPEAR. She was helped by the fact that what went in were the electrons and what came out were a quark and an antiquark, back to back to conserve momentum. These correlated jets showed up, barely but decisively, in the analysis. When Democritus and I were sitting in the CDF control room, needlelike bundles of ten or so hadrons, two jets 180 degrees apart, were flashed on the large screen every few minutes. There is no reason why there should be such a structure unless the jet is the offspring of a very high energy, very high momentum quark, which dresses itself before going out.

But the major discovery of the 1970s along these lines was made at the PETRA e^+ e^- machine in Hamburg, Germany. This machine, colliding at the total energy of 30 GeV, also showed, without need for analysis, the two-jet structure. Here one could almost see the quarks in the data. But something else was also seen.

One of the four detectors on-line at PETRA had its own acronym: TASSO, for Two-Armed Solenoidal Spectrometer. The TASSO group was looking for events in which *three* jets would appear. A consequence of QCD theory is that when e^+ and e^- annihilate to produce a quark and an antiquark, there is a reasonable probability that one of the outgoing quarks will radiate a messenger particle, a gluon. There is enough energy here to convert "virtual" gluon to real gluon. The gluons share the quarks' shyness and, like quarks, dress themselves before leaving the black box of the encounter domain. Therefore three jets of hadrons. But this takes more energy.

In 1978, runs of total energy of 13 and 17 GeV came out empty, but at 27 GeV, something happened. The analysis was pushed by another woman physicist, Sau Lan Wu, a professor at the University of Wisconsin. Wu's program soon uncovered more than forty events in which there were three jets of hadrons, each jet having three to ten tracks (hadrons). The array looked like the hood ornament of a Mercedes.

The other PETRA groups soon got on the bandwagon. Looking through their data, they also found the three-jet events. A year later, thousands had been collected. The gluon had thus been "seen." The pattern of tracks was calculated by theorist John Ellis at CERN using QCD, and one must credit his intervention in motivating the search. The announcement of the gluon's detection was made at a conference at Fermilab in the summer of 1979, and it was my job to go on the Phil Donahue television show in Chicago to explain the discovery. I put more energy into explaining that the Fermilab buffalo were not roaming the lab as early warning devices for dangerous radiation. But in physics, the real news was the gluons — the bosons, not the bisons.

So now we have all the messenger particles, or gauge bosons as they are more eruditely called. ("Gauge" came from gauge symmetry, and boson is derived from the Indian physicist S. N. Bose, who described the class of particles with integer values of spin.) Whereas the matter particles all have spin of ½ and are called fermions, the messenger particles all have spin 1 and are bosons. We've skipped over some details. The photon, for instance, was predicted by Einstein in 1905 and observed experimentally by Arthur Compton in 1923, using x-rays scattered from atomic electrons. Although neutral currents had been discovered in the mid-1970s, the W's and Z's were not directly observed until 1983–84, when they were detected in the CERN hadron collider. As mentioned, the gluons were pinned down by 1979.

In this long discussion of the strong force, we should note that we define it as the quark-quark force carried by gluons. But what about the "old" strong force between neutrons and protons? We now understand this as the residual effects of the gluons, sort of leaking out of the neutrons and protons that bind together in the nucleus. The old strong force that is well described by exchange of pions is now seen as a consequence of the complexities of quark-gluon processes.

END OF THE ROAD?

Entering the 1980s, we had figured out all the matter particles (quarks and leptons), and we had the messenger particles, or gauge bosons, of the three forces (excluding gravity) pretty much in hand. Adding the force particles to the matter particles, you have the complete standard model, or SM. Here, then, is the "secret of the universe":

MATTER

First generation	*Second generation*	*Third generation*
	QUARKS	
u	c	t?
d	s	b
	LEPTONS	
ν_e	ν_μ	ν_τ
e	μ	τ

FORCES

GAUGE BOSONS

electromagnetism	photon (γ)
weak force	W^- W^+ Z^0
strong force	eight gluons

Remember that the quarks come in three colors. So if one is nasty, one can count eighteen quarks, six leptons, and twelve gauge boson force carriers. There is also an anti table in which all the matter particles appear as antiparticles. That would give you sixty particles total. But who's counting? Stick to the above table; it's all you need to know. At last we believe we have Democritus's a-toms. They are the quarks and leptons. The three forces and their messenger particles account for his "constant violent motion."

It may seem arrogant to sum up our entire universe in a chart, albeit a messy one. Yet humans appear to be driven to construct such syntheses; "standard models" have been a recurrent theme in Western history. The current standard model wasn't given that name until the 1970s, and the term is peculiar to the recent modern history of physics. But certainly there have been other standard models through the centuries. The next page shows just a few of them.

THE STANDARD MODEL: an accelerated history

Architects	Date	PARTICLES	FORCES	Grade	Comments
Thales (Milesian)	600 B.C.	Water	Not mentioned	**B–**	He was the first to explain the world positing natural causes rather than gods. Replaced mythology with logic.
Empedocles (Acragan)	460 B.C.	Earth, air, fire, and water	Love and strife	**B+**	Came up with the idea of multiple "particles" that combine to make all kinds of matter.
Democritus (Abderan)	430 B.C.	The invisible, indivisible *atomos*, or a-tom	Constant violent motion	**A**	His model required too many particles, each with a different shape, but his basic idea of an uncuttable a-tom remains the definition of an elementary particle.
Isaac Newton (English)	1687	Hard, massy, impenetrable atoms	Gravity (for the cosmos) Unknown forces (for atoms)	**C**	He loved atoms but didn't advance their cause. His gravity is giving the big boys a major headache in the 1990s.
Roger J. Boscovich (Dalmatian)	1760	"Points of force," indivisible and without shape or dimension	Attractive and repulsive forces acting between points	**B+**	His theory was incom-plete, limited, but the idea of "zero radius" pointlike particles that create "fields of force" is essential to modern physics.
John Dalton (English)	1808	Atoms — the basic units of chemical elements: carbon, oxygen, etc.	Force of attraction between atoms	**C+**	He jumped the gun by resurrecting Democritus's term — Dalton's atom wasn't indivisible — but provided a clue when he said atoms differed by weight, not shape, as Democritus thought.
Michael Faraday (English)	1820	Electric charges	Electro-magnetism (plus gravity)	**B**	Applied atomism to electricity when he speculated that currents consisted of "corpuscles of electricity" — electrons.
Dmitri Mendeleev (Siberian)	1870	Fifty-plus atoms, arranged in the periodic table of the elements	Didn't speculate on forces	**B**	He took Dalton's concept and organized all the known chemical elements. His periodic table hinted strongly at a deeper, more meaningful structure.
Ernest Rutherford (New Zealander)	1911	Two particles: nucleus and electron	Nuclear (strong) force plus electromagne-tism, gravity	**A–**	By discovering the nucleus, he uncovered a new simplicity within all of Dalton's atoms.
Bjorken, Fermi, Friedman, Gell-Mann, Glashow, Kendall, Lederman, Perl, Richter, Schwartz, Steinberger, Taylor, Ting, plus a cast of thousands	1992	Six quarks and six leptons, plus their antiparticles. Quarks come in three colors.	Electromagnetism, the strong force, the weak force: twelve force-carrying particles — plus gravity	**In-com-plete**	"Γυφφαω." (laughter) — *Democritus of Abdera*

Why is our standard model incomplete? One obvious flaw is that the top quark hasn't yet been seen. Another is that one of the forces is missing: gravity. No one knows how to work this grand old force into the model. Another aesthetic flaw is that it's not simple enough — it should look more like Empedocles' earth, air, fire, and water, plus love and strife. There are too many parameters in the standard model, too many knobs to twiddle.

Which is not to say that the standard model is not one of the great accomplishments of science. It represents the work of a lot of guys (of both genders) who stayed up late at night. But in admiring its beauty and scope, one can't help feeling uneasy, and desirous of something simpler, a model that even an ancient Greek could love.

Listen: do you hear a laugh emanating from the void?

· 8 ·

THE GOD PARTICLE AT LAST

And the Lord looked upon Her world, and She marveled at its beauty — for so much beauty there was that She wept. It was a world of one kind of particle and one force carried by one messenger who was, with divine simplicity, also the one particle.

And the Lord looked upon the world She had created and She saw that it was also boring. So She computed and She smiled and She caused Her Universe to expand and to cool. And lo, it became cool enough to activate Her tried and true agent, the Higgs field, which before the cooling could not bear the incredible heat of creation. And in the influence of Higgs, the particles suckled energy from the field and absorbed this energy and grew massive. Each grew in its own way, but not all the same. Some grew incredibly massive, some only a little, and some not at all. And whereas before there was only one particle, now there were twelve, and whereas before the messenger and the particle were the same, now they were different, and whereas before there was only one force carrier and one force, now there were twelve carriers and four forces, and whereas before there was an endless, meaningless beauty, now there were Democrats and Republicans.

And the Lord looked upon the world She had created and She was convulsed with wholly uncontrolled laughter. And She summoned Higgs and, suppressing Her mirth, She dealt with him sternly and said:

"Wherefore hast thou destroyed the symmetry of the world?"

And Higgs, shattered by the faintest suggestion of disapproval, defended thusly:

"Oh, Boss, I have not destroyed the symmetry. I have merely caused it to be hidden by the artifice of energy consumption. And in so doing I have indeed made it a complicated world.

"Who could have foreseen that out of this dreary set of identical objects, we could have nuclei and atoms and molecules and planets and stars?

"Who could have predicted the sunsets and the oceans and the organic ooze formed by all those awful molecules agitated by lightning and heat? And who could have expected evolution and those physicists poking and probing and seeking to find out what I have, in Your service, so carefully hidden?"

And the Lord, hard put to stop Her laughter, signed forgiveness and a nice raise for Higgs.

— The Very New Testament 3:1

IT WILL BE OUR TASK in this chapter to convert the poetry(?) of the Very New Testament to the hard science of particle cosmology. But we cannot abandon our discussion of the standard model just yet. There are a few loose ends to tie up — and a few we can't tie up. Both sets are important in the story of the standard-model-and-beyond, and I must recount a few additional experimental triumphs that firmly established our current view of the microworld. These details provide a feeling for the model's power as well as its limitations.

There are two kinds of bothersome flaws in the standard model. The first has to do with its incompleteness. The top quark is still missing as of early 1993. One of the neutrinos (the tau) has not been directly detected, and many of the numbers we need are imprecisely known. For example, we don't know if the neutrinos have any rest mass. We need to know how CP symmetry violation — the process of the origin of matter — enters, and, most important, we need to introduce a new phenomenon, which we call the Higgs field, in order to preserve the mathematical consistency of the standard model. The second kind of flaw is a purely aesthetic one. The standard model is complicated enough to appear to many as only a way station toward a simpler view of the world. The Higgs idea, and its attendant particle, the Higgs boson, is relevant to all the issues we have just listed, so much so that we have named this book in its honor: the God Particle.

A FRAGMENT OF STANDARD-MODEL AGONY

Consider the neutrino.

"Which neutrino?"

Well, it doesn't matter. Let's take the electron neutrino — the garden-variety, first-generation neutrino — since it has the lowest mass. (Unless, of course, all neutrino masses are zero.)

"Okay, the electron neutrino."

It has no electric charge.

It has no strong or electromagnetic force.

It has no size, no spatial extent. Its radius is zero.

It may not have a mass.

Nothing has so few properties (deans and politicians excepted) as the neutrino. Its presence is less than a whisper.

As kids we recited:

> Little fly upon the wall
> Have you got no folks at all?
> No mother?
> No father?
> Pooey on you, ya bastard!

And now I recite:

> Little neutrino in the world
> With the speed of light you're hurled.
> No charge, no mass, no space dimension?
> Shame! You do defy convention.

Yet the neutrino exists. It has a sort of location — a trajectory, always heading in one direction with a velocity close (or equal) to that of light. The neutrino does have spin, although if you ask what it is that's spinning you expose yourself as one who has not yet been cleansed of impure prequantum thinking. Spin is intrinsic to the concept of "particle," and if the mass of the neutrino is indeed zero, its spin and its constant, undeviating velocity of light combine to give it a unique new attribute called chirality. This forever ties the direction of spin (clockwise or counterclockwise) to the direction of motion. It can have "right-handed" chirality, meaning that it advances with clockwise spin, or it can be left-handed, advancing with a counterclockwise spin. Therein lies a lovely symmetry. The gauge theory prefers all particles to have zero mass and universal chiral symmetry. There is that word again: symmetry.

Chiral symmetry is one of these elegant symmetries that describe the early universe — one pattern that repeats and repeats and repeats like wallpaper, but unrelieved by corridors, doors, or corners — unending. No wonder She found it boring and ordered in the Higgs field to give mass and break chiral symmetry. Why does mass break chiral symmetry? Once a particle has mass, it travels at speeds less than that of light. Now you, the observer, can go faster than the particle. Then, relative to you, the particle has reversed its direction of motion but not its spin, so a left-handed object to some observers becomes right-handed to others. But there are the neutrinos, survivors perhaps of the war on chiral symmetry. The neutrino is always left-handed, the antineutrino always right-handed. This handedness is one of the very few properties the poor little fellow has.

Oh yes, neutrinos have another property, the weak force. Neutrinos emerge from weak processes that take forever (sometimes microseconds) to happen. As we have seen, they can collide with another particle. This collision requires so close a touch, so deep an intimacy, as to be exceedingly rare. For a neutrino to collide hard in an inch-thick slab of steel would be as likely as finding a small gem buffeted randomly in the vastness of the Atlantic Ocean — that is, as likely as catching it in one cup of the Atlantic's water, randomly sampled. And yet for all its lack of properties, the neutrino has enormous influence on the course of events. For example, it is the outrush of huge numbers of neutrinos from the core that instigates the explosion of stars, scattering heavier elements, recently cooked in the doomed star, throughout space. The debris of such explosions eventually coalesces and accounts for the silicon and iron and other good stuff we find in our planets.

Recently, strenuous efforts have been made to detect the mass of the neutrino, if indeed it has any. The three neutrinos that are a part of our standard model are candidates for what astronomers call "dark matter," material that, they say, pervades the universe and dominates its gravitationally driven evolution. All we know so far is that neutrinos could have a small mass . . . or they could have zero mass. Zero is such a very special number that even the very slightest mass, say a millionth that of the electron, would be of great theoretical significance. As part of the standard model, neutrinos and their masses are an aspect of the open questions that lie therein.

HIDDEN SIMPLICITY: STANDARD-MODEL ECSTASY

When a scientist, say of the British persuasion, is really, really angry at someone and is driven to the extremes of expletives, he will say under his breath, "Bloody Aristotelian." Them's fightin' words, and a deadlier insult is hard to imagine. Aristotle is generally credited (probably unreasonably) with holding up the progress of physics for about 2,000 years — until Galileo had the courage and the conviction to call him out. He shamed Aristotle's acolytes in full view of the multitudes on the Piazza del Duomo, where today the Tower leans and the piazza is lined with souvenir sellers and ice cream stands.

We've reviewed the story of things falling from crooked towers — a feather floats down, a steel ball drops rapidly. That seemed like good stuff to Aristotle, who said, "Heavy falls fast, light falls slow." Perfectly intuitive. Also, if you roll a ball, it eventually comes to rest. Therefore, said Ari, rest is "natural and preferred, whereas motion requires a motive force keeping it moving." Eminently clear, confirmed by our everyday experience, and yet . . . wrong. Galileo saved his contempt, not for Aristotle, but for the generations of philosophers who worshiped at Aristotle's temple and accepted his views without question.

What Galileo saw was a profound simplicity in the laws of motion, provided we could remove complicating factors such as air resistance and friction, things that are very much a part of the real world but that hide the simplicity. Galileo saw mathematics — parabolas, quadratic equations — as the way the world must really be. Neil Armstrong, the first astronaut on the moon, dropped a feather and a hammer on the airless lunar surface, demonstrating the Tower experiment for all the world's viewers. With no resistance, the two objects dropped at the same rate. And a ball rolling on a horizontal surface would, in fact, roll forever if there were no friction. It rolls much farther on a highly polished table, and farther yet on an air track or on slippery ice. It takes some ability to think abstractly, to imagine motion without air, without rolling friction, but when you do, the reward is a new insight into the laws of motion, of space and time.

Since that heartwarming story, we have learned about hidden simplicity. It is nature's way to hide the symmetry, simplicity, and beauty that can be described by abstract mathematics. What we now see, in place of Galileo's air resistance and friction (and equivalent political obstructions), is our standard model. To track this idea to the 1990s, we have to pick up the story of the heavy messenger particles that carry the weak force.

STANDARD MODEL, 1980

The decade of the eighties opens with a large measure of theoretical smugness. The standard model sits there, with its pristine summary of three hundred years of particle physics, challenging the experimentalists to "fill in the blanks." The W^+, W^-, and Z^0 have not yet been observed, nor has the top quark. The tau neutrino requires a three-neutrino experiment, and such experiments have been proposed, but the arrangements are complicated, with small chance of success. They have not been approved. Experiments on the charged tau lepton strongly indicate that the tau neutrino must exist.

The top quark is the subject of research at all the machines, the electron-positron colliders as well as the proton machines. A brand-new machine, Tristan, is under construction in Japan (Tristan — what is the deep connection between Japanese culture and Teutonic mythology?). It is an $e^+ e^-$ machine that can produce top plus anti-top, $t\bar{t}$, if the mass of the top quark is no heavier than 35 GeV, or seven times heavier than its differently flavored cousin bottom, weighing in at 5 GeV. The experiment and the expectations of Tristan, at least insofar as top is concerned, are doomed. The top is heavy.

THE CHIMERA OF UNIFICATION

The search for the W was the all-out effort of the Europeans, determined to show the world that they had come into their own in this business. To find the W required a machine energetic enough to produce it. How much energy is required? This depends on how heavy the W is. Responding to the insistent and forceful arguments of Carlo Rubbia, CERN set out to build a proton-antiproton collider in 1978 based upon their 400 GeV proton machine.

By the late 1970s, the W and Z were estimated by the theorists to be "a hundred times heavier than the proton." (The rest mass of the proton, remember, is close enough to a convenient 1 GeV.) This estimate of the W and Z masses was made with such confidence that CERN was willing to invest $100 million or more on a "sure thing," an accelerator capable of delivering enough energy in a collision to make W's and Z's and a set of elaborate and expensive detectors to observe the collisions. What gave them this arrogant confidence?

There was a euphoria arising from the sense that a unified theory, the ultimate goal, was close at hand. Not a world model of six quarks and six leptons and four forces, but a model of perhaps only one class of particles and one grand — oh, so grand — unified force. This would surely be the realization of the ancient Greek view, the objective all along as we proceeded from water to air to earth to fire to all four.

Unification, the search for a simple and all-encompassing theory, is the Holy Grail. Einstein, as early as 1901 (at age twenty-two) wrote about the connections between molecular (electrical) forces and gravity. From 1925 to his death in 1955, he sought in vain for a unified electromagnetic-gravitational force. This huge effort by one of the greatest physicists of his, or any other, time failed. We now know that there are two other forces, the weak and the strong. Without these forces Einstein's efforts toward unification were doomed. The second major reason for Einstein's failure was his divorce from the central achievement of twentieth-century physics (to which he contributed strongly in its formative phases), the quantum theory. He never accepted this radical and revolutionary concept, which in fact provided the framework for unification of all the forces. By the 1960s three of the four forces had been formulated in terms of a quantum field theory and had been refined to the point where "unification" cried out.

All the deep theorists were after it. I remember a seminar at Columbia in the early fifties when Heisenberg and Pauli presented their new, unified theory of elementary particles. The seminar room (301 Pupin Hall) was densely crowded. In the front row were Niels Bohr, I. I. Rabi, Charles Townes, T. D. Lee, Polykarp Kusch, Willis Lamb, and James Rainwater — the present and future laureate contingent. Postdocs, if they had the clout to be invited, violated all the fire laws. Grad students hung from special hooks fastened to the rafters. It was crowded. The theory was over my head, but my not understanding it didn't mean it was correct. Pauli's final comment was an admission. "Yah, this is a crazy theory." Bohr's comment from the audience, which everyone remembers, went something like this: "The trouble with this theory is that it isn't crazy enough." Since the theory vanished like so many other valiant attempts, Bohr was right again.

A consistent theory of forces must meet two criteria: it must be a quantum field theory that incorporates the special theory of relativity and gauge symmetry. This latter feature and, as far as we know, only this guarantees that the theory is mathematically consistent, renormalizable. But there is much more; this gauge symmetry business has deep aesthetic appeal. Curiously, the idea comes from the one force that has not yet been formulated as a quantum field theory: gravity. Einstein's gravity (as opposed to Newton's) emerges from the desire to have the laws of physics be the same for all observers, those at rest as well as observers in accelerated systems and in the presence of gravitational fields, such as on the surface of the earth, which rotates at 1,000 miles per hour. In such a whirling laboratory, forces appear that make experiments come out quite differently than they would in smoothly moving — nonaccelerated — labs. Einstein sought laws that would look the same to *all* observers. This "invariance" requirement that Einstein placed on nature in his general theory of relativity (1915) logically implied the existence of the gravitational force. I say this so quickly, but I worked so hard to understand it! The theory of relativity contains a built-in symmetry that implies the existence of a force of nature — in this case, gravitation.

In an analogous way, gauge symmetry, implying a more abstract invariance imposed upon the relevant equations, also generates, in each case, the weak, the strong, and the elecromagnetic force.

THE GAUGE

We are on the threshold of the private driveway that leads to the God Particle. We must review several ideas. One has to do with the matter particles: quarks and leptons. They all have a spin of one half in the curious quantum units of spin. There are the force fields that can also be represented by particles: the quanta of the field. These particles all have integral spin — a spin of one unit. They are none other than the messenger particles and gauge bosons we have often discussed: the photons, the W's and the Z, and the gluons, all discovered and their masses measured. To make sense out of this array of matter particles and force carriers, let's reconsider the concepts of invariance and symmetry.

We've tap-danced around this gauge symmetry idea because it's hard, maybe impossible, to explain fully. The problem is that this book is in English, and the language of gauge theory is math. In English we must rely on metaphors. More tap-dancing, but perhaps it will help.

For example, a sphere has perfect symmetry in that we can rotate it through any angle about any axis without producing any change in the system. The act of rotation can be described mathematically; after the rotation the sphere can be described with an equation that is identical in every detail to the equation before rotation. The sphere's symmetry leads to the invariance of the equations describing the sphere to the rotation.

But who cares about spheres? Empty space is also rotationally invariant, like the sphere. Thus the equations of physics must be rotationally invariant. Mathematically, this means that if we rotate an x-y-z-coordinate system through any angle about any axis, that angle will not appear in the equation. We have discussed other such symmetries. For example, an object positioned on a flat infinite plane can be moved any distance in any direction, and again the system is identical (invariant) to the situation before the motion. This movement from point A to point B is called a translation, and we believe that space is also invariant to translation; that is, if we add 12 meters to all distances, the 12 will drop out of the equations. Thus, continuing the litany, the equations of physics must display invariance to translations. To complete this symmetry/conservation story, we have the law of conservation of energy. Curiously, the symmetry with which this is associated has to do with time, that is, with the fact that the laws of physics are invariant to translation in time. This means that in the equations of physics, if we add a constant interval of time, say 15 seconds, everywhere that time appears, the addition will wash out, leaving the equation invariant to this shift.

Now for the kicker. Symmetry reveals new features of the nature of space. I referred to Emmy Noether earlier in the book. Her 1918 contribution was the following: for every symmetry (showing up as the inability of the basic equations to notice, for example, space rotations and translations and time translation), there is a corresponding conservation law! Now conservation laws can be tested experimentally. Noether's work connected translation invariance to the well-tested law of conservation of momentum, rotation invariance to conservation of angular momentum, and time translation to conservation of energy. So these experimentally unassailable conservation laws (using the logic backward) tell us about the symmetries respected by time and space.

The parity conservation discussed in Interlude C is an example of a discrete symmetry that applies to the microscopic quantum domain. Mirror symmetry amounts to a literal reflection in a mirror of all coordinates of a physical system. Mathematically, it amounts to changing all z-coordinates to $-z$ where z points toward the mirror. As we saw, although the strong and electromagnetic forces respect this symmetry, the weak force doesn't, which of course gave us infinite joy back in 1957.

So far, most of this material is review and the class is doing well. (I feel it.) We saw in Chapter 7 that there can be more abstract symmetries not related to geometry, upon which our examples above have so far depended. Our best quantum field theory, QED, turns out to be invariant to what looks like a dramatic change in mathematical description — not a geometric rotation, translation, or reflection, but a much more abstract change in describing the field. The name of the change is gauge transformation, and any more detailed description is not worth the math anxiety it would induce. Suffice it to say that the equations of quantum electrodynamics (QED) are invariant to gauge transformation. This is a very powerful symmetry in that one can derive all the properties of the electromagnetic force from it alone. That's not the way it was done in history, but some graduate textbooks do it that way today. The symmetry ensures that the force carrier, the photon, is massless. Because the masslessness is connected to the gauge symmetry, the photon is called a "gauge boson." (Remember that "boson" describes particles, often messenger particles, that have integer spin.) And because it has been shown that QED, the strong force, and the weak force are described by equations that exhibit gauge symmetry, *all* the force carriers — photons, the W's and the Z, and gluons — are called gauge bosons.

Einstein's thirty years of fruitless effort to find a unified theory was bested in the late 1960s by Glashow, Weinberg, and Salam's successful unification of the weak force and the electromagnetic force. The major implication of the theory was the existence of a family of messenger particles: the photon, the W^+ and W^- and Z^0.

Now comes the God Particle theme. How do we have heavy W's and Z's in a gauge theory? How do such disparate objects as the zero-mass photon and the massive W's and Z's appear in the same family? Their huge mass differences account for the large differences in behavior between the electromagnetic and the weak force.

We will come back to this teasing introduction later; too much theory exhausts my spirit. And besides, before the theorists can go off to answer this question we must find the W. As if they wait.

FIND THE W

So CERN put down its money (or, more correctly, gave it to Carlo Rubbia), and the quest for the W was on. I should note that if the W is about 100 GeV in mass, one needs a good deal more than 100 GeV of collision energy available. A 400 GeV proton colliding with a proton at rest can't do it, for only 27 GeV is available for making new particles. The rest of the energy is used to conserve momentum. That is why Rubbia proposed the collider route. His idea was to make an antiproton source, using the injector to the CERN 400 GeV Super Proton Synchroton (SPS) to manufacture p-bars. When an adequate number had been accumulated, he'd put them into the SPS magnet ring more or less as we explained it back in Chapter 6.

Unlike the later Tevatron, the SPS was not a superconducting accelerator. This means that its maximum energy was limited. If both beams, protons and antiprotons, were accelerated to the full energy of the SPS, 400 GeV, you would have 800 GeV available — enormous. But the energy selected was 270 GeV in each beam. Why not 400 GeV? First, the magnets would then have to carry a high current for a long time — hours — during the collision time. CERN's magnets were not designed for this and would overheat. Second, remaining for any length of time at high field is expensive. The SPS magnets were designed to ramp their magnetic fields up to the full energy of 400 GeV, dwell for a few seconds while delivering beams to customers doing fixed-target experiments, and then reduce the field to zero. Rubbia's idea of colliding two beams was ingenious, but his basic problem was that his machine was not designed originally to be a collider.

The CERN authorities agreed with Rubbia that 270 GeV in each beam — making a total energy of 540 GeV — would probably be enough to make W's, which "weigh" only 100 GeV or so. The project was approved and an adequate number of Swiss francs were given in 1978. Rubbia assembled two teams. The first was a group of accelerator geniuses — French, Italian, Dutch, English, Norwegian, and an occasional visiting Yankee. Their language was broken English but flawless "acceleratorese." The second team, experimental physicists, had to build a massive detector, named UA-1 in a flight of poetic imagination, to observe the collisions between protons and antiprotons.

In the p-bar accelerator group, a Dutch engineer, Simon Van der Meer, had invented a method of compressing antiprotons into a small volume in the storage ring that accumulates these scarce objects. Called "stochastic cooling," this invention was the key to getting enough p-bars to make a respectable number of p/p-bar collisions, that is, about 50,000 per second. Rubbia, a superb technician, hurried his group, built his constituency, handled marketing, calls, and propaganda. His technique: have talk, will travel. His presentations are machine-gun style, with five transparencies projected per minute, an intimate mixture of blarney, bravado, bombast, and substance.

CARLO AND THE GORILLA

To many in physics, Carlo Rubbia is a scientist of heroic proportions. I once had the job of introducing him before he gave the banquet talk at a well-attended international meeting in Santa Fe. (This was after he won the Nobel Prize for finding the W and the Z.) I introduced him with a story.

At the Nobel ceremonies in Stockholm, King Olaf pulls Carlo aside and tells him there's a problem. Because of a screwup, the king explains, there's only one medal available this year. To determine which laureate gets the gold, the king has designed three heroic tasks, located in three tents on the field in full view of the assemblage. In the first tent, Carlo is told, he will find four liters of highly distilled slivovitz, the beverage that helped dissolve Bulgaria. The assigned time for drinking all this is 20 seconds! The second tent contains a gorilla, unfed for three days and suffering from an impacted wisdom tooth. The task: remove the offending tooth. The time: 40 seconds. The third tent hides the most accomplished courtesan of the Iraqi army. The task: satisfy her completely. The time: 60 seconds.

At the starter's gun, Carlo bounds into tent one. The gurgle is heard by all and, in 18.6 seconds, four drained liter bottles of slivovitz are triumphantly displayed.

Losing no time, the mythical Carlo staggers into the second tent, from which enormous, deafening roars are heard by all. Then silence. And in 39.1 seconds, Carlo stumbles out, wobbles to the microphone and pleads, "All right, where ish the gorilla with the toothache?"

The audience, perhaps because the conference wine was so generously served, roared with appreciation. I finally introduced Carlo, and as he passed me on his way to the lectern, he whispered, "I don't get it. Explain it later."

Rubbia did not suffer fools gladly, and his strong control stirred resentment. Sometime after his success, Gary Taubes wrote a book about him, *Nobel Dreams,* which was not flattering. Once, at a winter school with Carlo in the audience, I announced that the movie rights to the book had been sold and that Sydney Greenstreet, whose girth was roughly the same as Carlo's, had been signed to play him. Someone pointed out that Sydney Greenstreet was dead but would otherwise be a good choice. At another gathering, a summer conference on Long Island, someone put up a sign on the beach: "No Swimming. Carlo is using the ocean."

Rubbia drove hard on all fronts in the search for the W. He continually urged on the detector builders assembling the monster magnet that would detect and analyze events with fifty or sixty particles emerging from head-on collisions of 270 GeV protons and 270 GeV antiprotons. He was similarly knowledgeable about and active in the construction of the antiproton accumulator, or AA ring, the device that would put Van der Meer's idea to work and produce an intense source of antiprotons for insertion and acceleration in the SPS ring. The ring had to have radio-frequency cavities, enhanced water cooling, and a specially instrumented interaction hall where the UA-1 detector would be assembled. A competing detector, UA-2, natch, was approved by CERN authorities to keep Rubbia honest and buy some insurance. UA-2 was definitely the Avis of the situation, but the group building it was young and enthusiastic. Limited by a smaller budget, they designed a quite different detector.

Rubbia's third front was to keep the CERN authorities enthusiastic, roil the world community, and set the stage for the great W experiment. All of Europe was rooting for this, for it meant the coming of age of European science. One journalist claimed that a failure would crush "popes and prime ministers."

The experiment got under way in 1981. Everything was in place — UA-1, UA-2, the AA ring — tested and ready. The first runs, designed as checkout trials of everything in the complex system of collider plus detector, were reasonably fruitful. There were leaks, mistakes, accidents, but eventually, data! And all at a new level of complexity. The 1982 Rochester Conference was to be in Paris, and the CERN lab went all out to get results.

Ironically, UA-2, the afterthought detector, made the first splash by observing jets, the narrow bundles of hadrons that are the signatures of quarks. UA-1, still learning, missed this discovery. Whenever David beats Goliath, everyone except Goliath feels warm. In this case Rubbia, who hates to lose, recognized that the observation of jets was a real triumph for CERN — that all of the effort in machines, detectors, and software had paid off in a strong indicator. It all worked! If jets were seen, W's were soon.

A RIDE ON NO. 29

Perhaps a fantastic voyage can best illustrate the way detectors work. Here I will switch over to the CDF detector at Fermilab because it is more modern than UA-1, although the general idea of all the "four pi" detectors is the same. (Four pi — 4π — means that the detector completely surrounds the point of collision.) Remember that when a proton and an antiproton collide, a spray of particles comes off in all directions. On the average, one third are neutral, the rest charged. The task is to find out exactly where each particle goes and what it does. As with any physical observation, one is only partially successful.

Let's ride on one particle. Say it's track No. 29. It zips out at some angle to the line of the collision, encounters the thin metal wall of the vacuum vessel (the beam tube), zips through this, no sweat, and for the next twenty or so inches passes through a gas containing an immense number of very thin gold wires. Although there is no sign, this is Charpak territory. The particle may pass close to forty or fifty of these wires before reaching the end of the tracking chamber. If the particle is charged, each nearby wire records its passage, together with an estimate of how close it came. The accumulated information from the wires defines the particle's path. Since the wire chamber is in a strong magnetic field, the charged particle's path is curved, and a measurement of this curve, calculated by the on-board computer, gives the physicist the momentum of particle No. 29.

Next the particle passes through the cylindrical wall defining the magnetic wire chamber and passes into a "calorimeter sector," which measures particle energy. Now the particle's subsequent behavior depends on what it is. If it is an electron, it fragments on a series of closely spaced thin lead plates, giving up its entire energy to sensitive detectors that provide the meat for the lead sandwiches. The computer notes that the progress of No. 29 ceases after three or four inches of lead-scintillator calorimeter and concludes: electron! If, however, No. 29 is a hadron, it penetrates ten to twenty inches of calorimeter material before exhausting all of its energy. In both cases the energy is measured and cross-checked against the momentum measurement, determined by the particle's curvature in the magnet. But the computer graciously leaves it up to the physicist to draw a conclusion.

If No. 29 is a neutral particle, the tracking chamber doesn't record it at all. When it turns up in the calorimeter, its behavior is essentially the same as that of a charged particle. In both cases the particle produces nuclear collisions with calorimeter materials, and the debris produces further collisions until all the original energy is exhausted. So we can record and measure neutrals, but we can't chart the momentum, and we lose precision in the direction of motion since no track is left in the wire chamber. One neutral particle, the photon, can be easily identified by its relatively quick absorption by the lead, like the electron. Another neutral, the neutrino, leaves the detector entirely, carrying away its energy and its momentum, leaving behind not even a hint of its fragrance. Finally, the muon moves through the calorimeter leaving a small amount of energy (it has no strong nuclear collision). When it emerges, it finds some thirty to sixty inches of iron, through which it passes only to find a muon detector — wire chambers or scintillation counters. This is how muons are tagged.

One does all this for forty-seven particles, or whatever the number is, in this one particular event. The system stores the data, close to one million bits of information — equivalent to the amount of information in a hundred-page book — for each event. The data collection system must quickly decide whether this event is interesting or not; it must discard or record the event or pass the data into a "buffer" memory and clear all registers in order to be ready for the next event. This arrives on the average of a millionth of a second later if the machine is working very well. In the most recent full run at the Tevatron (1990–91), the total amount of information stored on magnetic tape of the CDF detector was equivalent to the text of one million novels or five thousand sets of *Encyclopaedia Britannica.*

Among the outgoing particles are some with very short lifetimes. These may move only a few tenths of an inch away from the collision point in the beam tube before spontaneously disintegrating. W's and Z's are so short lived that their flight distance is unmeasurable, and one must identify their existence from measurements on the particles to which they give rise. These are often hidden among the debris that typically flies out of each collision. Since the W is massive, the decay products have higher than average energy, which helps locate them. Such exotics as a top quark or a Higgs particle will have a set of expected decay modes that must be extracted from the mess of emerging particles.

The process of converting enormous numbers of electronic data bits to conclusions about the nature of the collisions takes impressive efforts. Tens of thousands of signals have to be checked and calibrated; tens of thousands of lines of code must be inspected and verified by looking at events that have to "make sense." Small wonder that it takes a battalion of highly skilled and motivated professionals (even though they may officially be classified as graduate students or postdocs) armed with powerful work stations and well-honed analysis codes two or three years to do justice to the data collected in a Tevatron collider run.

TRIUMPH!

At CERN, where collider physics was pioneered, it all worked, validating the design. In January 1983, Rubbia announced W's. The signal was five clear events that could be interpreted only as the production and subsequent disintegration of a W object.

A day or so later, UA-2 announced that it had four additional events. In both cases, the experimenters had to sort through about one million collisions that produced all manner of nuclear debris. How does one convince oneself as well as the multitude of skeptics? The particular W decay most conducive to discovery is $W^+ \rightarrow e^+ +$ neutrino, or $W^- \rightarrow e^- +$ antineutrino. In a detailed analysis of this kind of event one has to verify (1) that the single observed track is indeed an electron and not anything else, and (2) that the electron energy adds up to about half the mass of the W. The "missing momentum," which the invisible neutrino carries off, can be deduced by adding up all the momentum seen in the event and comparing it to "zero," which is the momentum of the initial state of colliding particles. The discovery was greatly facilitated by the lucky accident that W's are made almost at rest under the CERN collider parameters. To discover a particle, lots of constraints must be satisfied. An important condition is that all the candidate events yield the same value (within allowable measurement errors) for the W mass.

Rubbia was given the honor of presenting his results to the CERN community, and, uncharacteristically, he was nervous; eight years of work had been invested. His talk was spectacular. He had all the goods and the showmanship to display them with passionate logic(!). Even the Rubbia-haters cheered. Europe had its Nobel Prize, duly given to Rubbia and Van der Meer in 1985.

Some six months after the W success, the first evidence appeared for the existence of the neutral partner, the Z zero. With zero electric charge, it decays into, among many possibilities, an e^+ and an e^- (or a pair of muons, μ^+ and μ^-). Why? For those who fell asleep during the previous chapter, since the Z is neutral, the charges of its decay products must cancel each other out, so particles of opposite signs are logical decay products. Because both electron and muon pairs can be precisely measured, the Z^0 is an easier particle to recognize than the W. The trouble is that the Z^0 is heavier than the W, and fewer are made. Still, by late 1983, the Z^0 was established by both UA-1 and UA-2. With the discovery of the W's and the Z^0 and a determination that their masses are just what was predicted, the electroweak theory — which unified electromagnetism and the weak force — was solidly confirmed.

TOPPING OFF THE STANDARD MODEL

By 1992, tens of thousands of W's had been collected by UA-1 and UA-2, and the new kid, CDF, at the Fermilab Tevatron. The mass of the W is now known to be about 79.31 GeV. Some two million Z^0's were collected by CERN's "Z^0 factory," LEP (Large Electron-Positron Storage Ring), a seventeen-mile-around electron accelerator. The Z^0 mass is measured to be 91.175 GeV.

Some accelerators became particle factories. The first factories — in Los Alamos, Vancouver, and Zurich — produced pions. Canada is now designing a kaon factory. Spain wants a tau-charm factory. There are three or four proposals for beauty or bottom factories, and the CERN Z^0 factory is, in 1992, in full production. At SLAC a smaller Z^0 project might more properly be called a loft, or perhaps a boutique.

Why factories? The production process can be studied in great detail and, especially for the more massive particles, there are many decay modes. One wants samples of many thousands of events in each mode. In the case of the massive Z^0, there are a huge number of modes, from which one learns much about the weak and electroweak forces. One also learns from what isn't there. For example, if the mass of the top quark is less than half that of the Z^0, then we have (compulsory) $Z^0 \rightarrow$ top + antitop. That is, a Z zero can decay, albeit rarely, into a meson, composed of a top quark lashed to an antitop quark. The Z^0 is much more likely to decay into electron pairs or muon pairs or bottom-quark pairs, as mentioned. The success of the theory in accounting for these pairs encourages us to believe that the decay of Z^0 into top/antitop is predictable. We say it is compulsory because of the totalitarian rule of physics. If we make enough Z^0's, according to the probabilities of quantum theory, we should see evidence of the top quark. Yet in the millions of Z^0's produced at CERN, Fermilab, and elsewhere, we have never seen this particular decay. This tells us something important about the top quark. It must be heavier than half of the Z^0 mass. That's why the Z^0 can't produce it.

WHAT ARE WE TALKING ABOUT?

A very broad spectrum of hypothetical particles has been proposed by theorists following one trail or another toward unification. Usually the properties of these particles, except for the mass, are well specified by the model. Not seeing these "exotics" provides a lower limit for their mass, following the rule that the larger the mass the harder it is to produce.

Some theory is involved here. Theorist Lee says: a p/p-bar collision will produce a hypothetical particle — call it the Lee-on — if there is enough energy in the collision. However, the probability or relative frequency of producing the Lee-on depends on its mass. The heavier it is, the less frequently it is produced. The theorist hastens to supply a graph relating the number of Lee-ons produced per day to the particle's mass. For example: mass = 20 GeV, 1,000 Lee-ons (mind-numbing); 30 GeV, 2 Lee-ons; 50 GeV, one thousandth of a Lee-on. In the last case one would have to run the equipment for 1,000 days to get one event, and experimenters usually insist on at least ten events since they have additional problems with efficiency and background. So after a given run, say of 150 days (a year's run), in which no events are found, one looks at the curve, follows it down to where, say, ten events should have been produced — corresponding to a mass of, say, 40 GeV for the Lee-on. A conservative estimate is that some five events could have been missed. So the curve tells us that if the mass were 40 GeV, we would have seen a weak signal of a few events. But we saw nothing. Conclusion: the mass is heavier than 40 GeV.

What next? If the Lee-on or the top quark or the Higgs is worth the game, one has a choice of three strategies. First, run longer, but this is a tough way to improve. Second, get more collisions per second; that is, raise the luminosity. Right on! That is exactly what

Fermilab is doing in the 1990s, with the goal of improving the collision rate by about a hundredfold. As long as there is plenty of energy in the collision (1.8 TeV is plenty), raising the luminosity helps. The third strategy is to raise the energy of the machine, which increases the probability of producing all heavy particles. That's the Super Collider route.

With the discovery of the W and Z, we have identified six quarks, six leptons, and twelve gauge bosons (messenger particles). There is a bit more to the standard model that we have not yet fully confronted, but before we approach that mystery, we should beat on the model a bit. Writing it as three generations at least gives it a pattern. We note some other patterns, too. The higher generations are successively heavier, which means a lot in our cold world today but wouldn't have been very significant when the world was young and very hot. All the particles in the very young universe had enormous energies — billions and billions of TeV, so a little difference in rest mass between a bottom quark and an up quark wouldn't mean much. All quarks, leptons, and so on were once upon a time on an equal footing. For some reason She needed and loved them all. So we have to take them all seriously.

The Z^0 data at CERN suggest another conclusion: it is very unlikely that we have a fourth or fifth generation of particles. How is that for a conclusion? How could these scientists working in Switzerland, lured by the snow-capped mountains, deep, icy lakes, and magnificent restaurants, come to such a limiting conclusion?

It's a neat argument. The Z^0 has plenty of decay modes, and each mode, each possibility for decay, shortens its life a bit. If there are a lot of diseases, enemies, and hazards, human life is also shortened. But that is a sick analogy. Each opportunity to decay opens a channel or a route for the Z^0 to shake this mortal coil. The sum total of all routes determines the lifetime. Let's note that not all Z^0's have the same mass. Quantum theory tells us that if a particle is unstable — doesn't live forever — its mass must be somewhat indeterminate. The Heisenberg relations tell us how the lifetime affects the mass distribution: long lifetime, narrow width; short lifetime, broad width. In other words, the shorter the lifetime, the less determinate the mass and the broader the range of masses. The theorists can happily supply us a formula for the connection. The distribution width is easy to measure if you have a lot of Z^0's and a hundred million Swiss francs to build a detector.

The number of produced Z^0's is zero if the sum of the e^+ and the e^- energies at the collision is substantially less than the average Z^0 mass of 91.175 GeV. The operator raises the energy of the machine until a low yield of Z^0's is recorded by each of the detectors. Increase the machine energy, and the yield increases. It is a repeat of the J/psi experiment at SLAC, but here the width is about 2.5 GeV; that is, one finds a peak yield at 91.175, which decreases to about half on either side, at 89.9 GeV and 92.4 GeV. (If you'll recall, the J/psi width was much narrower: about 0.05 MeV.) The bell-shaped curve gives us a width, which is in effect a lifetime. Every possible Z^0 decay mode decreases its lifetime and increases the width by about 0.20 GeV.

What has this to do with a fourth generation? We note that each of the three generations has a low-mass (or zero-mass) neutrino. If there is a fourth generation with a low-mass neutrino, then the Z^0 must include, as one of its decay modes, the neutrino ν_x and its antiparticle, $\bar{\nu}_x$, of this new generation. This possibility would add 0.17 GeV to the width. So the width of the Z^0 mass distribution was carefully studied. And it turned out to be exactly what the three-generation standard model had predicted. The data on the width of the Z^0 excludes the existence of a low-mass fourth-generation neutrino. All four LEP experiments chimed in to agree that their data allowed only three neutrino pairs. A fourth generation with the same structure as the other three, including a low- or zero-mass neutrino, is excluded by the Z^0 production data.

Incidentally, the same remarkable conclusion had been claimed by cosmologists years earlier. They based their conclusions on the way neutrons and protons combined to form the chemical elements during an early phase of the expansion and cooling of the universe after that humongous bang. The amount of hydrogen compared to the amount of helium depends (I won't explain) on how many neutrino species there are, and the data on abundances strongly suggested three species. So the LEP research is relevant to our understanding of the evolution of the universe.

Well, here we are with an almost complete standard model. Only the top quark is missing. The tau neutrino is too, but that is not nearly so serious, as we have seen. Gravity must be postponed until the theorists understand it better, and, of course, the Higgs is missing, the God Particle.

SEARCH FOR TOP

A *NOVA* TV program called "Race for the Top" was shown in 1990 when CERN's p-bar/p collider and Fermilab's CDF were both running. CDF had the advantage of three times higher energy, 1.8 TeV against CERN's 620 GeV. CERN, by cooling their copper coils a bit better, had succeeded in raising their beam energies from 270 GeV to 310 GeV, squeezing every bit of energy they could in order to be competitive. Still, a factor of three hurts. CERN's advantage was nine years of experience, software development, and know-how in data analysis. Also they had redone the antiproton source, using some of Fermilab's ideas, and their collision rate was slightly better than ours. In 1989–90, the UA-1 detector was retired. Rubbia was now director general of CERN with an eye to the future of his laboratory, so UA-2 was given the task of finding top. An ancillary goal was to measure the mass of the W more precisely, for this was a crucial parameter of the standard model.

At the time the *NOVA* program was put to bed, neither group had found any evidence for top. In fact, by the time the program aired, the "race" was over, in that CERN was just about out of the picture. Each group had analyzed the absence of a signal in terms of top's unknown mass. As we have seen, not finding a particle tells you something about its mass. The theorists knew everything about the production of top and about certain decay channels — everything but the mass. The production probability depends critically on the unknown mass. Fermilab and CERN both set the same limits: the mass of the top quark was greater than 60 GeV.

Fermilab's CDF continued to run, and slowly the machine energy began to pay off. By the time the collider run was over, CDF had run for eleven months and had seen more than 100 billion (10^{11}) collisions — but no top. The analysis gave a limit of 91 GeV for the mass, making the top at least eighteen times heavier than the bottom quark. This surprising result disturbed many theorists working on unified theories, especially in the electroweak pattern. In these models the top quark should be much lower in mass, and this led some theorists to view top with special interest. The mass concept is somehow tied in with Higgs. Is the heaviness of the top quark a special clue? Until we find top, measure its mass, and in general subject it to the experimental third degree, we won't know.

The theorists went back to their calculations. The standard model was actually still intact. It could accommodate a top quark as heavy as 250 GeV, the theorists figured, but anything heavier would indicate a fundamental problem with the standard model. Experimenters were reinvigorated in their determination to pursue the top quark. But with top's mass greater than 91 GeV, CERN dropped out. The $e^+ e^-$ machines are too low in energy and therefore useless; of the world's inventory, only Fermilab's Tevatron can make top. What is needed is at least five to fifty times the present number of collisions. This is the challenge for the 1990s.

THE STANDARD MODEL IS A SHAKY PLATFORM

I have a favorite slide that pictures a white-gowned deity, with halo, staring at a "Universe Machine." It has twenty levers, each one designed to be set at some number, and a plunger labeled "Push to create universe." (I got this idea from a sign a student put up on the bathroom hand drier: "Push to get a message from the dean.") The idea is that twenty or so numbers must be specified in order to begin the universe. What are these numbers (or parameters, as they are called in the physics world)? Well, we need twelve numbers to specify the masses of the quarks and leptons. We need three numbers to specify the strengths of the forces. (The fourth, gravity, really isn't a part of the standard model, at least not yet.) We need some numbers to show how one force relates to another. Then we need a number for how the CP-symmetry violation enters, and a mass for the Higgs particle, and a few other handy items.

If we have these basic numbers, all other parameters are derived therefrom — for example, the 2 in the inverse-square law, the mass of the proton, the size of the hydrogen atom, the structure of H_2O and the double helix (DNA), the freezing temperature of water, and the GNP of Albania in 1995. I wouldn't have any idea how to obtain most of the derived numbers, but we do have these enormous computers . . .

The drive for simplicity leads us to be very sarcastic about having to specify twenty parameters. It's not the way any self-respecting God would organize a machine to create universes. One parameter — or two, maybe. An alternative way of saying this is that our experience with the natural world leads us to expect a more elegant organization. So this, as we have already complained, is the real problem with the standard model. Of course we still have an enormous amount of work to do to pinpoint these parameters accurately. The problem is the aesthetics — six quarks, six leptons, and twelve force-carrying gauge particles, and the quarks come in three colors, and then there are the antiparticles. And gravity waiting in the wings. Where is Thales now that we need him?

Why *is* gravity left out? Because no one has yet succeeded in forcing gravity — the general theory of relativity — to conform to the quantum theory. The subject, quantum gravity, is one of the theoretical frontiers of the 1990s. In describing the universe in its present grand scale, we don't need quantum theory. But once upon a time the entire universe was no bigger than an atom; in fact, it was a good deal smaller. The extraordinarily weak force of gravity was enhanced by the enormous energy of the particles that made all the planets, stars, galaxies of billions of stars, all that mass compressed to a pinhead on a pinhead, a size tiny compared to an atom. The rules of quantum physics must apply here in this primal gravitational maelstrom, and we don't know how to do it! Among theorists the marriage of general relativity and quantum theory is the central problem of contemporary physics. Theoretical efforts along these lines are called "super gravity" or "supersymmetry" or "superstrings" or the "Theory of Everything" (TOE).

Here we have exotic mathematics that curls the eyebrows of some of the best mathematicians in the world. They talk about ten dimensions: nine space and one time dimension. We live in four dimensions: three space dimensions (east-west, north-south, and up-down) and one time dimension. We can't possibly intuit more than three space dimensions. "No problem." The superfluous six dimensions have been "compactified," curled up to an unimaginably small size so as not to be evident in the world we know.

Today's theorists have a bold objective: they're searching for a theory that describes a pristine simplicity in the intense heat of the very early universe, a theory with no parameters. Everything must emerge from the basic equation; all the parameters must come out of the theory. The trouble is, the only candidate theory has no connection with the world of observation — not yet anyway. It has a brief instant of applicability at the imaginary domain that the experts call the "Planck mass," a domain where all the particles in the universe have energies of 1,000 trillion times the energy of the Super Collider. The time interval of this greater glory lasted for a trillionth of a trillionth of a trillionth of a second. Shortly thereafter, the theory gets confused — too many possibilities, no clear road indicating that we the people and planets and galaxies are indeed a prediction.

In the middle 1980s, TOE had a tremendous appeal for young physicists of the theoretical persuasion. In spite of the risk of long years of investment for small returns, they followed the leaders (like lemmings, some would say) to the Planck mass. We who stayed home at Fermilab and CERN received no postcards, no faxes. But disillusion began to set in. Some of the more stellar recruits to TOE quit, and pretty soon, buses began arriving back from the Planck mass with frustrated theorists looking for something real to calculate. The entire adventure is still not over, but it has slowed to a quieter pace, while the more traditional roads to unification are tried.

These more popular roads toward a complete, overarching principle have groovy names: grand unification, constituent models, supersymmetry, Technicolor, to name a few. They all share one problem: there are no data! These theories made a rich stew of predictions. For example, supersymmetry (affectionately shortened to "Susy"), probably the most popular theory, if theorists voted (and they don't), predicts nothing less than a doubling of the number of particles. As I've explained, the quarks and leptons, collectively called fermions, all have one half unit of spin, whereas the messenger particles, collectively called bosons, all have one full unit of spin. In Susy this asymmetry is repaired by postulating a boson partner for every fermion and a fermion partner for every boson. The naming is terrific. The Susy partner of the electron is called "selectron," and the partners of all the leptons are collectively called "sleptons." The quark partners are "squarks." The spin-one-half partners of the spin-one bosons are given a suffix "ino" so that gluons are joined by "gluinos," photons couple with "photinos," and we have "winos" (partner of the W) and "zinos." Cute doesn't make a theory, but this one is popular.

The search for squarks and winos will go on as the Tevatron increases its power through the 1990s and the machines of the year 2000 come on-line. The Super Collider being built in Texas will enable exploration of the "mass domain" up to about 2 TeV. The definition of mass domain is very loose and depends on the details of the reaction that makes a new particle. However, a sign of the power of the Super Collider is that if no Susy particles are found in this machine, most Susy protagonists have agreed to abandon the theory in a public ceremony in which they break all their wooden pencils.

But the SSC has a more immediate goal, a quarry more pressing than the squarks and sleptons. As a compact summary of everything we know, the standard model has two major defects, one aesthetic, one concrete. Our aesthetic sense tells us that there are too many particles, too many forces. Worse, the many particles are distinguished by the seemingly random masses assigned to quarks and leptons. Even the forces differ largely because of the masses of the messenger particles. The concrete problem is one of inconsistency. When the force-field theories, in impressive agreement with all of the data, are asked to predict the results of experiments carried out at very high energies, they churn out physical absurdities. Both problems can be illuminated and possibly solved by an object (and a force) that must be added gingerly to the standard model. The object and the force go by the same name: Higgs.

AT LAST . . .

> All visible objects, man, are but as pasteboard masks. But in each event . . . some unknown but still reasoning thing puts forth the mouldings of its features from behind the unreasoning mask. If man will strike, strike through the mask!
>
> — Captain Ahab

One of the finest novels in American literature is Herman Melville's *Moby Dick*. It is also one of the most disappointing — at least for the captain. For hundreds of pages we hear about Ahab's quest to find and harpoon a large white oceangoing mammal named Moby Dick. Ahab is pissed. This whale has bitten off his leg, and he wants revenge. Some critics suggest that the whale bit off a lot more than leg, which would explain more adequately the good captain's pique. Ahab explains to his first mate, Starbuck, that Moby Dick is more than a whale. He is a pasteboard mask; he represents a deeper force in nature that Ahab must confront. So for hundreds of pages Ahab and his men scurry furiously around the ocean, having adventures and misadventures, killing lots of smaller whales of various masses. Finally, thar she blows: the great white whale. And then, in quick succession, the whale drowns Ahab, kills all the other harpooners, then sinks the ship for good measure. End of story. Bummer. Perhaps Ahab needed a bigger harpoon, one denied by nineteenth-century budgetary restraints. Let's not let that happen to us. Moby Particle is within striking distance.

•

We have to ask this question about our standard model: is it simply a pasteboard mask? How can a theory be in accordance with all the data at low energy and predict nonsensical effects at high energy? The answer is to suggest that the theory is leaving something out, some new phenomenon which, when installed in the theory, will contribute negligibly to the data at, say, Fermilab energies and therefore will not spoil agreement with experimental data. Examples of what's left out might be a new particle or a change in the behavior of a force. These postulated new phenomena must contribute negligibly at low energy but massively at Super Collider or higher energy. When a theory does not include these terms (because we don't know about them) we get mathematically inconsistent results at these high energies.

This is somewhat like Newtonian physics, which works very successfully for ordinary phenomena but predicts that we can accelerate an object to infinite velocity; this implausible consequence is totally contradicted when Einstein's special theory of relativity is installed. Relativity theory has infinitesimally tiny effects at the velocities of bullets and rockets. However, as the velocities approach that of light, a new effect appears: the masses of the speeding objects begin to increase, and infinite velocities become impossible. What happens is that special relativity merges into Newtonian results at velocities that are small compared to the velocity of light. The weakness of this example is that whereas the concept of infinite velocity may have been disturbing to Newtonians, it was not nearly as traumatic as what happens to the standard model at high energies. We'll return to this soon.

THE MASS CRISIS

I have hinted at the function of the Higgs particle in giving mass to massless particles and thereby disguising the true symmetry of the world. This is a new and bizarre idea. Heretofore, as we have seen in our myth-history, simplicity was gained by finding substructures — the Democritan idea of *atomos.* And so we went from molecules to chemical atoms to nuclei to protons and neutrons (and their numerous Greek relatives) to quarks. History would lead one to expect that now we reveal the little people inside the quark, and indeed this may still happen. But we really don't think that is the way the long-awaited complete theory of the world will come out. Perhaps it's more like the kaleidoscope I referred to earlier, in which some split mirrors convert a few bits of colored glass into a myriad of seemingly complex designs. Higgs's ultimate purpose (this isn't science, it's philosophy) may be to create a more amusing, more complex world as suggested in the parable that started this chapter.

The new idea is that all of space contains a field, the Higgs field, which permeates the vacuum and is the same everywhere. This means that when you look up at the stars on a clear night you are looking through the Higgs field. Particles, influenced by this field, acquire mass. This by itself is not remarkable since particles can acquire energy from the (gauge) fields we have discussed, the gravitational field or the electromagnetic field. For example, if you carry a lead block to the top of the Eiffel Tower, the block acquires potential energy because of its altered position in the earth's gravitational field. Since $E = mc^2$, this increase in potential energy is equivalent to an increment in mass, in this case the mass of the earth–lead-block system. Here we have to gently add a small complexity to Einstein's hoary equation. The mass, m, actually has two parts. One is the rest mass, m_0, which is what is measured in the laboratory when the particle is at rest. The other part of the mass is "acquired" by the particle by virtue of its motion (like the protons in the Tevatron) or by virtue of its potential energy in a field. We see a similar dynamic in atomic nuclei. For example, if you separate the proton and neutron that make up the deuterium nucleus, the sum of the masses increases.

But the potential energy acquired from the Higgs field differs in several ways from the action of the more familiar fields. The Higgs-acquired mass is actually rest mass. In fact, in what may be the most intriguing version of the Higgs theory, *all* rest mass is generated by the Higgs field. Another difference is that the amount of mass soaked up from the field differs for various particles. Theorists say that the masses of the particles in our standard model are a measure of how strongly they are coupled to the Higgs field.

The Higgs influence on the masses of quarks and leptons reminds one of Pieter Zeeman's discovery, in 1896, of the splitting of the energy levels of an electron in an atom when a magnetic field is applied to the atom. The field (playing the metaphoric role of Higgs) breaks the symmetry of space that the electron had enjoyed. For example, one energy level, influenced by the magnet, splits into three; level A gains energy from the field, level B loses energy, and level C doesn't change at all. Of course, we now understand completely how all of this happens. It is simple quantum electromagnetism.

So far we have no idea what the rules are that control the Higgs-generated mass increments. But the question nags: why only these masses — the masses of the W^+, W^-, and Z^0, and the up, down, charm, strange, top, and bottom, as well as the leptons — which form no obvious pattern? The masses vary from that of the electron, at .0005 GeV, to the top quark's, which must be greater than 91 GeV. We should recall that this bizarre idea — Higgs — was used with great success in formulating the electroweak theory. There the Higgs field was proposed as a way of hiding the unity of the electromagnetic and the weak force. In unity there are four massless messenger particles — the W^+, W^-, Z^0, and the photon — that carry the electroweak force. Along comes the Higgs field, and presto, the W's and Z soak up the essence of Higgs and grow heavy; the photon is untouched. The electroweak shatters into the weak (weak because the messengers are so fat) and the electromagnetic force, whose properties are determined by the massless photon. The symmetry is spontaneously broken, the theorists say. I prefer the description that Higgs hides the symmetry by its mass-giving power. The masses of the W's and the Z were successfully predicted from the parameters of the electroweak theory. And the relaxed smiles of the theorists remind us that 't Hooft and Veltman established that this whole theory has no infinities.

I dwell on this issue of mass in part because it has been with me all during my professional life. In the 1940s the issue seemed well focused. We had two particles that exemplified the puzzle of mass: the electron and the muon. They seemed to be in all respects identical except that the muon weighed two hundred times more than its puny cousin. The fact that these were leptons, ignoring the strong force, made it more intriguing. I became obsessed with the problem and made the muon my favorite object of study. The aim was to try to find some difference, other than mass, in the behavior of the muon and the electron as a clue to the mechanism of mass differences.

The electron is occasionally captured by a nucleus, giving rise to a neutrino and a recoiling nucleus. Can the muon do this? We measured the process of muon capture — bingo, same process! A high-energy electron beam scatters protons. (This reaction was studied at Stanford.) We measured the same reaction at Brookhaven with muons. A small difference in rates enticed us for years, but nothing came of it. We even discovered that the electron and the muon have separate neutrino partners. And we have already discussed the superprecise g minus 2 experiment, in which the magnetism of the muon was measured and compared to that of the electron. Except for the extra mass effect, they were the same.

All efforts to find a clue to the origin of mass failed. Along the way, Feynman wrote his famous inquiry: "Why does the muon weigh?" Now, at least, we have a partial, by no means complete, answer. A stentorian voice says, "Higgs!" For fifty or so years we have been puzzling about the origin of mass, and now the Higgs field presents the problem in a new context; it is not only the muon. It provides, at the least, a common source for all masses. The new Feynmanian question could be: how does the Higgs field determine the sequence of seemingly patternless masses that is given to the matter particles?

The variation of mass with state of motion, the change of mass with system configuration, and the fact that some particles — the photon surely and the neutrinos possibly — have zero rest mass all challenge the concept of mass as a *fundamental* attribute of matter. Then we must recall the calculation of mass that came out infinite, which we never solved — just "renormalized" away. This is the background with which we face the problem of the quarks, leptons, and force carriers, which are differentiated by masses. It makes our Higgs story tenable — that mass is not an intrinsic property of particles but a property acquired by the interaction of particles and their environment. The idea that mass is not intrinsic like charge or spin is made even more plausible by the idyllic notion of zero mass for all quarks and leptons. In this case, they would obey a satisfying symmetry, chiral symmetry, in which their spins would forever be associated with their direction of motion. But that idyll is hidden by the Higgs phenomenon.

Oh, one more thing. We talked about gauge bosons and their one-unit spin; we also discussed fermion matter particles (spin of one half unit). What breed of cat is the Higgs? It is a spin-zero boson. *Spin* implies directionality in space, but the Higgs field gives mass to objects at every location and with no directionality. Higgs is sometimes called a "scalar [no direction] boson" for that reason.

THE UNITARITY CRISIS

Much as we are intrigued by the mass-endowing attributes of this new field, one of my favorite theorists, Tini Veltman, rates that job of the Higgs far below its major obligation, which is nothing less than making our standard model consistent. Without Higgs, the model fails a simple test of consistency.

Here's what I mean. We have talked a lot about collisions. Let's aim one hundred particles at a specific target, say a piece of iron with one square inch of area. A theorist of modest ability can calculate the probability (remember, quantum theory permits us to predict only probability) that there will be a scattering. For example, the theory may predict that ten particles will scatter out of the one hundred that we direct at our target, for a probability of 10 percent. Now many theories predict that the probability of scattering depends on the energy of the beam we are using. At low energy all of the force theories we know — strong, weak, and electromagnetic — predict probabilities that are in agreement with the actual experiments. However, it is known that for the weak force the probability increases with energy. For example, at medium energy the scattering probability may increase to 40 percent. If the theory predicts that the scattering probability is greater than 100 percent, then clearly the theory ceases to be valid. Something is wrong, since a probability of more than 100 percent makes no sense. It literally means that more particles are scattered than were in the beam in the first place. When this happens we say the theory violates unitarity (exceeds unit probability).

In our history, the puzzle was that the theory of the weak force was in good agreement with the experimental data at low energy but predicted nonsense at high energy. This crisis was discovered when the energy at which disaster was predicted was outside the energy reach of the existing accelerators. But the failure of the theory indicated that something was being left out, some new process, some new particle perhaps, which, if we only knew what it was, would have the effect of preventing the increase of probability to nonsense values. The weak force, you will remember, was invented by Fermi to describe the radioactive decay of nuclei. These decays are basically low-energy phenomena, and as the Fermi theory evolved, it became very accurate at predicting a huge number of processes in the 100 MeV energy domain. One motivation of the two-neutrino experiment was to test the theory at higher energies, because the predictions were that a unitarity crisis would occur at about 300 GeV. Our experiment, carried out at a few GeV, confirmed that the theory was heading toward a crisis. This turned out to be an indicator that the theorists had left out of the theory a W particle of approximately 100 GeV mass. The original Fermi theory, which did not include W's, was mathematically equivalent to using an infinitely massive force carrier, and 100 GeV is so extremely large compared to the early experiments (below 100 MeV) that the old theory worked well. But when we asked the theory what 100 GeV neutrinos would do, the 100 GeV W had to be included to avoid a unitarity crisis — but more is needed.

Well, this review is simply to explain that our standard model suffers from a unitarity disease in its most virulent form. The disaster now strikes at an energy of about 1 TeV. The object that would avoid disaster if . . . if it existed is a neutral heavy particle with special properties that we call — you guessed it — a Higgs particle. (Earlier we referred to the Higgs field, but we should remember that the quanta of a field are a set of particles.) It might be the very same object that creates the diversity of masses or it might be a similar object. There might be one Higgs particle or there might be a family of Higgs particles.

THE HIGGS CRISIS

Lots of questions must be answered. What are the properties of the Higgs particles and, most important, what is their mass? How will we recognize one if we meet it in a collision? How many types are there? Does Higgs generate all masses or only some increment to masses? And how do we learn more about it? Since it is Her particle, we can wait, and if we lead an exemplary life, we'll find out when we ascend to Her kingdom. Or we can spend $8 billion and build us a Super Collider in Waxahachie, Texas, which has been designed to produce the Higgs particle.

The cosmologists are also fascinated by the Higgs idea, since they sort of stumbled on the need for scalar fields to participate in the complex process of expanding the universe, thus adding to the burden Higgs must bear. More about this in Chapter 9.

The Higgs field as it is now contrived can be destroyed by high energy (or high temperatures). These generate quantum fluctuations that can neutralize the Higgs field. Thus the joint particle-cosmology picture of an early universe, pure and with dazzling symmetry, is too hot for Higgs. But as temperature/energy drops below 10^{15} degrees Kelvin or 100 GeV, the Higgs acts up and does its mass-generating thing. So, for example, before Higgs we have massless W's, Z's, and photons and a unified electroweak force. The universe expands and cools and along comes the Higgs — making the W and Z fat, for some reason ignoring the photon — and this results in breaking the electroweak symmetry. We get a weak force, mediated by massive force carriers W^+, W^-, Z^0, and we get a separate electromagnetic force, carried by photons. It is as if to some particles the Higgs field is like a heavy oil through which they move sluggishly, seeming to be massive. To other particles the Higgs is like water, and to still others, such as photons and perhaps neutrinos, it is invisible.

I should probably review the origin of the Higgs idea, since I've been a bit coy about letting the cat out of the bag. It is also called hidden symmetry or "spontaneous symmetry breaking." The idea was introduced into particle physics by Peter Higgs of the University of Edinburgh. It was used by theorists Steven Weinberg and Abdus Salam, working independently, to understand the conversion of a unified and symmetric electroweak force, transmitted by a happy family of four zero-mass messenger particles, into two very different forces: QED with its massless photon and the weak force with massive W^+, W^-, and Z^0's. Weinberg and Salam built on the earlier work of Sheldon Glashow, who, following Julian Schwinger, just knew that there was a consistent, unified electroweak theory but didn't put all the details together. And there were Jeffrey Goldstone and Martinus Veltman and Gerard 't Hooft. And there are others who should be mentioned, but that's life. Besides, how many theorists does it take to light up a light bulb?

Another way of looking at Higgs is from the point of view of symmetry. At high temperatures the symmetry is exposed — regal, pure simplicity. At lower temperatures the symmetry is broken. Time for some more metaphors.

Consider a magnet. A magnet is a magnet because, at low temperatures, its atomic magnets are aligned. A magnet has a special direction, its north-south axis. Thus it has lost the symmetry of a piece of nonmagnetic iron in which all spatial directions are equivalent. We can "fix" the magnet. By raising the temperature, we go from magnetic iron to nonmagnetic iron. The heat generates molecular agitation, which eventually destroys the alignment, and we have a purer symmetry. Another popular metaphor is the Mexican hat: a symmetric dome surrounded by a symmetric turned-up brim. A marble is perched on the top of the dome. Perfect rotational symmetry, but no stability. When the marble falls to a more stable (lower-energy) position, somewhere on the brim, the symmetry is destroyed even though the basic structure is symmetric.

In another metaphor we imagine a perfect sphere filled with water vapor at very high temperature. The symmetry is perfect. If we let the system cool, eventually we get a pool of water with some ice floating in it and residual water vapor above. The symmetry has been totally destroyed by the simple act of cooling, which in this metaphor allows the gravitational field to exert itself. However, paradise can be regained by simply heating up the system.

So: before Higgs, symmetry and boredom; after Higgs, complexity and excitement. When you next look out at the night sky you should be aware that all of space is filled with this mysterious Higgs influence, which is responsible, so this theory holds, for the complexity of the world we know and love.

Now picture the formulas (ugh!) that give correct predictions and postdictions of the properties of particles and forces we measure at Fermilab and in our accelerator labs of the 1990s. When we plug in reactions to be carried out at much higher energies, the formulas churn out nonsense. Aha, but if we include the Higgs field, then we modify the theory and get a consistent theory even at energies of 1 TeV. Higgs saves the day, saves the standard model with all its virtues. Does all this prove that it is correct? Not at all. It's only the best the theorists can do. Perhaps She is even more clever.

A DIGRESSION ON NOTHING

Back in the days of Maxwell, physicists felt that they needed a medium that would pervade all space and through which light and other electromagnetic waves could travel. They called it an aether and established properties so that it could do its job. Aether also provided an absolute coordinate system that enabled measurement of the velocity of light. Einstein's flash of insight showed that aether was an unnecessary burden on space. Here one is tampering with a venerable concept, none other than the "void" invented (or discovered) by Democritus. Today the void, or more precisely, the "vacuum state," is front and center.

The vacuum state consists of those regions of the universe where all matter has been removed and no energy or momentum exists. It is "nothing at all." James Bjorken, in talking about this state, said that he was tempted to do for particle physics what John Cage did for music: a four-minute-and-twenty-two-second . . . nothing. Only fear of the conference chairman dissuaded him. Bjorken, expert as he is on the properties of the vacuum state, doesn't compare to 't Hooft, who understands nothing at all much better.

The sad part of the story is that the pristine absoluteness of the vacuum state (as a concept) has been so polluted (wait until the Sierra Club finds out!) by twentieth-century theorists that it is vastly more complicated than the discarded nineteenth-century aether. What replaces the aether, in addition to all the ghostly virtual particles, is the Higgs field, whose full dimensions we do not yet know. To do its job, there must exist, and experiments must reveal, at least one Higgs particle, electrically neutral. This may be only the tip of the iceberg; a zoo of Higgs boson quanta may be needed to completely describe the new aether. Clearly there are new forces here and new processes. We can summarize the little we know: at least some of the particles that represent the Higgs aether must have zero spin, must be intimately and mysteriously connected to mass, and must manifest themselves at temperatures equivalent to an energy of less than 1 TeV. There is controversy also about the Higgs structure. One school says it's a fundamental particle. Another idea is that it is composed of new, quarklike objects, which could eventually be seen in the laboratory. A third camp is intrigued by the huge mass of the top quark and believes that Higgs is a bound state of top and antitop. Only data will tell. Meanwhile, it's a miracle that we can see the stars at all.

The new aether is then a reference frame for energy, in this case potential energy. And Higgs alone doesn't explain the other debris and theoretical garbage that is dumped in the vacuum state. The gauge theories deposit their requirements, the cosmologists exploit "false" vacuum energy, and in the evolution of the universe, the vacuum can stretch and expand.

One longs for a new Einstein who will, in a flash of insight, give us back our lovely nothingness.

FIND THE HIGGS!

So Higgs is great. Why, then, hasn't it been universally embraced? Peter Higgs, who loaned his name to the concept (not willingly), works on other things. Veltman, one of the Higgs architects, calls it a rug under which we sweep our ignorance. Glashow is less kind, calling it a toilet in which we flush away the inconsistencies of our present theories. And the other overriding objection is that there isn't a single shred of experimental evidence.

How does one prove the existence of this field? Higgs, just like QED, QCD, or the weak force, has its own messenger particle, the

Higgs boson. Prove Higgs exists? Just find the particle. The standard model is strong enough to tell us that the Higgs particle with the lowest mass (there may be many) must "weigh" less than 1 TeV. Why? If it is more than 1 TeV, the standard model becomes inconsistent, and we have the unitarity crisis.

The Higgs field, the standard model, and our picture of how God made the universe depend on finding the Higgs boson. There is no accelerator on earth, unfortunately, that has the energy to create a particle as heavy as 1 TeV.

You could, however, build one.

THE DESERTRON

In 1981 we at Fermilab were deeply involved in building the Tevatron and the p-bar/p collider. We were, of course, paying some attention to what was going on in the world and especially to the CERN quest for the W. By late spring of that year we were getting confident that superconducting magnets could work and could be mass-produced with the required stringent specifications. We were convinced, or at least 90 percent convinced, that the 1 TeV mass scale, the terra incognita of particle physics, could be reached at relatively modest cost.

Thus it made sense to start thinking of the "next machine" (whatever would follow the Tevatron), as an even bigger ring of superconducting magnets. But in 1981 the future of particle research in this country was mortgaged to a machine struggling to survive at the Brookhaven lab. This was the Isabelle project, a proton-proton collider of modest energy that should have been working by 1980 but had been delayed by technical problems. In the interval the physics frontier had moved on.

At the annual Fermilab users' meeting in May of 1981, after duly reporting on the State of the Laboratory, I ventured a guess about the future of the field, especially "the energy frontier at 1 TeV." I remarked that Carlo Rubbia, already a dominating influence at CERN, would soon "pave the LEP tunnel with superconducting magnets." The LEP ring, about seventeen miles in circumference, contained conventional magnets for its $e^+ e^-$ collider. LEP needed that huge radius to reduce the energy lost by the electrons. These radiate energy when they are constrained into a circular orbit by magnets. (The smaller the radius, remember, the more the radiation.) So CERN's LEP machine used weak fields and a large radius. This also made it ideal for accelerating protons, which because of their much larger mass don't radiate very much energy. The farsighted LEP designers surely had this in mind as an eventual application of the big tunnel. Such a machine with superconducting magnets could easily go to about 5 TeV in each ring, or 10 TeV in the collision. And all the United States had to offer in competition beyond the Tevatron at 2 TeV was the ailing Isabelle, a 400 GeV collider (0.8 TeV in total), although it did have a very high collision rate.

By the summer of 1982, both the Fermilab superconducting-magnet program and the CERN proton-antiproton collider looked as if they would be successful. When American high-energy physicists gathered at Snowmass, Colorado, in August to discuss the status and the future of the field, I made my move. In a talk entitled "The Machine-in-the-Desert," I proposed that the community seriously consider making its number-one priority the building of a huge new accelerator based on the "proven" technology of supermagnets and forge ahead to the 1 TeV mass domain. Let's recall that to produce particles that might have a mass of 1 TeV, the quarks participating in the collision must contribute at least this amount of energy. The protons, carrying the quarks and gluons, must have much higher energy. My guess in 1982 was 10 TeV in each beam. I made a wild guesstimate at the cost and rested my case solidly on the premise that the lure of the Higgs was too attractive to pass up.

There was a moderately lively debate at Snowmass over the Desertron, as it was initially called. The name was based on the idea that a machine so large could be built only in a place devoid of people and land value and hills and valleys. What was wrong about that idea was that I, a New York City kid, practically raised in the subways, had completely forgotten the power of deep tunneling. History rubbed it in. The German machine HERA goes under the densely populated city of Hamburg. CERN's LEP tunnel burrows under the Jura Mountains.

I was attempting to forge a coalition of all the American labs to back this idea. SLAC was always looking toward electron acceleration; Brookhaven was struggling to keep Isabelle alive; and a lively and very talented gang at Cornell were trying to upgrade their electron machine to a status they called CESR II. I dubbed my Desertron lab "Slermihaven II" to dramatize the union of all the fiercely competitive labs behind the new venture.

I won't belabor the politics of science, but after a year full of trauma, the U.S. particle-physics community formally recommended abandoning Isabelle (renamed CBA for Colliding Beam Accelerator) in favor of the Desertron. Now called the Superconducting Super Collider, it was to have 20 TeV in each beam. At the same time — July 1983 — Fermilab's new accelerator hit the front pages as a success, accelerating protons to a record of 512 GeV. This was soon followed by further successes, and about a year later the Tevatron went to 900 GeV.

PRESIDENT REAGAN AND THE SUPER COLLIDER: A TRUE STORY

By 1986, the SSC proposal was ready to be submitted to President Reagan for approval. As director of Fermilab, I was asked by an assistant secretary of the DOE if we could make a short video for the president. He thought a ten-minute exposure to high-energy physics would be useful when the proposal was discussed at a Cabinet meeting. How do you teach a president high-energy physics in ten minutes? More important, how do you teach *this* president? After considerable agony, we hit on the idea of having some high school kids visit the lab, be taken on a tour of the machinery, ask a lot of questions, and receive answers designed for them. The president would see and hear all this and maybe get a notion of what high-energy physics is all about. So we invited kids from a nearby school. We coached a few just a bit and let the rest be spontaneous. We filmed about thirty minutes and cut it down to the best fourteen minutes. Our Washington contact warned us: no more than ten minutes! Something about attention span. So we cut more and shipped him ten lucid minutes of high-energy physics for high school sophomores. In a few days we had our reaction. "Way too complicated! Not even close."

What to do? We redid the soundtrack, wiping out the kids' questions. Some of them, after all, were pretty tough. A voice-over expert then related the kinds of questions the kids might have asked (written out by me), and gave the answers while the action remained the same: the scientist guides pointing, the kids gawking. This time we made it crystal clear and very simple. We tested it on nontechnical people. Then we sent it in. Our DOE guy was getting impatient.

Again he was underwhelmed. "Well, it's better, but it's still too complicated."

I began to get a little nervous. Not only was the SSC in danger, but my job was at stake. That night I awoke at 3 A.M. with a brilliant idea. The next video would go this way: a Mercedes pulls up to the lab entrance, and a distinguished gentleman of fifty-five or so emerges. The voice-over says: "Meet Judge Sylvester Matthews of the Fourteenth Federal District Court, who is visiting a large government research lab." The "judge" explains to his hosts, three handsome young physicists (one female), that he has moved into the neighborhood and drives past the lab on his way to court every day. He reads about our work in the *Chicago Tribune*, knows we are dealing with "volts" and "atoms," and, since he never studied physics, is curious about what goes on. He enters the building, thanking the physicists for taking time with him this morning.

My idea was that the president would identify with an intelligent layperson who is self-assured enough to say that he doesn't understand. In the subsequent eight and a half minutes, the judge frequently interrupts the physicists to insist that they go slower and clarify this and that point. At nine-plus minutes, the judge shoots his cuff, looks at his Rolex, and thanks the young scientists graciously. Then, with a shy smile: "You know I really didn't understand most of the things you told me, but I do get a sense of your enthusiasm, of the grandeur of the quest. It somehow reminds me of what it must have been like to explore the West . . . man alone on horseback with a vast, unexplored land . . ." (Yes, I wrote that.)

When the video got to Washington, the assistant secretary was ecstatic. "You've done it! It's terrific. Just right! It will be shown at Camp David over the weekend."

Greatly relieved, I went to bed smiling, but I woke up at 4 A.M. in a cold sweat. Something was wrong. Then I knew. I hadn't told the assistant secretary that the "judge" was an actor, hired from the Chicago Actors' Bureau. This was around the time the president was having trouble finding a confirmable appointee to the Supreme Court. Suppose he . . . I tossed and sweated until it was 8 A.M. in Washington. With my third call I got him.

"Say, about that video . . ."

"I told you it was great."

"But I have to tell — "

"It's good, don't worry. It's on its way to Camp David."

"Wait!" I screamed. "Listen! The judge. It's not a real judge. He's an actor, and the president may want to talk to him, interview him. He looks so intelligent. Suppose he . . ." [Long pause]

"The Supreme Court?"

"Yeah."

[Pause, then snickering] "Look, if I tell the president he's an actor, he'll *surely* appoint him to the Supreme Court."

Not long afterward the president approved the SSC. According to a column by George Will, the discussion about the proposal had been brief. During a Cabinet meeting the president listened to his secretaries, who were about evenly divided on the merits of the SSC. He then quoted a well-known quarterback: "Throw deep." By which everyone assumed he meant "Let's do it." The Super Collider became national policy.

Over the next year a lively search for a site for the SSC engaged communities all around the nation and in Canada. Something about the project seemed to excite people. Imagine a machine that could cause the mayor of Waxahachie, Texas, to stand up in public and conclude a fiery speech with "And this nation must be the first to find the Higgs scalar boson!" Even "Dallas" featured the Super Collider in a subplot in which J. R. Ewing and others attempted to buy up land around the SSC site.

When I referred to the mayor's comment at a meeting of the National Conference of Governors, in one of the several million talks I gave while selling the SSC, I was interrupted by the governor of Texas. He corrected my pronunciation of Waxahachie. Apparently I had deviated by more than the normal difference between Texan and New Yorkese. I couldn't resist. "Sir, I really tried," I assured the governor. "I went there, stopped at a restaurant, and asked the waitress to tell me where I was, clearly and distinctly. 'B-U-R-G-E-R — K-I-N-G,' she enunciated." Most of the governors laughed. Not the Texan.

The year 1987 was the year of three supers. First, there was the supernova that flared in the Large Magellanic Cloud about 160,000 years ago and finally got its signal to our planet so that neutrinos from outside our solar system were detected for the first time. Then there was the discovery of high-temperature superconductivity, which excited the world with its possible technological benefits. Early on there was hope that we would soon have room-temperature superconductors. Visions arose of reduced power costs, levitated trains, a myriad of modern miracles, and, for science, much-reduced costs of building the SSC. Now it's clear that we were too optimistic. In 1993 high-temperature superconductors are still a lively frontier for research and for a deeper understanding of the nature of material, but the commercial and practical applications are still a long way off.

The third super was the search for the site of the Super Collider. Fermilab was one of the contenders largely because the Tevatron could be used as an injector to the SSC main ring, an oval track with a circumference of fifty-three miles. But after weighing all considerations, the DOE's select committee picked the Waxahachie site. The decision was announced in October 1988, a few weeks after I had entertained a huge meeting of the Fermilab staff with my Nobel jokes. Now we had quite a different meeting as the gloomy staff gathered to hear the news and wonder about the future of the laboratory.

In 1993 the SSC is under construction, with a probable completion date of 2000, give or take a year or two. Fermilab is aggressively upgrading its facility in order to increase the number of p-bar/p collisions, to improve its chances of finding top, and to explore the lower levels of the great mountain the SSC is designed to scale.

Of course, the Europeans are not sitting on their hands. After a period of vigorous debate, study, design reports, and committee meetings, Carlo Rubbia, as CERN's director general, decided to "pave the LEP tunnel with superconducting magnets." The energy of an accelerator, you will recall, is determined by the combination of its ring diameter and the strength of its magnets. Constrained by the seventeen-mile circumference of the tunnel, the CERN designers were forced to strive for the highest magnetic field that they could technologically visualize. This was 10 tesla, about 60 percent stronger than the SSC's magnets and two and a half times stronger than the Tevatron's. Meeting this formidable challenge will require a new level of sophistication in superconducting technology. If it succeeds, it will give the proposed European machine an energy of 17 TeV compared to the SSC's 40 TeV.

The total investment in financial and human resources, if both of these new machines are actually built, is enormous. And the stakes are very high. What if the Higgs idea turns out to be wrong? Even if it is, the drive to make observations "in the 1 TeV mass domain" is just as strong; our standard model must be either modified or discarded. It's like Columbus setting out for the East Indies. If he doesn't reach it, thought the true believers, he will find something else, perhaps something even more interesting.

· 9 ·

INNER SPACE, OUTER SPACE, AND THE TIME BEFORE TIME

> You walk down Piccadilly
> With a poppy or a lily
> In your medieval hand —
> And everyone will say
> As you walk your mystic way,
> If this young man expresses himself
> In terms too deep for me,
> Why, what a singularly deep young man
> This deep young man must be.
>
> — Gilbert and Sullivan, *Patience*

IN HIS "DEFENSE OF POETRY," the English romantic poet Percy Bysshe Shelley contended that one of the sacred tasks of the artist is to "absorb the new knowledge of the sciences and assimilate it to human needs, color it with human passions, transform it into the blood and bone of human nature."

Not many romantic poets rushed to accept Shelley's challenge, which may explain the present sorry state of our nation and planet. If we had Byron and Keats and Shelley and their French, Italian, and Urdu equivalents explaining science, the science literacy of the general public would be far higher than it is now. This, of course, excludes you — not "dear reader" anymore, but friend and colleague who has fought with me through to Chapter 9 and is, by royal edict, a fully qualified, literate reader.

People who measure science literacy assure us that only one in three can define a molecule or name a single living scientist. I used to characterize these dismal statistics by adding, "Did you know that only sixty percent of the residents of Liverpool understand non-Abelian gauge theory?" Of twenty-three graduates randomly selected at Harvard's 1987 commencement ceremonies, only two could explain why it's hotter in summer than in winter. The answer, by the way, is not "because the sun is closer in summer." It isn't closer. The earth's axis of rotation is tilted, so when the northern hemisphere is tilted toward the sun, the rays are closer to being perpendicular to the surface, and that half of the globe enjoys summer. The other hemisphere gets oblique rays — winter. Six months later the situation is reversed.

The sad part about the ignorance of the twenty-one out of twenty-three Harvard grads — Harvard, by God! — who couldn't answer the question is what they are missing. They have gone through life without understanding a seminal human experience: the seasons. Of course, there are those bright moments when people surprise you. Several years ago, on Manhattan's IRT subway, an elderly man sweating over an elementary calculus problem in his textbook turned in desperation to the stranger sitting next to him, asking if he knew any calculus. The stranger nodded yes, and proceeded to solve the man's problem for him. Of course, it's not every day that an old man studies calculus next to the Nobel Prize–winning theoretical physicist T. D. Lee on the subway.

I had a similar train experience, but with a different ending. I was sitting on a crowded commuter train coming out of Chicago when a nurse boarded, leading a group of patients from the local mental hospital. They arranged themselves around me as the nurse began counting: "One, two, three — " She looked at me. "Who are you?"

"I'm Leon Lederman," I answered, "Nobel Prize winner and director of Fermilab."

She pointed at me and sadly continued: "Yes, four, five, six . . ."

But seriously, the concern over science illiteracy is legitimate, among other reasons because of the ever-increasing linkage of science, technology, and public welfare. Then, too, there is the great pity of missing out on the world view I have tried to present in these pages. Though still incomplete, it has grandeur, beauty, and an emerging simplicity. As Jacob Bronowski said:

> The progress of science is the discovery at each step of a new order which gives unity to what had long seemed unlike. Faraday did this when he closed the link between electricity and magnetism. Clerk Maxwell did it when he linked both with light. Einstein linked time with space, mass with energy, and the path of light past the sun with the flight of a bullet; and spent his dying years in trying to add to these likenesses another, which would find a single imaginative order

between the equations of Clerk Maxwell and his own geometry of gravitation.

When Coleridge tried to define beauty, he returned always to one deep thought: beauty he said, is "unity in variety." Science is nothing else than the search to discover unity in the wild variety of nature — or more exactly, in the variety of our experience.

INNER SPACE/OUTER SPACE

To see this edifice in its proper context, we now make an excursion to astrophysics, and I must explain why particle physics and astrophysics have in recent times been fused to a new level of intimacy, which I once called the inner space/outer space connection.

While the inner-space jocks were building ever more powerful microscope-accelerators to see down into the subnuclear domain, our outer-space colleagues were synthesizing data from telescopes of ever greater power, supplied with new technologies for increasing sensitivity and the ability to see fine detail. Another breakthrough came with space-based observatories carrying instruments to detect infrared, ultraviolet, x-rays, gamma rays — in short, the entire range of the electromagnetic spectrum, much of which had been blocked by our opaque and shimmering atmosphere.

The synthesis of the past one hundred years in cosmology is the "standard cosmological model." It holds that the universe began as a hot, dense, compact state about 15 billion years ago. Then the universe was infinitely or almost infinitely dense, infinitely or almost infinitely hot. The "infinite" description sits uncomfortably with physicists; all the qualifiers are the result of the fuzzy influence of quantum theory. For reasons we may never know, the universe exploded and has been expanding and cooling ever since.

Now how in the world did cosmologists find that out? The Big Bang model arose in the 1930s after the discovery that the galaxies, collections of 100 billion or so stars, were all flying away from one Edwin Hubble, who happened to be measuring their velocities in 1929. Hubble had to collect enough light from distant galaxies to resolve the spectral lines and compare them to lines of the same elements on earth. He noted that all of the lines shifted systematically toward the red. It was known that a source of light moving away from an observer would do just that. The "red shift" was in fact a measure of the relative velocity of source and observer. Over the years

Hubble found that all galaxies were moving away from him in all directions. Hubble showered regularly, and there was nothing personal in all this; it was simply a manifestation of the expansion of space. Because space is expanding the distances between all galaxies, astronomer Hedwina Knubble, observing from the planet Twilo in Andromeda, would see the same phenomenon — galaxies flying away from her. Indeed, the more distant the object, the faster it is moving. This is the essence of Hubble's law. It implies that if you ran the film backward, the most distant galaxies, moving faster, would close in on the nearer objects, and finally the whole mess would rush together and coalesce into a very, very small volume at a time presently estimated as about 15 billion years ago.

The most famous metaphor in science asks you to imagine yourself a two-dimensional creature, a Flatlander. You know from east-west and north-south, but up and down *do not exist*. Cast up-down out of your ken. You live on the surface of a balloon that is expanding. All over the surface are the residences of observers — planets and stars, clustered into galaxies all over the sphere. All two-dimensional. From any vantage point, all objects are moving away as the surface continually expands. The distance between any two points in this universe increases. That is how it is in our three-dimensional world. The other virtue of this metaphor is that, as in our own universe, there is no special place. All points on the surface are democratically equivalent to all other points. No center. No edge. No danger of falling off the universe. Since our expanding-universe metaphor (the balloon surface) is all we know, it is not a case of stars rushing out *into* space. It is space carrying along the whole kaboodle, which is expanding. It isn't easy to visualize an expansion that is happening everywhere in the universe. There is no outside, no inside. There is only this universe, expanding. Into what is it expanding? Think again of your life as a Flatlander on the surface of a balloon. The surface is all that exists in our metaphor.

Two major additional consequences of the Big Bang theory finally wore down the opposition, and now a fair consensus holds. One is the prediction that the light of the original incandescence — assuming it was hot, hot — would still be around as remnant radiation. Recall that light consists of photons, and the energy of photons is related inversely to their wavelength. A consequence of the universe's expansion is that all lengths are expanded. The wavelengths, originally infinitesimal, as befits high-energy photons, were thus predicted to have grown to the microwave region of a few millimeters. In 1965 the embers of the Big Bang, the microwave radiation, were discovered. The entire universe is awash with these photons, moving in all possible directions. Photons that started a journey billions of years ago when the universe was much smaller and hotter ended up on a Bell Laboratories antenna in New Jersey. What a fate!

After this discovery it became crucial to measure the distribution of wavelengths (here, please reread Chapter 5 with the book turned upside down), which was eventually done. Using the Planck equation, this measurement gives you the average temperature of the stuff (space, stars, dust, an occasional beeping satellite that escaped) that has been bathed in these photons. According to the latest (1991) NASA measurements from the COBE satellite, this is 2.73 degrees above absolute zero (2.73 degrees Kelvin). This remnant radiation is also strong evidence of the hot Big Bang theory.

While we are listing successes, we should also point out difficulties, all of which were eventually overcome. Astrophysicists have been carefully examining the microwave radiation in order to measure temperatures in different parts of the sky. The fact that these temperatures matched up with extraordinary precision (better than .01 percent) was a cause for some concern. Why? Because when two objects have exactly the same temperature, it is plausible to assume that they were once in contact. Yet the experts were sure that the different regions having precisely the same temperatures were never in contact. Not "hardly ever," but *never.*

Astrophysicists are allowed to speak so categorically because they have calculated how far apart two regions of the sky were at the time the microwave radiation observed by COBE was emitted. That time was 300,000 years after the Big Bang, not as early as one would like but as close as we can get. It turns out that these separations were so large that even with the velocity of light there was no time for the two regions to communicate. Yet they have the same temperature, or very close to it. Our Big Bang theory couldn't explain this. A failure? Another miracle? It became known as the causality, or isotropy, crisis. *Causality* because there seemed to be a causal connection between sky regions that never should have been in contact. *Isotropy* because everywhere you look on the grand scale you see pretty much the same pattern of stars, galaxies, clusters, and dust. One could live with this in a Big Bang model by saying that the similarity of the billions of pieces of the universe that had never been in contact was a pure accident. But we don't like "accidents." Miracles are okay if you invest in the lottery or are a Chicago Cubs fan, but not in science. When they appear, we suspect that something grander is lurking in the shadows. More on this later.

AN ACCELERATOR WITH AN UNLIMITED BUDGET

A second major success of the Big Bang model has to do with the composition of our universe. You think of the world as being made of air, earth, water (I'll leave out fire), and billboards. But if we look up and measure with our spectroscopic telescopes, we find mostly hydrogen, then helium. These account for 98 percent of the universe. The rest is composed of the other ninety or so elements. We know by our spectroscopic telescopes the relative amounts of the lighter elements — and lo! — the BB theorists say that these abundances are precisely what one would expect. Here is how we know.

The prenatal universe had in it all the matter in the presently observed universe — that is, about 100 billion galaxies, each with its 100 billion suns (can you hear Carl Sagan?). Everything we can see today was squeezed into a volume vastly smaller than the head of a pin. Talk about overcrowding! The temperature was high — about 10^{32} degrees Kelvin, a lot hotter than our current 3 degrees or so. And consequently matter was decomposed into its most primordial components. A plausible picture is of a "hot soup," or plasma, of quarks and leptons (or whatever is inside, if anything) smashing into each other with energies like 10^{19} GeV, or a trillion times the energy of the biggest collider a post-SSC physicist can imagine building. Gravity roared in as a powerful (but presently little understood) influence at this microscopic scale.

After this fanciful beginning, there was expansion and cooling. As the universe cooled, the collisions became less violent. The quarks, in intimate contact with one another as part of the dense glob that was the baby universe, began to coagulate into protons, neutrons, and the other hadrons. Earlier, any such union would have come apart in the ensuing violent collisions, but the cooling was relentless, the collisions kinder and gentler. By age three minutes, the temperatures had fallen enough to allow protons and neutrons to combine and, where earlier these would quickly have come apart, now stable nuclei formed. This was the nucleosynthesis period, and since we know a lot of nuclear physics, we can calculate the relative abundances of the chemical elements that did form. They are the nuclei of very light elements; the heavier elements require slow "cooking" in stars. Of course, atoms (nuclei plus electrons) didn't form until the temperature fell enough to allow electrons to organize themselves around nuclei. The right temperature arrived at about 300,000 years. Before that time we had no atoms, and we needed no chemists. Once neutral atoms formed, photons could move freely, and that is why we got our microwave photon information late.

Nucleosynthesis was a success: the calculated and the measured abundances agreed. Wow! Since the calculations are an intimate mix of nuclear physics, weak-force reactions, and early universe conditions, this agreement is very strong support for the Big Bang theory.

In the course of telling this story I have also explained the inner space/outer space connection. The early universe was nothing more than an accelerator lab with a totally unconstrained budget. Our astrophysicists need to know all about quarks and leptons and the forces in order to model evolution. And, as we pointed out in Chapter 6, particle physicists are provided data from Her One Great Experiment. Of course, at times earlier than 10^{-13} seconds, we are much less sure of the physics laws.

Nevertheless, we continue to make progress in our understanding of the Big Bang domain and the evolution of the universe. Our observations are made 15 billion years after the fact. Information that has been rattling around the universe for almost that amount of time occasionally stumbles into our observatories. We are also aided by the standard model and by the accelerator data that support it and try to extend it. But theorists are impatient; the hard accelerator data give out at energies equivalent to a universe that has lived 10^{-13} seconds. Astrophysicists need to know the operative laws at much earlier times, so they goad the particle theorists to roll up their sleeves and contribute to the torrent of papers: Higgs, unification, compositeness (what's inside the quark), and a host of speculative theories that venture beyond the standard model to build a bridge to a more perfect description of nature and a road to the Big Bang.

THERE ARE THEORIES, AND THEN THERE ARE THEORIES

It is 1:15 A.M. in my study. Several hundred yards away, the Fermilab machine is colliding protons on antiprotons. Two massive detectors are receiving data. The battle-hardened CDF group of 342 scientists and students are busy checking out the new pieces of their 5,000-ton detector. Not all of them, of course. On the average, at this time, a dozen people will be in the control room. Partway around the ring the new D-Zero detector, with its 321 collaborators, is being tuned up. The run, a month old, had the usual shaky start, but data taking will go on for about sixteen months, with a break for phasing in a new piece of the accelerator designed to increase the collision rate. Although the main thrust is to find the top quark, testing and extending the standard model is an essential part of the drive.

About 5,000 miles away, our CERN colleagues are also working hard to test a variety of theoretical ideas about how to extend the standard model. But while this good, clean work is going on, theoretical physicists are working, too, and I propose to give here a very brief, plumber's version of three of the most intriguing theories: GUTs, supersymmetry, and superstrings. This will be a superficial treatment. Some of these speculations are truly profound and can be appreciated only by the creators, their mothers, and a few close friends.

But first a comment on the word "theory," which lends itself to popular misconceptions. "That's *your* theory" is a popular sneer. Or "That's only a theory." Our fault for sloppy use. The quantum theory and the Newtonian theory are well-established, well-verified components of our world view. They are not in doubt. It's a matter of derivation. Once upon a time it was Newton's (as yet unverified) "theory." Then it was verified, but the name stuck. "Newton's theory" it will always be. On the other hand, superstrings and GUTs are *speculative* efforts to extend current understanding, building on what we know. The better theories are verifiable. Once upon a time that was the sine qua non of any theory. Nowadays, addressing events at the Big Bang, we face, perhaps for the first time, a situation in which a theory may never be experimentally tested.

GUTs

I have described the unification of the weak and electromagnetic forces into the electroweak force, carried by a quartet of particles: W^+, W^-, Z^0, and the photon. I have also described QCD — quantum chromodynamics — which deals with the behavior of quarks, in three colors, and gluons. These forces are now both described by quantum field theories obeying gauge symmetry.

Attempts to join QCD with the electroweak force are known collectively as grand unification theories (GUTs). The electroweak unification becomes evident in a world whose temperature exceeds 100 GeV (roughly the mass of the W, or 10^{15} degrees K). As chronicled in Chapter 8, we can achieve this temperature in the lab. GUTs unification, on the other hand, requires a temperature of 10^{15} GeV, which puts it out of the range of even the most megalomaniacal accelerator builder. The estimate is derived by looking at three parameters that measure the strengths of the weak, electromagnetic, and strong forces. There is evidence that these parameters in fact change with energy, the strong forces getting weaker and the electroweak forces stronger. The merger of all three numbers occurs at an energy of 10^{15} GeV. This is the grand unification regime, a place where the symmetry of the laws of nature is at a higher level. Again, this is a theory yet to be verified, but the trend of the measured strengths does indicate a convergence near this energy.

There are a number of grand unified theories, a *large* number, and they all have their ups and downs. For example, an early entrant to the GUT contest predicted that the proton was unstable and would decay into a neutral pion and a positron. The lifetime of a proton in this theory is 10^{30} years. Since the age of the universe is considerably less — somewhat over 10^{10} years — not too many protons have decayed. The decay of a proton would be a spectacular event. Remember, we considered the proton to be a stable hadron — and a good thing, too, because a reasonably stable proton is very important to the future of the universe and the economy. Yet in spite of the very low expected rate of decay, the experiment is doable. For example, if the lifetime is indeed 10^{30} years, and we watch a single proton for one year, we have only 1 divided by 10^{30} as our chance of seeing the decay — 10^{-30}. Instead, we can watch lots of protons. In 10,000 tons of water there are about 10^{33} protons (trust me). This would mean that 1,000 protons should decay in a year.

So enterprising physicists went underground — into a salt mine under Lake Erie in Ohio, into a lead mine under Mount Toyama in Japan, and into the Mont Blanc tunnel that connects France and Italy — to be shielded from the background of cosmic radiation. In these tunnels and deep mines they placed huge, clear plastic containers of pure water, about 10,000 tons worth. That would be a cube roughly 70 feet on each side. The water was stared at by hundreds of large, sensitive photomultiplier tubes, which would detect the bursts of energy released by the decay of a proton. So far no proton decays have been observed. This doesn't mean that these ambitious experiments have not proved valuable, for they have established a new measure of the proton's lifetime. Allowing for inefficiencies, the proton lifetime, if indeed the particle is unstable, must be *longer* than 10^{32} years.

Interestingly, the long and unsuccessful wait for protons to decay was enlivened by unexpected excitement. I have already told about the supernova explosion of February 1987. Simultaneously a burst of neutrino events was seen by the Lake Erie and Mount Toyama underground detectors. The combination of light and neutrinos was in disgustingly good agreement with models of stellar explosion. You should have seen the astronomers preen! But the protons just don't decay.

GUTs have a hard time but, ever resilient, GUT theorists continue to be enthusiastic. One doesn't have to build a GUT accelerator to test the theory. GUT theories have testable consequences in addition to proton decay. For example, SU(5), one of the grand unified theories, makes the postdiction that the electric charge of particles is quantized, and must come in multiples of one third the charge of the electron. (Remember the quark charges?) Very satisfying. Another consequence is the consolidation of the quarks and leptons in one family. In this theory, quarks (inside the proton) can be converted to leptons and vice versa.

GUTs predict the existence of supermassive particles (X bosons) that are one thousand trillion times heavier than protons. The mere possibility that these exist and can appear as virtual particles does have tiny, tiny consequences, such as the rare decay of protons. Incidentally, the prediction of this decay has practical, if very far-out, implications. If the nucleus of hydrogen (a single proton), for example, could be converted to pure radiation, it would provide a source of energy one hundred times more efficient than fusion energy. A few tons of water could provide all the energy needed by the United States in a day. Of course, right now we'd have to heat the water to GUT temperatures, but perhaps some kid now being turned off to science by an insensitive kindergarten teacher might have the idea that would make this more practical. So, help the teacher!

At the temperatures of the GUT scale (10^{28} degrees Kelvin) symmetry and simplicity have reached the point where there is only one kind of matter (lepto-quark?) and one force with an array of force-carrying particles and, oh yes, gravity, dangling there.

SUSY

Supersymmetry, or Susy, is the favorite of the betting theorists. We were introduced to Susy earlier. This theory unifies the matter particles (quarks and leptons) and the force carriers (gluons, W's . . .). It makes a huge number of experimental predictions, not one of which has (yet) been observed. But what fun!

We have gravitinos and winos and gluinos and photinos — the matterlike partners of gravitons, W's, and the rest. We have supersymmetric partners of quarks and leptons: squarks and sleptons, respectively. The burden on this theory is to show why these partners, one for every known particle, have not been seen. Oh, say the theorists, remember antimatter. Until the 1930s no one dreamed that every particle would have its twin antiparticle. And remember that symmetries are created to be broken (like mirrors?). The partner particles haven't been seen because they are heavy. Build a big enough machine and they will all appear.

Mathematical theorists assure the rest of us that the theory has a splendid symmetry in spite of its obscene proliferation of particles. Susy also promises to lead us to a true quantum theory of gravity. Attempts to quantize the general theory of relativity — our theory of gravity — have been beset with infinities up the wazoo in a way that could not be renormalized. Susy promises to lead us to a beautiful quantum theory of gravity.

Susy also civilizes the Higgs particle, which, lacking this symmetry, could not do the job it was invented to do. The Higgs particle, being a scalar (zero-spin) boson, is particularly sensitive to the busy vacuum around it. Its mass is influenced by the virtual particles of all masses that fleetingly occupy its space, each one contributing energy and, therefore, mass until the poor Higgs would grow far, far too obese to save the electroweak theory. What happens with supersymmetry is that the Susy partners influence the Higgs mass with their opposite signs. That is, the W particle makes the Higgs heavier, while wino cancels the effect, so the theory allows the Higgs to have a useful mass. Still, all this doesn't prove that Susy is right. It's just beautiful.

The issue is far from settled. Buzz words appear: supergravity, the geometry of superspace — elegant mathematics, dauntingly complex. But one experimentally intriguing consequence is that Susy willingly and generously supplies candidates for dark matter, stable neutral particles that could be massive enough to account for this pervasive material that haunts the observable universe. Susy particles were presumably made in the Big Bang era, and the lightest of the predicted particles — perhaps the photino, the higgsino, or the gravitino — could survive as stable remnants to constitute the dark matter and satisfy the astronomers. The next generation of machines must either confirm or deny Susy . . . but, oh, oh, oh what a gal!

SUPERSTRINGS

I believe it was *Time* magazine that forever embellished the lexicon of particle physics by trumpeting this as the Theory of Everything, or TOE. A recent book put it even better: *Superstrings, Theory of Everything?* (This is read with a rising inflection.) String theory promises a unified description of all forces, even gravity, all particles, space and time, free of arbitrary parameters and infinities. In short: everything. The basic premise replaces point particles by short segments of string. String theory is characterized by a structure that pushes the frontiers of mathematics (as physics has very occasionally done in the past) and the conceptual limitations of the human imagination to the extremes. The creation of this theory has its own history and its own heroes: Gabrielle Veneziano, John Schwarz, André Neveu, Pierre Ramond, Jeff Harvey, Joel Sherk, Michael Green, David Gross, and a gifted pied piper by the name of Edward Witten. Four of the prominent theorists worked together at an obscure institution in New Jersey and have become known as the Princeton String Quartet.

String theory is a theory about a very distant place, almost as far away as Atlantis or Oz. We are talking about the Planck domain, and if it ever existed (like Oz), it would have been in the very earliest flicker of Big Bang cosmology. There is no way we can imagine experimental data from that epoch. That doesn't mean we shouldn't persevere. Suppose one finds a mathematically consistent (no infinities) theory that somehow describes Oz and has as its low, low energy consequence our standard model? If it is also unique — that is, has no competitors that do the same thing — then we will all rejoice and lay down our pencils and trowels. Uniqueness is what superstrings doesn't enjoy. Within the major assumptions of superstrings are an enormous number of possible paths to the world of data. Let's see what else characterizes this stuff without pretending to explain it. Oh yes, as mentioned in Chapter 8, it requires ten dimensions: nine space dimensions and one time.

Now we all know there are only three space dimensions, although we have warmed up to the issue by imagining living in a two-dimensional world. So why not nine? "Where are they?" you rightly ask. Curled up. Curled up? Well, the theory started with gravity, which is based upon geometry, so one can visualize that six of the dimensions got curled up into a tiny ball. The size of the ball is typical of the Planck regime, 10^{-33} centimeters, about the size of the string that replaces the point particle. The particles we know emerge as vibrations of these strings. A stretched string (or wire) has an infinite number of vibration modes. That is the basis of the violin — or the lute, if you remember way back when we met Galileo's old man. The vibrations of real strings are classified in terms of a fundamental note and its harmonies or frequency modes. The mathematics of microstrings is similar. Our particles come from the lowest-frequency modes.

There is no way I can describe what has excited the leaders of this theory. Ed Witten gave a fantastic, gripping lecture about all this at Fermilab some years ago. For the first time in my experience, when he concluded there was almost ten seconds of silence (that's a lot!) before the applause. I rushed over to my lab to explain what I had learned to my colleagues on shift, but by the time I got there I had lost most of it. The artful lecturer makes you think you understand it.

As the theory met increasingly more difficult mathematics and a proliferation of possible directions, the progress and the intensity surrounding superstrings dropped to a more sensible level, and now we can only wait. There continues to be interest on the part of very capable theorists, but I suspect it will be a long time before TOE reaches the standard model.

FLATNESS AND DARK MATTER

Waiting for a theory rescue, Big Bang still has puzzles. Let me select one more problem that has confounded physicists even as it has led us — experimenters and theorists alike — to some tantalizing notions about the Very Beginning. It is known as the flatness problem, and it has a very human content — the morbid interest in whether the universe will continue to expand forever or whether it will slow down and reverse to a period of universal contraction. The issue is how much gravitational mass there is in the universe. If there is enough, the expansion will be reversed and we will have the Big Crunch. This is known as the closed universe. If there isn't enough, the universe will continue to expand forever, growing colder — an open universe. Between these two regimes is a "critical mass" universe, one that has just enough mass to slow the expansion but not enough to reverse it — a so-called flat universe.

Time for a metaphor. Think about sending a rocket up off the surface of the earth. If we give the rocket too small a velocity, it falls back to earth (closed universe). The earth's gravitation is too strong. If we give it a huge velocity, it can escape the earth's gravitation and soar into the solar system (open universe). Obviously there is a critical velocity such that ever so slightly less speed results in fallback, and ever so slightly more results in escape. Flatness occurs when the velocity is right on. The rocket escapes, but with ever-decreasing velocity. For rockets on our earth the critical velocity is 11.3 kilometers per second. Now, following the example, think of a fixed-velocity rocket (the Big Bang) and ask how heavy a planet (total mass density of the universe) results in escape or fallback.

One can estimate the gravitational mass of the universe by counting the stars. People have done this, and, taken alone, the number is too small to halt the expansion; it predicts an open universe, by a very wide margin. However, there is very strong evidence for the existence of a distribution of nonradiating matter, "dark matter," pervading the universe. When observed matter and estimated dark matter are combined, measurements indicate that the mass in the universe is close to — not less than 10 percent of nor more than two times — the critical mass. Thus it is still an open question whether the universe will continue to expand or will contract eventually.

There are many speculative candidates for dark matter. Most of them are particles, of course, with fancy names — axions, photinos — given by loving theorist-inventors. One of the most fascinating possibilities for dark matter is one or more of the standard-model neutrinos. There should be an enormous density of these elusive objects left over from the Big Bang era. They would be ideal candidates if . . . if they had a finite rest mass. We already know that the electron neutrino is too light, leaving two candidates, of which the tau neutrino is the favorite. Two reasons: (1) it exists, and (2) we know almost nothing about its mass.

Not long ago at Fermilab we carried out an ingenious and subtle experiment designed to detect whether the tau neutrino has a finite mass that would serve to close the universe. (Here cosmological needs drove an accelerator experiment, an indication of the particle-cosmology union.)

Imagine a graduate student on shift on a bleak winter's night, imprisoned in a small electronics hut on the wind-swept prairie of Illinois. Data have been accumulating for eight months. He checks the progress of the experiment, and as part of his routine he examines the data on the neutrino mass effect. (You don't measure the mass directly, but an influence the mass would have on some reactions.) He runs the entire sample of data through the calculation.

"What's this?" He becomes instantly alert. He can't believe the screen. "Oh, my God!" He runs computer checks. All are positive. There it is — mass! Enough to close the universe. This twenty-two-year-old graduate student experiences the incredible, breath-stopping conviction that he alone on the planet, among 5.32 billion of his fellow *sapiens,* knows the future of the universe. Talk about a Eureka moment!

Well, it's a nice story to think about. The part about the graduate student was true, but the experiment failed to detect any mass. That particular experiment just wasn't good enough, but it could have been, and . . . perhaps someday it will be. Colleague reader, please read this to your uncertain teenager *con brio!* Tell him or her that (1) experiments often fail, and (2) they don't always fail.

CHARLTON, GOLDA, AND GUTH

But even if we don't yet understand *how* the universe contains the critical mass needed for a flat universe, we're pretty sure that it does. We'll see why. Of all the masses nature could have chosen for Her universe (say 10^6 times critical mass or 10^{-16} times critical mass), She chose something nearly flat. In fact, it's worse than that. It appears to be a miracle that the universe has survived the two opposing fates — immediate runaway expansion or immediate crunch — for 15 billion years. It turns out that the flatness at age one second had to be close to perfect. If it deviated by ever so little, on one side we would have had the Big Crunch even before we made a single nucleus; if the deviation were on the other side, the expansion of the universe would have progressed by this time to a stone-cold dead thing. Again a miracle! Much as scientists may envision the Wise One, der Alte, a Charlton Heston type with fake long flowing beard and a strange laser-induced glow, or (as in my own view) a Margaret Mead or

Golda Meir or Margaret Thatcher type of deity, the contract clearly says that the laws of nature are not to be amended, that they are what they are. Thus the flatness problem is too much of a miracle and one seeks causes to make the flatness "natural." That's why my graduate student was freezing his ass off trying to determine whether neutrinos are dark matter or not. Infinite expansion or Big Crunch. He wanted to know. So do we.

The problem of flatness, the problem of the uniform 3-degree radiation, and several other problems of the Big Bang model were solved, at least theoretically, in 1980 by Alan Guth, an MIT particle theorist. His improvement is known as the Inflationary Big Bang model.

INFLATION AND THE SCALAR PARTICLE

In this brief history of the past 15 billion years I forgot to mention that the evolution of the universe is pretty much all contained in Einstein's equations of general relativity. Once the universe cools to a temperature of 10^{32} degrees Kelvin, classical (nonquantum) relativity prevails, and the subsequent events are indeed consequences of Einstein's theory. Unfortunately, the great power of the theory of relativity was discovered, not by the master but by his followers. In 1916, before Hubble and Knubble, the universe was thought to be a much more sedate, static object, and Einstein in his self-proclaimed "greatest blunder" added a term to his equation to prevent the expansion that the equation predicted. Since this is not a book on cosmology (and there are some excellent ones around), we will hardly do justice to the concepts, many of which are above my salary level.

What Guth discovered was a process, allowed by the Einstein equations, that generated an explosive force so huge as to produce a runaway expansion; the universe inflated from a size much smaller than that of a proton (10^{-15} meters) to the size of a golf ball in a time interval of 10^{-33} seconds or so. This inflationary phase arose through the influence of a new field, a nondirectional (scalar) field — a field that looks and acts and smells like . . . Higgs!

It is Higgs! The astrophysicists have discovered a Higgs thing in a wholly new context. What is the role of the Higgs field in promoting this bizarre pre-expanding-universe event that we call inflation?

We have noted that the Higgs field is closely tied to the concept of mass. What induces the wild inflation is the assumption that the preinflationary universe is suffused with a Higgs field whose energy content is so large that it drives a very rapid expansion. So "In the beginning there was a Higgs field" may not be too far from the truth. The Higgs field, which is constant throughout space, changes over time — in accordance with the laws of physics. These laws (added to the Einstein equations) generate the inflationary phase, which occupies the enormous time interval of 10^{-35} seconds to 10^{-33} seconds after Creation. Theoretical cosmologists describe the initial state as a "false vacuum" because of the energy content of the Higgs field. The ultimate transition to a true vacuum releases this energy to create the particles and the radiation, all at the enormous temperature of the Beginning. Following this, the more familiar Big Bang phase of relatively serene expansion and cooling begins. The universe is confirmed at the age of 10^{-33} seconds. "Today I am a universe," one intones at this phase.

Having donated all of its energy to the creation of particles, the Higgs field retires temporarily, reappearing several times in various disguises in order to keep the mathematics consistent, suppress infinities, and supervise the increasing complexity as the forces and particles continue to differentiate. Here is the God Particle in all its splendor.

Now wait. I didn't make any of this up. The originator of the theory, Alan Guth, was a young particle physicist trying to solve what appeared to be a totally different problem: the standard Big Bang model predicted the existence of magnetic monopoles — isolated single poles. North and south would then be related as matter and antimatter are. Looking for monopoles was a favorite game of particle hunters, and every new machine had its monopole search. But all proved unsuccessful. So at least monopoles are very rare, in spite of the absurd cosmological prediction that there should be enormous numbers of them. Guth, an amateur cosmologist, hit on the idea of inflation as a way of modifying the Big Bang cosmology to eliminate monopoles; then he discovered that by improving his inflation idea, he could solve all the other defects of that cosmology. Guth later commented on how lucky he was to make this discovery because all the components were known — a comment on the virtue of innocence in the creative act. Wolfgang Pauli once complained about his loss of creativity, "Ach, I know too much."

To complete this final tribute to Higgs, I should briefly explain how this rapid expansion solves the isotropy, or causality, and flatness crises. The inflation, which takes place at speeds vastly greater than the speed of light (the theory of relativity sets no limit on how fast space can expand), is just what we need. In the beginning, small regions of the universe were in intimate contact. Inflation vastly expanded these regions, separating their parts into causally disconnected regions. After inflation the expansion was slow compared to light velocity, so we continually discover new regions of the universe as their light finally reaches us. "Ah," the cosmic voice says, "we meet again." Now it is not a shock to realize that they are just like us: isotropy!

Flatness? The inflationary universe makes a clear statement: the universe is at critical mass; the expansion will continue to slow forever, but it will never reverse. Flatness: in Einstein's general theory of relativity, all is geometry. The presence of mass causes space to be curved; the more the mass, the greater the curvature. A flat universe is a critical condition between two opposing types of curvature. Large mass generates inward curvature of space, like the surface of a sphere. This is attractive and tends to a closed universe. Small mass produces an outward curvature, like the surface of a saddle. This tends to an open universe. Flatness represents a universe with a critical mass, "in between" inward and outward curvature. Inflation has the effect of stretching a tiny amount of curved space to so huge a domain as to make it effectively flat — very flat. The prediction of exact flatness, a universe that is critically poised between expansion and contraction, can be tested by identifying the dark matter and continuing the process of measuring the mass density. This, we are assured by the astros, will be done.

Other successes of the inflationary model have given it wide acceptance. For example, one of the "minor" annoyances of Big Bang cosmology is that it doesn't explain the lumpiness of the universe — the existence of galaxies, stars, and the rest. Qualitatively that lumpiness seems okay. By chance fluctuation, some matter clumps together out of a smooth plasma. The slight extra gravitational attraction draws other stuff to it, making the gravity even stronger. The process continues, and sooner or later we have a galaxy. But the details show that the process is far too slow if it is dependent on "chance fluctuations," so the seeds of galaxy formation must have been implanted during the inflationary phase.

Theorists who have thought about these seeds imagine them as small (less than 0.1 percent) density variations in the initial distribution of matter. Where did these seeds come from? Guth's inflation provides a very attractive explanation. One has to go to the quantum phase of the universe's history, in which spooky quantum mechanical fluctuations during inflation can lead to the irregularities. Inflation enlarges these microscopic fluctuations to a scale commensurate with galaxies. Recent observations (announced in April 1992) by the COBE satellite of ever so small differences in the temperature of the microwave background radiation in different directions are delightfully consistent with the inflationary scenario.

What COBE saw reflects conditions when the universe was young — only 300,000 years old — and stamped with the imprint of the inflation-induced distributions that made the background radiation hotter where it was less dense, cooler where it was more dense. The observed temperature differences thus provide experimental evidence for the existence of the necessary seeds for galaxy formation. No wonder the news made headlines all over the world. The temperature differences were only a few millionths of a degree and required extraordinary experimental care, but what a payout! One could detect, in the homogenized goop, evidence of the clumpiness that presaged the galaxies, suns, planets, and us. "It was like seeing the face of God," said exuberant astronomer George Smoot.

Heinz Pagels stressed the philosophical point that the inflationary phase is the ultimate Tower of Babel device, effectively cutting us off from whatever went before. It stretched and diluted all the structures that preexisted. So although we have an exciting story about the beginning, from time 10^{-33} seconds to time 10^{17} seconds (now), there are still those pesky kids out there who say, Yes, but the universe exists and how did it start?

In 1987 we had a "face of God" sort of conference at Fermilab when a group of astro/cosmo/theorists gathered to discuss how the universe began. The official title of the conference was Quantum Cosmology, and it was called so that the experts could commiserate about the domain of ignorance. No satisfactory theory of quantum gravity exists, and until one does, there will be no way of coping with the physical situation of the universe at the earliest moments.

The conference roster was a Who's Who of this exotic discipline: Stephen Hawking, Murray Gell-Mann, Yakov Zeldovich, André Linde, Jim Hartle, Mike Turner, Rocky Kolb, and David Schramm, among others. The arguments were abstract, mathematical, and very lively. Most of it was over my head. What I enjoyed most was Hawking's summary talk on the origin of the universe, given Sunday morning at about the time when 16,427 other sermons on roughly the same subject were being delivered from 16,427 pulpits around the nation. Except. Except that Hawking's talk was delivered through a voice synthesizer, giving it just that extra authenticity. As usual, he had a lot of interesting and complicated things to say, but he expressed the most profound thought quite simply. "The universe is what it is because it was what it was," he intoned.

Hawking was saying that the application of quantum theory to cosmology has as a task the specifications of initial conditions that must have existed at the very moment of creation. His premise assumes that the proper laws of nature — which, we hope, will be formulated by some genius now in third grade — will then take over and describe the subsequent evolution. The new great theory must integrate a description of the universe's initial conditions with a perfect understanding of the laws of nature and so explain all cosmological observations. It must also have as a consequence the standard model of the 1990s. If, before this breakthrough, we have achieved, via data from the Super Collider, a new standard model with a much more concise accounting for all of the data since Pisa, so much the better. Our sarcastic Pauli once drew an empty rectangle and claimed he had replicated the finest work of Titian — only the details were missing. Indeed, our painting "The Birth and Evolution of the Universe" requires a few more brushstrokes. But the frame is beautiful.

BEFORE TIME BEGAN

Let's go back to the prenatal universe again. We live in a universe about which we know a great deal. Like the paleontologist who reconstructs a mastodon from a fragment of a shinbone, or an archeologist who can visualize a long-defunct city from a few ancient stones, we are aided by the laws of physics emerging from the laboratories of the world. We are convinced (though we cannot prove this) that only one sequence of events, played backward, can lead via the laws of nature from our observed universe to the beginning and "before." The laws of nature must have existed before even time began in order for the beginning to happen. We say this, we believe it, but can we prove it? No. And what about "before time began"? Now we have left physics and are in philosophy.

The concept of time is tied to the appearance of events. A happening marks a point in time. Two happenings define an interval. A regular sequence of happenings can define a "clock" — a heartbeat, the swing of a pendulum, sunrise/sunset. Now imagine a situation where nothing ever happens. No tick-tock, no meals, no happening. The very concept of time in this sterile world has no meaning. Such may have been the state of the universe "before." The Great Event, the Big Bang, was a formidable happening that created, among other things, time.

What I am saying is that if we cannot define a clock, we cannot give a meaning to time. Consider the quantum idea of the decay of a particle, say our old friend the pion. Until it decays, there is no way of determining time in the universe of the pion. Nothing about it changes. Its structure, if we understand anything, is identical and unchanging until it decays in its own personal version of the Big Bang. Contrast this with our human experience of the decay of a *homo sapiens*. Believe me, there are plenty of signs that the decay is progressing or even imminent! In the quantum world, however, there is no meaning to the questions "When will the pion decay?" or "When did the Big Bang take place?" We can, on the other hand, ask the question "How long ago did the Big Bang take place?"

We can try to imagine the pre–Big Bang universe: timeless, featureless, but in some unimaginable way beholden to the laws of physics. These give the universe, like a doomed pion, a finite probability of exploding, changing, undergoing a transition, a change of state. Here we can improve on the metaphor used to start the book. Again we compare the universe in the Very Beginning to a huge boulder on top of a towering cliff, but now it is sitting in a trough. This would make it stable according to classical physics. Quantum physics, however, permits tunneling — one of the weird effects we examined in Chapter 5 — and the first event is that the boulder appears outside of the trough and, oops, goes over the edge of the cliff, falling to release its potential energy and create the universe as we know it. In very speculative models, our dear, dear Higgs field plays the role of the metaphoric cliff.

It is comforting to visualize the disappearance of space and time as we run the universe backward toward the beginning. What happens as space and time tend toward zero is that the equations we use to explain the universe break down and become meaningless. At this point we are just plumb out of science. Perhaps it is just as well that space and time cease to have meaning; it gives us the possibility that the vanishing of the concept takes place smoothly. What remains? What remains must be the laws of physics.

When dealing with all the elegant new theories about space, time, and the beginning, an obvious frustration sets in. As opposed to almost all other periods in science — certainly since the 1500s — there seems to be no way for experiments and observations to help out, at least not in the next few days. Even in Aristotle's time, one could (at risk) count the teeth in a horse's mouth in order to enter the debate on the number of teeth the horse has. Now our colleagues are debating a subject that has only one piece of data: the existence of a universe. This of course brings us to the whimsical subtitle of our book: the universe is the answer, but damned if we know the question.

RETURN OF THE GREEK

It was almost 5 A.M. I had dozed over the last pages of Chapter 9. My deadline was (long) past, and I had no inspiration. Suddenly I heard a commotion outside our old farmhouse in Batavia. The horses in the stable were milling around and kicking. I walked out to see this guy in a toga and a pair of brand-new sandals coming out of the barn.

LEDERMAN: Democritus! What are you doing here?

DEMOCRITUS: Call those horses? You should see the Egyptian chariot horses I raised in Abdera. Seventeen hands and up. They could *fly!*

LEDERMAN: Yes, well, how are you?

DEMOCRITUS: Do you have an hour? I've been invited to the control room of the Wake Field Accelerator that just turned on in Teheran on January 12, 2020.

LEDERMAN: Yeow! Can I come?

DEMOCRITUS: Sure, if you behave. Here, hold my hand and say Πλανχκ Μασσ. [Planck mass]

LEDERMAN: Πλανχκ Μασσ.

DEMOCRITUS: Louder!

LEDERMAN: Πλανχκ Μασσ!

Suddenly we were in a surprisingly small room that looked totally different from what I had expected — the command deck of Star Ship *Enterprise*. There were a few multicolored screens with very sharp images (high-definition TV). But the banks of oscilloscopes and dials were gone. Over in one corner a group of young men and women were engaged in an animated discussion. A technician standing next to me was punching buttons on a palm-sized box and watching one of the screens. Another technician was speaking Persian into a microphone.

LEDERMAN: Why Teheran?

DEMOCRITUS: Oh, some years after world peace, the UN decided to locate the New World Accelerator at the ancient crossroads of the world. The government here is one of the most stable, and they also made the best case for good geology, proximity to cheap power, water, and skilled labor, and the best shishkebab south of Abdera.

LEDERMAN: What's going on?

DEMOCRITUS: The machine is colliding 500 TeV protons against 500 TeV antiprotons. Ever since 2005, when the Super Collider discovered the Higgs at a mass of 422 GeV, there was this urgent need to explore the "Higgs sector" to see if there are more kinds of Higgs.

LEDERMAN: They found the Higgs?

DEMOCRITUS: One of them. They think there is a whole family of Higgses.

LEDERMAN: Anything else?

DEMOCRITUS: Oh, hell, yes. You should have been here when the on-line data showed this crazy event with six jets and eight electron pairs. By now they have seen several squarks, gluinos, as well as the photino . . .

LEDERMAN: Supersymmetry?

DEMOCRITUS: Yes, as soon as the machine energies went above 20 TeV, these little guys poured out.

Democritus called to someone in heavily accented Persian, and we were soon handed mugs of steaming fresh yak milk. When I asked for a display screen to see events, someone clamped a virtual-reality helmet over my head, and events, constructed from the data by God-knows-what-kind of computer, flashed before my eyes. I noticed that these 2020 physicists (the preschool kids of my era) still needed to be pictorially spoon-fed the information. A tall, young black woman with a spectacular Afro hairdo, carrying what looked like a computer notepad, sauntered over. Ignoring Democritus, she looked me over with some amusement. "Blue jeans, just like my grandfather used to wear. With that outfit you must be from UN headquarters. Are you inspecting us?"

"No," I said. "I'm from Fermilab, and I've been out of the business for a few years. What's going on?"

The next hour passed in a dazing blur of explanations of neural networks, jet algorithms, top quark and Higgs calibration points, vacuum-deposited diamond semiconductors, femtobytes, and — worse — twenty-five years of experimental progress. She was from Michigan, a product of the prestigious Detroit High School of Science. Her husband, a Kazakhstani postdoc, was employed by the University of Quito. She explained that the machine had a radius of only one hundred miles, this modest size made possible by a 1997 breakthrough in room-temperature superconductors. Her name was Mercedes.

MERCEDES: Yeah, the Super Collider R&D group stumbled on these new materials while they were tracking down some weird effects in the niobium alloys. One thing led to another, and suddenly we had this cruciferous material that begins superconducting at 50 degrees Fahrenheit, about the temperature of a cool day in autumn.

LEDERMAN: What is the critical field?

MERCEDES: Fifty tesla! If I remember my history, your Fermilab machine was at four tesla. Today there are twenty-five companies making or growing the stuff. The economic impact in FY 2019 is about three hundred billion dollars. The super-train, which floats between New York and Los Angeles, cruises at two thousand miles per hour. Huge clumps of steel wool, energized by the new stuff, now provide pure water to most of the cities of the world. Every week we read about some new application.

Democritus, sitting quietly up until now, bored in on the central question.

DEMOCRITUS: Have you seen anything inside quarks?

MERCEDES: [*shaking her head, smiling*] That was my Ph.D. thesis. The best measurement came out of the last Super Collider experiment. The radius of the quark is less than an incredibly small 10^{-21} centimeters. As far as we can tell, quarks and leptons are as good an approximation to points as you can get.

DEMOCRITUS: [*jumping up and down, clapping, laughing hysterically*] *Atomos!* Finally!

LEDERMAN: Any surprises?

MERCEDES: Well, with Susy and the Higgs, a young theorist from CUNY — a guy named Pedro Monteagudo — has written a new Susy-GUT equation that successfully predicts the Higgs-generated masses of all the quarks and leptons. Just as Bohr explained the energy levels in the hydrogen atom.

LEDERMAN: Yeow! Really?

MERCEDES: Yeah, the Monteagudo equation has taken over from Dirac, Schrödinger, and all points west. Look at my T-shirt.

As if I needed such an invitation. But as I focused on the curious hieroglyphic displayed there, I felt a fuzzy, earthquake-like dizziness, and it all faded.

•

"Shit." I was back home, groggily lifting my head off my papers. I noticed one photocopy of a news headline: CONGRESSIONAL FUNDING FOR THE SUPER COLLIDER IN DOUBT. My computer modem was beeping, and an E-mail message was "inviting" me to Washington for a Senate hearing on the SSC.

GOOD-BYE

You and I, dear colleague, have come a long way from Miletus. We have traversed the road of science from then and there to here and now. Regretfully we have sped past many of the milestones, major and minor. But we have paused at a few of the important sights: at Newton and Faraday, Dalton and Rutherford, and, of course, at McDonald's for a hamburger. We see a new synergy between inner and outer space, and like a driver on a forested winding road, we see occasional glimpses, obscured by trees and fog, of a towering edifice: an intellectual construct 2,500 years in the making.

Along the way I have tried to insert some irreverent details about the scientists. It is important to distinguish between the scientists and the science. Scientists, more often than not, are people, and as such they span the enormous range of variability that makes people so . . . so interesting. Scientists are serene and ambitious; they are driven by curiosity and ego; they exhibit angelic virtue and immense greed; they are wise beyond measure and childish well into their dotage; intense, obsessed, laid-back. Among the subset of humans called scientists, there are atheists, agnostics, the militantly apathetic, the deeply religious, and those who view the Creator as a personal deity, either all-wise or somewhat bumbling, like Frank Morgan in *The Wizard of Oz*.

The range of abilities among scientists is also huge. This is okay because science needs the mixers of cement as well as the master architects. We count among us minds of awesome power, those who are only monstrously clever, those possessed of magic hands, uncanny intuition, and that most vital of all scientific attributes: luck. We also have jerks, assholes, and those who are just dumb . . . *dumb!*

"You mean relative to you others," my mother once protested.

"No, Mom, dumb like anyone is dumb."

"So how did he get a Ph.D.?" she challenged.

"Sitzfleisch, Mom." Sitzfleisch: the ability to sit through any task, to do it again and again until the job is somehow done. Those who give out Ph.D.'s are human too — sooner or later they give in.

Now, if there is any unifier to this collection of human beings we call scientists, it is the pride and reverence with which each of us adds our contribution to that intellectual edifice: our science. It may be a brick, fitted meticulously and cemented into place, or it may be a magnificent lintel (to stress out the metaphor) gracing columns placed there by our masters. We build with a sense of awe, heavily tinged with skepticism, guided by what we found when we arrived, bringing all our human variables, coming to this effort from all directions, each carrying our own cultural dress and language, but somehow finding instant communication, instant understanding, and empathy in the common task of building the tower of science.

It is time to let you go back to your real life. For the past three years I have been yearning for a time when this would be over. Now I admit that I will miss you, colleague reader. You have been my constant companion on airplanes, in very quiet, late-night writing sessions. I have pictured you as retired history teacher, turf accountant, college student, wine merchant, motorcycle mechanic, high school sophomore, and, when I need cheering up, an incredibly beautiful contessa who wants to run her hands through my hair. Like a reader finishing a novel, reluctant to leave the characters behind, I will miss you.

THE END OF PHYSICS?

Before I go, I have a statement to make on this ultimate T-shirt business. I may have given the impression that the God Particle, once understood, will provide the ultimate revelation: how the universe works. This is the domain of the really-deep-thinkers, the particle theorists who are paid to really think deep. Some of them believe that

The Road to reductionism will come to an end; we will essentially know it all. Science will then concentrate on complexity: super bucky-balls, viruses, the morning traffic jam, a cure for hatred and violence . . . all good stuff.

There is another view — that we are like children (in the metaphor of Bentley Glass) playing on the shore of a vast ocean. This view allows for the truly endless frontier. Behind the God Particle is revealed a world of splendid, blinding beauty, but one to which our mind's eye will adapt. Soon we will perceive that we do not have all the answers; what is inside the electron, quark, and black hole will draw us ever on.

I think I favor the optimists (or are they pessimists giving up job security?), those theorists who believe we will "know it all," but the experimentalist in me prevents summoning up the requisite arrogance. The experimental road to Oz, the Planck mass, to that epoch less than 10^{-40} seconds after The Event makes our total voyage from Miletus to Waxahachie look like a pleasure cruise on Lake Winnebago. I think not only of accelerators girdling the solar system and detector edifices to match, not only of the billions and billions of hours of sleep my students and theirs will lose, but I worry about the necessary sense of optimism that our society must summon if this quest is to continue.

What we really do know and will know much better in a decade or so can be measured by the SSC energy: 40 trillion volts. But important things must also happen at energies so high as to make our forthcoming SSC collisions seem docile. There are still boundless possibilities for complete surprises. Operating under new laws of nature as unimaginable today as quantum theory (or the cesium atomic clock) would have been to Galileo, we could find ancient civilizations existing inside quarks. Gasp! Before the men in white coats arrive, let me switch to another frequently raised question.

It is astonishing how often otherwise competent scientists forget the lessons of history, namely, that the major impacts of science on society have always come from the kind of research that drives the quest for the a-tom. Without taking anything away from genetic engineering, materials science, or controlled fusion, the quest for the a-tom has paid for itself many millionfold, and there is no sign so far that this has changed. The investment in abstract research, at less than one percent of the budgets of industrial societies, has performed much better than the Dow Jones average has for over three hundred years. Yet from time to time we are terrorized by frustrated policy makers who want to focus science on the *immediate* needs of society, forgetting or perhaps never understanding that most of the major advances in technology that have influenced the quality and quantity of human life have come out of pure, abstract, curiosity-driven research. Amen.

OBLIGATORY GOD ENDING

Looking for inspiration on how to wind up this book, I studied the endings of a few dozen science books written for a general audience. They are always philosophical, and the Creator almost always appears in the favorite image of the author or in the image of the author's favorite author. I have noticed two kinds of closing summaries in popular science books. One kind is characterized by humility. The downgrading of humankind usually begins by reminding the reader that we are many times removed from centrality: our planet is not the center of the solar system, and the solar system is not the center of our galaxy, nor is our galaxy anything special as galaxies go. If this isn't enough to discourage even a Harvard man, we learn that the very material we and the things around us are made of consists of only a small sample of the fundamental objects in the universe. Then these authors note that humankind and all of its institutions and monuments matter very little to the continued evolution of the cosmos. The master of the humbling assessment may be Bertrand Russell:

> Such, in outline, but even more purposeless, more void of meaning, is the world which Science presents for our belief. Amid such a world, if anywhere, our ideals henceforward must find a home. That man is the product of causes which had no prevision of the end they were achieving; that his origin, his growth, his hopes and fears, his loves and his beliefs, are but the outcome of accidental collocations of atoms; that no fire, no heroism, no intensity of thought and feeling, can preserve an individual life beyond the grave; that all the labours of the ages, all the devotion, all the inspirations, all the noonday brightness of human genius, are destined to extinction in the vast death of the solar system, and that the whole temple of Man's achievement must inevitably be buried beneath the debris of a universe in ruins — all these things, if not quite beyond dispute, are yet so nearly certain, that no philosophy which rejects them can hope to stand. Only within the scaffolding of these truths, only on the firm foundation of unyielding despair, can the soul's habitation henceforth be safely built.

> Brief and powerless is Man's life, on him and all his race the slow, sure doom falls pitiless and dark. . . .

To which I say softly, Wow! The guy has a point. Steven Weinberg put it more succinctly: "The more the universe seems comprehensible, the more it seems pointless." Now we are surely humbled.

There are also those who go all the way in the other direction, who view the effort to understand the universe as not at all humbling but exalting. This group yearns to "know the mind of God" and says that by so doing we become a crucial part of the whole process. Thrillingly, we are restored to our rightful place at the center of the universe. Some philosophers of this ilk go so far as to say that the world is a product of the human mind's constructions; others, a bit more modest, say that our mind's very existence, even on the infinitesimal speck of an ordinary planet, must be a crucial part of the Grand Plan. To which I say, very softly now, that it's nice to be needed.

But I prefer a combination of the two approaches, and if we're going to work God in here somewhere, let's call on the folks who have given us so many memorable images of Her. So here is the script for the last scene in Hollywood's loving transmutation of this book.

•

The hero is the president of the Astrophysics Society, the only person ever to win three Nobel Prizes. He stands at night on the beach, legs planted wide, shaking his fist at the jeweled blackness of the sky. Anointed by his humanity, aware of mankind's most powerful achievements, he shouts at the universe above the sound of crashing waves. "I have created you. You are the product of my mind — my vision and my invention. It is I who provide you with reason, with purpose, with beauty. Of what use are you but for my consciousness and my constructions, which have revealed you?"

A fuzzy swirling light appears in the sky, and a beam of radiance illuminates our man-on-the-beach. To the solemn and climactic chords of the Bach B Minor Mass, or perhaps the piccolo solo of Stravinsky's "Rites," the light in the sky slowly configures itself into Her Face, smiling, but with an expression of infinite sweet sadness.

Fade to black. Roll credits.

ACKNOWLEDGMENTS

We believe it was Anthony Burgess (or was it Burgess Meredith?) who proposed an amendment to the Constitution that would prohibit an author from including in his acknowledgments a thank you to his wife for typing the manuscript. Our wives don't do typing, so you are spared that here. There are thanks to be given, however.

Michael Turner, a theorist and cosmologist, pored over the manuscript for subtle errors in theory (and some not so subtle); he caught many, fixed them, and steered us back on course. Given the experimental bias of the book, it was as if Martin Luther had asked the pope to proofread his ninety-five theses. Mike, if there are residual errors, blame the editors.

The Fermi National Accelerator Laboratory (and its patron saint in Washington, the U.S. Department of Energy) provided much of the inspiration and not a little of the mechanical support.

Willis Bridegam, librarian of Amherst College, made special resources of the Robert Frost Library and the Five College system available to us. Karen Fox provided creative research.

We suspect that Peg Anderson, our manuscript editor, became so embroiled in the subject that she asked all the right questions, and in so doing, she earned her battlefield commission as an honorary M.S. in physics.

Kathleen Stein, *Omni*'s incomparable interview editor, assigned the interview from which sprang the germ of the book. (Or was it a virus?)

Lynn Nesbit had more faith in the project than we did.

And John Sterling, our editor, sweated the whole thing out. We hope that whenever he sits down in a warm bath, he'll think of us, and scream something appropriate.

Leon M. Lederman
Dick Teresi

A NOTE ON HISTORY AND SOURCES

When scientists talk about history, one must be alert. It isn't history as a professional, scholarly historian of science would write it. One could call it "fake history." The physicist Richard Feynman called it conventionalized myth-history. Why? Scientists (certainly this scientist) use history as part of pedagogy. "See, here is a sequence of scientific events. First there was Galileo, then Newton and this apple . . ." Of course, that isn't the way it happens. There are crowds of others who help and hinder. The evolution of a new concept in science can be enormously complicated — and was even in the days before faxes. A quill pen can do plenty of damage.

In Newton's time there was a dense literature of published articles, books, correspondence, lectures. Priority battles (who gets the credit for being the first to make a discovery) go back long before Newton. Historians sort all of this out and create a vast and rich literature about the people and concepts. However, from the point of view of storytelling, myth-history has the great virtue of filtering out the noise of real life.

As for sources, when one sums up the knowledge gained over five decades working in physics, it is difficult to pin down the precise source of every fact, quote, or piece of information. There may, in fact, be no source for some of the best stories in science, but they have become such a part of the collective consciousness of scientists that they are "true," whether or not they ever happened. Still, we hit some books, and for the benefit of the reader, here are some of the better ones. This is by no means a complete list, nor do we mean to imply that the following publications are the original or best sources for the information cited. I list them in no particular order, except the whim of an experimentalist . . .

I profited from several biographies of Newton, especially the version by John Maynard Keynes and *Never at Rest* by Richard Westfall (Cambridge: Cambridge University Press, 1981). Abraham Pais's *Inward Bound: Of Matter and Forces in the Physical World* (New York: Oxford University Press, 1986) was an invaluable source, as was the classic *A History of Science* by Sir William Dampier (Cambridge: Cambridge University Press, 1948). The recent biographies *Schrödinger: Life and Thought* by Walter Moore (Cambridge: Cambridge University Press, 1989) and *Uncertainty: The Life and Science of Werner Heisenberg* by David Cassidy (New York: W. H. Freeman, 1991) were also of great help, as were *The Life and Times of Tycho Brahe* by John Allyne Gade (Princeton: Princeton University Press, 1947), *Galileo at Work: His Scientific Biography* by Stillman Drake (Chicago: University of Chicago Press, 1978), *Galileo Heretic* by Pietro Redondi (Princeton: Princeton University Press, 1987), and *Enrico Fermi, Physicist* by Emilio Segré (Chicago: University of Chicago Press, 1970). We are indebted to Heinz Pagels for two books: *The Cosmic Code* (New York: Simon & Schuster, 1982) and *Perfect Symmetry* (New York: Simon & Schuster, 1985), and to Paul Davies for *Superforce* (New York: Simon & Schuster, 1984).

Some books by nonscientists provided anecdotes, quotes, and other valuable information — most notably *Scientific Temperaments* by Philip J. Hilts (New York: Simon & Schuster, 1982) and *The Second Creation: Makers of the Revolution in Twentieth-Century Physics* by Robert P. Crease and Charles C. Mann (New York: Macmillan, 1986).

The Very Beginning scenario, as mentioned in the text, is more philosophy than physics. University of Chicago theorist/cosmologist Michael Turner says this is a reasonable guess. Charles C. Mann supplied some nice details on the remarkable number 137 in his *Omni* magazine article, entitled, oddly enough, "137." We consulted a number of sources for the beliefs of Democritus, Leucippus, Empedocles, and the other pre-Socratic philosophers: Bertrand Russell's *A History of Western Philosophy* (New York: Touchstone, 1972); W. K. C. Guthrie's *The Greek Philosophers: From Thales to Aristotle* (New York: Harper & Brothers, 1960) and *A History of Greek Philosophy* (Cambridge: Cambridge University Press, 1978); Frederick Copleston's *A History of Philosophy: Greece & Rome* (New York: Doubleday, 1960); and *The Portable Greek Reader,* edited by W. H. Auden (Viking Press, 1948).

Many dates and details were checked with *The Dictionary of Scientific Biography*, edited by Charles C. Gillispie (New York: Scribner's, 1981), a multivolume set that can cost one many enjoyable hours in the library.

Miscellaneous sources include *Johann Kepler* (Baltimore: Williams & Wilkins, 1931), which is a series of papers, and *Chemical Atomism in the Nineteenth Century* by Alan J. Rocke (Columbus: Ohio State University Press, 1984). Bertrand Russell's gloomy quote in Chapter 9 was taken from *A Free Man's Worship* (1923).

INDEX